Springer Series in Statistics

Springer
New York
Berlin
Heidelberg
Barcelona
Hong Kong
London
Milan
Paris
Singapore
Tokyo

Springer Series in Statistics

Andersen/Borgan/Gill/Keiding: Statistical Models Based on Counting Processes.
Andrews/Herzberg: Data: A Collection of Problems from Many Fields for the Student and Research Worker.
Anscombe: Computing in Statistical Science through APL.
Berger: Statistical Decision Theory and Bayesian Analysis, 2nd edition.
Bolfarine/Zacks: Prediction Theory for Finite Populations.
Borg/Groenen: Modern Multidimensional Scaling: Theory and Applications
Brémaud: Point Processes and Queues: Martingale Dynamics.
Brockwell/Davis: Time Series: Theory and Methods, 2nd edition.
Daley/Vere-Jones: An Introduction to the Theory of Point Processes.
Dzhaparidze: Parameter Estimation and Hypothesis Testing in Spectral Analysis of Stationary Time Series.
Fahrmeir/Tutz: Multivariate Statistical Modelling Based on Generalized Linear Models.
Farebrother: Fitting Linear Relationships: A History of the Calculus of Observations 1750 - 1900.
Farrell: Multivariate Calculation.
Federer: Statistical Design and Analysis for Intercropping Experiments, Volume I: Two Crops.
Federer: Statistical Design and Ánalysis for Intercropping Experiments, Volume II: Three or More Crops.
Fienberg/Hoaglin/Kruskal/Tanur (Eds.): A Statistical Model: Frederick Mosteller's Contributions to Statistics, Science and Public Policy.
Fisher/Sen: The Collected Works of Wassily Hoeffding.
Good: Permutation Tests: A Practical Guide to Resampling Methods for Testing Hypotheses.
Goodman/Kruskal: Measures of Association for Cross Classifications.
Gouriéroux: ARCH Models and Financial Applications.
Grandell: Aspects of Risk Theory.
Haberman: Advanced Statistics, Volume I: Description of Populations.
Hall: The Bootstrap and Edgeworth Expansion.
Härdle: Smoothing Techniques: With Implementation in S.
Hart: Nonparametric Smoothing and Lack-of-Fit Tests.
Hartigan: Bayes Theory.
Hedayat/Sloane/Stufken: Orthogonal Arrays: Theory and Applications.
Heyde: Quasi-Likelihood and its Application: A General Approach to Optimal Parameter Estimation.
Heyer: Theory of Statistical Experiments.
Huet/Bouvier/Gruet/Jolivet: Statistical Tools for Nonlinear Regression: A Practical Guide with S-PLUS Examples.
Jolliffe: Principal Component Analysis.
Kolen/Brennan: Test Equating: Methods and Practices.
Kotz/Johnson (Eds.): Breakthroughs in Statistics Volume I.

(continued after index)

Michael L. Stein

Interpolation of Spatial Data

Some Theory for Kriging

With 27 Illustrations

Springer

Michael L. Stein
Department of Statistics
University of Chicago
Chicago, IL 60637
USA

Library of Congress Cataloging-in-Publication Data
Stein, Michael Leonard.
Interpolation of spatial data : some theory for kriging / Michael L. Stein.
p. cm. — (Springer series in statistics)
Includes bibliographical references and index.
ISBN 0-387-98629-4 (alk. paper)
1. Kriging. I. Title. II. Series.
TN272.7.S74 1999
622'.1'015195—dc21 98-44772

Printed on acid-free paper.

Production managed by Timothy Taylor; manufacturing supervised by Thomas King.
Photocomposed copy prepared from the author's TEX files.
Printed and bound by Edwards Brothers, Inc., Ann Arbor, MI.
Printed in the United States of America.

9 8 7 6 5 4 3 2 1

ISBN 0-387-98629-4 Springer-Verlag New York Berlin Heidelberg SPIN 10693415

TO LAURIE

Preface

This monograph is an attempt to provide a mathematical treatment for the procedure known as kriging, which is a popular method for interpolating spatial data. Kriging is superficially just a special case of optimal linear prediction applied to random processes in space or random fields. However, optimal linear prediction requires knowing the covariance structure of the random field. When, as is generally the case in practice, the covariance structure is unknown, what is usually done is to estimate this covariance structure using the same data that will be used for interpolation. The properties of interpolants based on an estimated covariance structure are not well understood and it is common practice to ignore the effect of the uncertainty in the covariance structure on subsequent predictions. My goal in this monograph is to develop the mathematical tools that I believe are necessary to provide a satisfactory theory of interpolation when the covariance structure is at least partially unknown. This work uses these tools to prove a number of results, many of them new, that provide some insight into the problem of interpolating with an unknown covariance structure. However, I am unable to provide a complete mathematical treatment of kriging with estimated covariance structures. One of my hopes in writing this book is that it will spur other researchers to take on some of the unresolved problems raised here.

I would like to give a bit of personal history to help explain my devotion to the mathematical approach to kriging I take here. It has long been recognized that when interpolating observations from a random field possessing a semivariogram, the behavior of the semivariogram near the origin plays a crucial role (see, for example, Matheron (1971, Section 2-5)). In the mid 1980s I was seeking a way to obtain an asymptotic theory to support this general understanding. The asymptotic framework I had in mind was to take more and more observations in a fixed and bounded region of space, which I call fixed-domain asymptotics. Using this approach, I suspected that it should generally be the case that only the behavior of the semivariogram near the origin matters asymptotically for determining the properties of kriging predictors. Unfortunately, I had no idea how to prove such a result except in a few very special cases. However, I did know of an example in which behavior away from the origin of the semivariogram could have an asymptotically nonnegligible impact on the properties of kriging predictors. Specifically, as described in 3.5, the semivariograms corresponding to exponential and triangular autocovariance functions have the same behavior near the origin, but optimal linear interpolants under the two models do not necessarily have similar asymptotic behavior. I believed that there should be some mathematical formulation of the problem that would exclude the "pathological" triangular autocovariance function and would allow me to obtain a general theorem on asymptotic properties of kriging predictors. Soon after arriving at the University of Chicago in the fall of 1985, I was browsing through the library and happened upon *Gaussian Random Processes* by Ibragimov and Rozanov (1978). I leafed through the book and my initial reaction was to dismiss it as being too difficult for me to read and in any case irrelevant to my research interests. Fortunately, sitting among all the lemmas and theorems and corollaries in this book was a single figure on page 100 showing plots of an exponential and triangular autocovariance function. The surrounding text explained how Gaussian processes corresponding to these two autocovariance functions could have orthogonal measures, which did not make an immediate impression on me. However, the figure showing the two autocovariance functions stuck in my mind and the next day I went back to the library and checked out the book. I soon recognized that equivalence and orthogonality of Gaussian measures was the key mathematical concept I needed to prove results connecting the behavior of the semivariogram at the origin to the properties of kriging predictors. Having devoted a great amount of effort to this topic in subsequent years, I am now more firmly convinced than ever that the confluence of fixed-domain asymptotics and equivalence and orthogonality of Gaussian measures provides the best mathematical approach for the study of kriging based on estimated covariance structures. I would like to thank Ibragimov and Rozanov for including that single figure in their work.

This monograph represents a synthesis of my present understanding of the connections between the behavior of semivariograms at the origin, the properties of kriging predictors and the equivalence and orthogonality of Gaussian measures. Without an understanding of these connections, I believe it is not possible to develop a full appreciation of kriging. Although there is a lot of mathematics here, I frequently discuss the repercussions of the mathematical results on the practice of kriging. Readers whose main interests are in the practice of kriging should consider skipping most of the proofs on a first reading and focus on the statements of results and the related discussions. Readers who find even the statements of the theorems difficult to digest should carefully study the numerical results in Chapters 3 and 6 before concluding that they can ignore the implications of this work. For those readers who do plan to study at least some of the proofs, a background in probability theory at the level of, say, Billingsley (1995) and some familiarity with Fourier analysis and Hilbert spaces should be sufficient. The necessary second-order theory of random fields is developed in Chapter 2 and results on equivalence and orthogonality of Gaussian measures in Chapter 4. Section 1.3 provides a brief summary of the essential results on Hilbert spaces needed here.

In selecting topics for inclusion, I have tried to stick to topics pertinent to kriging about which I felt I had something worthwhile to say. As a consequence, for example, there is little here about nonlinear prediction and nothing about estimation for non-Gaussian processes, despite the importance of these problems. In addition, no mention is made of splines as a way of interpolating spatial data, even though splines and kriging are closely related and an extensive literature exists on the use of splines in statistics (Wahba 1990). Thus, this monograph is not a comprehensive guide to statistical approaches to spatial interpolation. Part I of Cressie (1993) comes much closer to providing a broad overview of kriging.

This work is quite critical of some aspects of how kriging is commonly practiced at present. In particular, I criticize some frequently used classes of models for semivariograms and describe ways in which empirical semivariograms can be a misleading tool for making inferences about semivariograms. Some of this criticism is based on considering what happens when the underlying random field is differentiable and measurement errors are negligible. In some areas of application, nondifferentiable random fields and substantial measurement errors may be common, in which case, one could argue that my criticisms are not so relevant to those areas. However, what I am seeking to accomplish here is not to put forward a set of methodologies that will be sufficient in some circumscribed set of applications, but to suggest a general framework for thinking about kriging that makes sense no matter how smooth or rough is the underlying random field and whether there is nonnegligible measurement error. Furthermore, I contend that the common assumption that the semivariogram of the underlying random field behaves linearly in a neighborhood of the origin (which implies the random field is not differentiable), is often made out of habit or ignorance and not because it is justified.

For those who want to know what is new in this monograph, I provide a summary here. All of 3.6 and 3.7, which study the behavior of predictions with evenly spaced observations in one dimension as the spacing between neighboring observations tends to 0, are new. Section 4.3 mixes old and new results on the asymptotic optimality of best linear predictors under an incorrect model. Theorem 10 in 4.3, which shows such results apply to triangular arrays of observations and not just a sequence of observations, is new. So are Corollaries 9 and 13, which extend these results to cases in which observations include measurement error of known variance. The quantitative formulations of Jeffreys's law in 4.4 and the plausible approximations in 6.8 giving asymptotic frequentist versions of Jeffreys's law are published here for the first time, although some of these ideas appeared in an NSF grant proposal of mine many years ago. Section 6.3, which points out an important error in Matheron (1971), is new, as is 6.7 on the asymptotic behavior of the Fisher information matrix for a periodic version of the Matérn model. Finally, the extensive numerical results in 3.5, 6.6 and 6.8 and the simulated example in 6.9 are new.

This work grew out of notes for a quarter-long graduate class in spatial statistics I have taught sporadically at the University of Chicago. However, this book now covers many more topics than could reasonably be addressed in a quarter or even a semester for all but the most highly prepared students. It would be a mistake not to get to Chapter 6, which has a much greater focus on practical aspects of kriging than the preceding chapters. I would recommend not skipping any sections entirely but instead judicially omitting proofs of some of the more technical results. The proofs in 3.6 and 6.7 depend critically on evenly spaced observations and do not provide much statistical insight; they are good candidates for omission. Other candidates for omission include the proofs of Theorem 1 and Theorems 10–12 in Chapter 4 and all proofs in 5.3 and 5.4. There are exercises at the end of most sections of highly varying difficulty. Many ask the reader to fill in details of proofs. Others consider special cases of more general results or address points not raised in the text. Several ask the reader to do numerical calculations similar to those done in the text. All numerical work reported on here, unless noted otherwise, was done in S-Plus.

There are many people to thank for their help with this work. Terry Speed pointed out the connection between my work and Jeffreys's law (see 4.4) and Wing Wong formulated the Bayesian version of this law described in 4.4. Mark Handcock calculated the predictive densities given in 6.10 using programs reported on in Handcock and Wallis (1994). Numerous people have read parts of the text and provided valuable feedback including Stephen Stigler, Mark Handcock, Jian Zhang, Seongjoo Song, Zhengyuan Zhu, Ji Meng Loh, and several anonymous reviewers. Michael Wichura provided frequent and invaluable advice on using TEX; all figures in this text were produced using his PICTEX macros (Wichura 1987). Mitzi Nakatsuka typed the first draft of much of this work; her expertise and dedication are gratefully acknowledged. Finally, I would like to gratefully acknowledge the support of the National Science Foundation (most recently, through NSF Grant DMS 95-04470) for supporting my research on kriging throughout my research career.

I intend to maintain a Web page containing comments and corrections regarding this book. This page can be reached by clicking on the book's title in my home page `http://galton.uchicago.edu/faculty/stein.html`.

Chicago, Illinois
December 1998

Michael L. Stein

Contents

1
Linear Prediction

1.1 Introduction

This book investigates prediction of a spatially varying quantity based on observations of that quantity at some set of locations. Although the notion of prediction sometimes suggests the assessment of something that has not yet happened, here I take it to mean the assessment of any random quantity that is presently not known exactly. This work focuses on quantities that vary continuously in space and for which observations are made without error, although Sections 3.7, 4.2, 4.3, 6.6 and 6.8 do address some issues regarding measurement errors. Our goals are to obtain accurate predictions and to obtain reasonable assessments of the uncertainty in these predictions. The approach to prediction I take is to consider the spatially varying quantity to be a realization of a real-valued random field, that is, a family of random variables whose index set is $\mathbb{R}^d$.

Much of this work focuses on the properties of predictors that are linear functions of the observations, although 1.4 describes a cautionary example on the potential inefficiencies of "optimal" linear predictors. Section 1.2 defines and derives best linear prediction of random fields based on a finite number of observations. Section 1.3 briefly reviews some properties of Hilbert spaces, which are a powerful tool for studying general linear prediction problems. Section 1.5 considers best linear unbiased prediction, which applies when the mean function of the random field is known up to a vector of linear parameters. Best linear unbiased prediction is frequently used in spatial statistics where it is commonly called universal kriging. Section

1.6 summarizes some basic themes of this work and briefly considers how these themes relate to practical issues in the prediction of random fields. Section 1.7 succinctly states my main recommendations for the practice of predicting random fields. Readers who can only spare 30 seconds on this book might want to skip directly to 1.7.

Chapter 2 provides a detailed discussion of properties of random fields relevant to this work. For now, let us introduce some essential definitions and notation. For a random variable X, I use $E(X)$ to indicate its expected value and $\mathrm{var}(X)$ for its variance. For random variables X and Y, $\mathrm{cov}(X,Y) = E(XY) - E(X)E(Y)$ is the covariance of X and Y. Suppose $\{Z(\mathbf{x}) : \mathbf{x} \in \mathbb{R}^d\}$ is a real-valued random field on $\mathbb{R}^d$ and $\mathbf{x}, \mathbf{y} \in \mathbb{R}^d$. The mean function of Z is $EZ(\mathbf{x})$, which I often denote by $m(\mathbf{x})$. The covariance function is $\mathrm{cov}\{Z(\mathbf{x}), Z(\mathbf{y})\}$, which I often denote by $K(\mathbf{x},\mathbf{y})$. Finally, a random field is Gaussian if all of its finite-dimensional distributions are Gaussian (multivariate normal). See Appendix A for a brief summary of results on multivariate normal distributions.

1.2 Best linear prediction

Suppose we observe a random field Z on $\mathbb{R}^d$ at $\mathbf{x}_1, \ldots, \mathbf{x}_n$ and wish to predict $Z(\mathbf{x}_0)$. I call the quantity to be predicted the predictand. If the law of Z is known, then inference about $Z(\mathbf{x}_0)$ should be based upon the conditional distribution of $Z(\mathbf{x}_0)$ given the observed values of $Z(\mathbf{x}_1), \ldots, Z(\mathbf{x}_n)$. In practice, specifying the law of a random field can be a daunting task. Furthermore, even if we are willing to believe that we know the law of Z, calculating this conditional distribution may be extremely difficult. For these reasons, it is common to restrict attention to linear predictors.

Suppose Z has mean function $m(\mathbf{x})$ and covariance function $K(\mathbf{x},\mathbf{y})$. If m and K are known, then we can obtain the mean and variance of any linear combination of observations of Z. For random vectors $\mathbf{X}$ and $\mathbf{Y}$, define $\mathrm{cov}(\mathbf{X}, \mathbf{Y}^T) = E\{(\mathbf{X} - E\mathbf{X})(\mathbf{Y} - E\mathbf{Y})^T\}$, where the expected value of a random matrix is just the matrix of expected values and $\mathbf{Y}^T$ is the transpose of $\mathbf{Y}$. Suppose we observe $\mathbf{Z} = (Z(\mathbf{x}_1), \ldots, Z(\mathbf{x}_n))^T$ and wish to predict $Z(\mathbf{x}_0)$ using a predictor of the form $\lambda_0 + \boldsymbol{\lambda}^T\mathbf{Z}$. The mean squared error (mse) of this predictor is just the squared mean of the prediction error plus its variance and is given by

$$E\{Z(\mathbf{x}_0) - \lambda_0 - \boldsymbol{\lambda}^T\mathbf{Z}\}^2 = \{m(\mathbf{x}_0) - \lambda_0 - \boldsymbol{\lambda}^T\mathbf{m}\}^2 + k_0 - 2\boldsymbol{\lambda}^T\mathbf{k} + \boldsymbol{\lambda}^T\mathbf{K}\boldsymbol{\lambda},$$

where $\mathbf{m} = E\mathbf{Z}$, $k_0 = K(\mathbf{x}_0, \mathbf{x}_0)$, $\mathbf{k} = \mathrm{cov}\,\{\mathbf{Z}, Z(\mathbf{x}_0)\}$ and $\mathbf{K} = \mathrm{cov}\left(\mathbf{Z}, \mathbf{Z}^T\right)$. It is apparent that for any choice of $\boldsymbol{\lambda}$, we can make the squared mean term 0 by taking $\lambda_0 = m(\mathbf{x}_0) - \boldsymbol{\lambda}^T\mathbf{m}$, so consider choosing $\boldsymbol{\lambda}$ to minimize the variance. For any $\boldsymbol{\lambda}, \boldsymbol{\nu} \in \mathbb{R}^n$,

$$\mathrm{var}\{Z(\mathbf{x}_0) - (\boldsymbol{\lambda} + \boldsymbol{\nu})^T\mathbf{Z}\}$$

$$= k_0 - 2(\boldsymbol{\lambda} + \boldsymbol{\nu})^T \mathbf{k} + (\boldsymbol{\lambda} + \boldsymbol{\nu})^T \mathbf{K}(\boldsymbol{\lambda} + \boldsymbol{\nu})$$
$$= k_0 - 2\boldsymbol{\lambda}^T \mathbf{k} + \boldsymbol{\lambda}^T \mathbf{K}\boldsymbol{\lambda} + \boldsymbol{\nu}^T \mathbf{K}\boldsymbol{\nu} + 2(\mathbf{K}\boldsymbol{\lambda} - \mathbf{k})^T \boldsymbol{\nu}. \quad (1)$$

Let us next show $\mathbf{k}$ is in $C(\mathbf{K})$, the column space of $\mathbf{K}$. Consider $\boldsymbol{\mu} \in \mathbb{R}^n$ such that $\mathrm{var}(\boldsymbol{\mu}^T \mathbf{Z}) = \boldsymbol{\mu}^T \mathbf{K}\boldsymbol{\mu} = 0$. Then $0 = \mathrm{cov}(\boldsymbol{\gamma}^T \mathbf{Z}, \boldsymbol{\mu}^T \mathbf{Z}) = \boldsymbol{\gamma}^T \mathbf{K}\boldsymbol{\mu}$ for all $\boldsymbol{\gamma} \in \mathbb{R}^n$, so $\mathbf{K}\boldsymbol{\mu} = \mathbf{0}$. In addition, $0 = \mathrm{cov}\{Z(\mathbf{x}_0), \boldsymbol{\mu}^T \mathbf{Z}\} = \mathbf{k}^T \boldsymbol{\mu}$, so that $\mathbf{K}\boldsymbol{\mu} = \mathbf{0}$ implies $\mathbf{k}^T \boldsymbol{\mu} = 0$. Thus, $\mathbf{k}$ is orthogonal to the null space of $\mathbf{K}$ and hence $\mathbf{k} \in C(\mathbf{K}^T) = C(\mathbf{K})$, as required. Consequently, there exists $\boldsymbol{\lambda}$ such that $\mathbf{K}\boldsymbol{\lambda} = \mathbf{k}$, and for such $\boldsymbol{\lambda}$,

$$\begin{aligned} \mathrm{var}\{Z(\mathbf{x}_0) - (\boldsymbol{\lambda} + \boldsymbol{\nu})^T \mathbf{Z}\} &= k_0 - 2\boldsymbol{\lambda}^T \mathbf{k} + \boldsymbol{\lambda}^T \mathbf{K}\boldsymbol{\lambda} + \boldsymbol{\nu}^T \mathbf{K}\boldsymbol{\nu} \\ &\geq k_0 - 2\boldsymbol{\lambda}^T \mathbf{k} + \boldsymbol{\lambda}^T \mathbf{K}\boldsymbol{\lambda} \end{aligned}$$

for all $\boldsymbol{\nu}$ since $\boldsymbol{\nu}^T \mathbf{K}\boldsymbol{\nu} = \mathrm{var}(\boldsymbol{\nu}^T \mathbf{Z}) \geq 0$. Thus, since $\boldsymbol{\lambda}^T \mathbf{Z}$ achieves this lower bound, it necessarily minimizes the variance of the prediction error. We call any linear predictor that minimizes the mean squared error among all linear predictors the best linear predictor or BLP. The preceding argument proves that the BLP always exists. Exercise 1 asks you to show it is essentially unique. If $\mathbf{K}$ is invertible, the values of λ_0 and $\boldsymbol{\lambda}$ that give the BLP are

$$\begin{aligned} \boldsymbol{\lambda} &= \mathbf{K}^{-1}\mathbf{k} \qquad \text{and} \\ \lambda_0 &= m(\mathbf{x}_0) - \mathbf{k}^T \mathbf{K}^{-1}\mathbf{m} \end{aligned} \quad (2)$$

and the resulting mse is $k_0 - \mathbf{k}^T \mathbf{K}^{-1}\mathbf{k}$.

If Z is Gaussian, then (see Appendix A) we have the much stronger result that the conditional distribution of $Z(\mathbf{x}_0)$ given $\mathbf{Z} = \mathbf{z}$ is $N\big(\lambda_0 + \boldsymbol{\lambda}^T \mathbf{z}, k_0 - \mathbf{k}^T \mathbf{K}^{-1}\mathbf{k}\big)$, where λ_0 and $\boldsymbol{\lambda}$ are given by (2) and $N(\mu, \sigma^2)$ is the univariate normal distribution with mean μ and variance σ^2. Thus, for a Gaussian random field with known mean and covariance functions, finding the conditional distribution of the process at $\mathbf{x}_0$ is straightforward.

We see that for Gaussian Z, the BLP gives the conditional expectation, so that the BLP is the best predictor (in terms of minimizing mse), linear or nonlinear (Rice 1995, p. 140). Thus, there is a temptation to believe that BLPs work well for processes that are not too far from Gaussian. However, as the example in 1.4 demonstrates, it is important to be careful about what one means by a process being close to Gaussian.

Exercises

1 Show that the BLP is unique in the sense that if $\lambda_0 + \boldsymbol{\lambda}^T \mathbf{Z}$ and $\mu_0 + \boldsymbol{\mu}^T \mathbf{Z}$ are both BLPs for $Z(\mathbf{x}_0)$, then $E(\lambda_0 + \boldsymbol{\lambda}^T \mathbf{Z} - \mu_0 - \boldsymbol{\mu}^T \mathbf{Z})^2 = 0$.

2 Suppose X_0, X_1 and X_2 are random variables with mean 0 and variance 1, $\mathrm{cov}(X_0, X_1) = \mathrm{cov}(X_1, X_2) = \rho$ with $|\rho| \leq 2^{-1/2}$ and $\mathrm{cov}(X_0, X_2) = 0$. Find the BLP of X_0 based on X_1 and X_2. Find the mse of the BLP. Note that unless $\rho = 0$, X_2 plays a role in the BLP despite the fact that it is uncorrelated with X_0. Why is there the restriction $|\rho| \leq 2^{-1/2}$?

3 Suppose $X_0, X_1, \ldots, X_n$ are random variables with mean 0, variance 1 and $\text{cov}(X_i, X_j) = \rho$ for $i \neq j$. Find the BLP of X_0 based on $X_1, \ldots, X_n$. Find the mse of the BLP.

1.3 Hilbert spaces and prediction

A classical problem in stochastic processes is to predict the future of a process based on having observed it up to the present. More specifically, for a process Z on $\mathbb{R}$ with finite second moments, consider finding the BLP of $Z(t)$, $t > 0$ based on observing $Z(s)$ for all $s \leq 0$, so that $s = 0$ is the present time. Wiener (1949) and Kolmogorov (1941) studied this problem for weakly stationary processes. Linear algebra, which worked fine when there were only a finite number of observations, is not an adequate tool in this setting. The right approach is to view the set of possible linear predictors as a Hilbert space. For background material on Hilbert spaces, see, for example, Akhiezer and Glazman (1981), although Section 5.6 of Cramér and Leadbetter (1967) contains pretty much everything you will need to know about Hilbert spaces to read this work.

Very briefly, a Hilbert space is a complete inner product space, or a linear space possessing an inner product and containing all of its limit points under the metric defined by the inner product. A linear space $\mathcal{L}$ is a set of elements $\mathbf{x}, \mathbf{y}, \ldots$ satisfying the conditions:

(a) there is an operation called addition and denoted by $+$ such that $\mathcal{L}$ is an Abelian (commutative) group with respect to addition;

(b) multiplication of elements of $\mathcal{L}$ by (real or complex) scalars $a, b, \ldots$ is defined and satisfies $a(\mathbf{x}+\mathbf{y}) = a\mathbf{x} + a\mathbf{y}$, $(a+b)\mathbf{x} = a\mathbf{x} + b\mathbf{x}$, $a(b\mathbf{x}) = (ab)\mathbf{x}$, $1\mathbf{x} = \mathbf{x}$ and $0\mathbf{x} = \mathbf{0}$, where $\mathbf{0}$ is the zero element of the group.

A linear space is an inner product space if for each $\mathbf{x}, \mathbf{y} \in \mathcal{L}$ there is a (real or complex) number $(\mathbf{x}, \mathbf{y})$ such that

(c) $(\mathbf{x}, \mathbf{y}) = \overline{(\mathbf{y}, \mathbf{x})}$, where, for a complex number z, $\bar{z}$ is its complex conjugate,

(d) $(a\mathbf{x} + b\mathbf{y}, \mathbf{z}) = a(\mathbf{x}, \mathbf{z}) + b(\mathbf{y}, \mathbf{z})$ and

(e) $(\mathbf{x}, \mathbf{x}) \geq 0$ with equality if and only if $\mathbf{x} = \mathbf{0}$.

We say $\mathbf{x}$ is orthogonal to $\mathbf{y}$, written $\mathbf{x} \perp \mathbf{y}$, if $(\mathbf{x}, \mathbf{y}) = 0$.

For any $\mathbf{x} \in \mathcal{L}$, define its norm, written $\|\mathbf{x}\|$, by the positive square root $(\mathbf{x}, \mathbf{x})^{1/2}$. If we define $\|\mathbf{x} - \mathbf{y}\|$ as the distance between $\mathbf{x}$ and $\mathbf{y}$, the inner product space is a metric space. The inner product space is complete and hence a Hilbert space if for any sequence $\mathbf{x}_1, \mathbf{x}_2, \ldots$ such that $\lim_{m,n\to\infty} \|\mathbf{x}_m - \mathbf{x}_n\| = 0$ there exists $\mathbf{x} \in \mathcal{L}$ such that $\lim_{n\to\infty} \|\mathbf{x}_n - \mathbf{x}\| = 0$. The Hilbert space is called separable if it has a countable dense subset.

We are mostly concerned with Hilbert spaces for which scalar multiplication is restricted to reals and the inner product is real.

For any subset $\mathcal{X}$ of a Hilbert space $\mathcal{H}$, the linear manifold spanned by $\mathcal{X}$, denoted by $\mathcal{M}_0$, is the set of all linear combinations $a_1\mathbf{x}_1 + \cdots a_n\mathbf{x}_n$ with n finite and $\mathbf{x}_1, \ldots, \mathbf{x}_n \in \mathcal{X}$. The closed linear manifold spanned by $\mathcal{X}$, denoted by $\mathcal{M}$, is just $\mathcal{M}_0$ together with its limit points under the metric defined by the inner product. Note that $\mathcal{M}$ is itself necessarily a Hilbert space. Any set whose closed linear manifold is $\mathcal{M}$ is called a basis for $\mathcal{M}$, so that $\mathcal{X}$ is automatically one basis for $\mathcal{M}$. In this work, we generally only consider Hilbert spaces with finite or countable bases. Every separable Hilbert space possesses a finite or countable basis (Exercise 4).

For studying prediction, the crucial concept is that of projection of an element of a Hilbert space onto a subspace. Suppose $\mathcal{H}$ is a Hilbert space and $\mathcal{G}$ a subspace. Given $h \in \mathcal{H}$, there exists a unique element $g \in \mathcal{G}$ such that

$$\|h - g\| = \inf_{g' \in \mathcal{G}} \|h - g'\| \tag{3}$$

(Exercise 5). We call g the projection of h onto $\mathcal{G}$. An important property of the projection g is that it is the unique element in $\mathcal{G}$ satisfying $h - g \perp g'$ for all $g' \in \mathcal{G}$ (Exercise 6). That is, the error of approximation is orthogonal to all elements of $\mathcal{G}$.

The Hilbert spaces we encounter most frequently in this work are those generated by a random field Z on some set R. More specifically, consider a random field Z on a set $R \subset \mathbb{R}^d$ with mean function m and covariance function K. Let $\mathcal{H}_R^0$ be the real linear manifold of $\{Z(\mathbf{x}) : \mathbf{x} \in R\}$ for some $R \subset \mathbb{R}^d$. For g and h in $\mathcal{H}_R^0$, define the inner product $(g, h) = E(gh)$. The closure of $\mathcal{H}_R^0$ with respect to this inner product is a Hilbert space, which I denote by $\mathcal{H}_R(m, K)$.

To characterize the BLP in terms of such a Hilbert space, we need to make sure the constant term is in the space of possible predictors. Specifically, letting Q be the set on which Z is observed, all linear predictors of $h \in \mathcal{H}_R(m, K)$ are of the form $c + g$, where c is a scalar and $g \in \mathcal{H}_Q(m, K)$. Let $g(h)$ be the unique element in $\mathcal{H}_Q(m, K)$ satisfying $\operatorname{cov}\{h - g(h), g'\} = 0$ for all $g' \in \mathcal{H}_Q(m, K)$ (see Exercises 5 and 6) and set $c(h) = Eh - Eg(h)$. Then $c(h) + g(h)$ is the BLP of h, which follows from $E[\{h - c(h) - g(h)\}(c' + g')] = 0$ for all real c' and all $g' \in \mathcal{H}_Q(m, K)$ (Exercise 7). We use this characterization in the next section to verify that a particular linear predictor is the BLP.

Exercises

4 Show that every separable Hilbert space has a finite or countable basis.

5 For a Hilbert space $\mathcal{H}$, a subspace $\mathcal{G}$ and $h \in \mathcal{H}$, show that there is a unique element $g \in \mathcal{G}$ such that (3) holds.

6 (Continuation of 5). Show that g is the unique element in $\mathcal{G}$ satisfying $h - g \perp g'$ for all $g' \in \mathcal{G}$.

7 Using the definitions in the last paragraph of this section, show that $E[\{h - c(h) - g(h)\}(c' + g')] = 0$ for all real c' and all $g' \in \mathcal{H}_Q(m, K)$. Use this in conjunction with Exercise 3 to verify that $c(h) + g(h)$ is the BLP of h.

1.4 An example of a poor BLP

Although much of this work focuses on linear prediction, it is important to keep in mind that "best" linear predictors can sometimes be highly inefficient compared to the best nonlinear predictors when the underlying random field is not Gaussian. This section presents an example showing that a stochastic process can in one sense be nearly Gaussian and yet a BLP performs infinitely worse than the best nonlinear predictor under this model. Thus, in reading Chapters 3–5, where we study properties of linear predictors in some depth, keep in mind that these results are largely irrelevant for some non-Gaussian random fields.

Suppose N is a Poisson process with constant intensity λ on $\mathbb{R}$, so that for a Borel set A, $N(A)$ is the number of events of the process in A and $E\{N(A)\}$ is λ times the Lebesgue measure of A. Define $Z(t) = N((t-1, t+1])$. Then $m(t) = 2\lambda$ and $K(s,t) = (2 - |s-t|)^+\lambda$, where t^+ means the positive part of t (Exercise 8). Observe Z on $R = [-2,-1] \cup [1,2]$ and consider predicting $Z(0)$. A partial realization of N, where the $\times$s represent the locations of events of N, and the corresponding values of Z on R is given in Figure 1. It is possible to show that with probability 1,

$$Z(0) = \{\#\text{ positive jumps of } Z \text{ on } [-2,-1]\} \\ + \{\#\text{ negative jumps of } Z \text{ on } [1,2]\}, \tag{4}$$

so that the mse of the best predictor is 0 (Exercise 9). For the realization shown in Figure 1, $Z(0) = 3$ and we see there is 1 positive jump on $[-2,-1]$ and 2 negative jumps on $[1,2]$. This optimal predictor of $Z(0)$ is decidedly nonlinear. On the other hand, the BLP of $Z(0)$ is

$$\tfrac{2}{3}\lambda + \tfrac{2}{3}\{Z(1) + Z(-1)\} - \tfrac{1}{3}\{Z(2) + Z(-2)\}, \tag{5}$$

which follows by showing that the error of the BLP has mean 0 and is uncorrelated with $Z(t)$ for all $t \in R$ (Exercise 10). The mse of the BLP is $\frac{2}{3}\lambda$, so the ratio of the mse of the BLP to the mse of the best nonlinear predictor is infinite for all λ. This is despite the fact that as $\lambda \to \infty$, $\{Z(\cdot) - 2\lambda\}/\lambda^{1/2}$ converges weakly (Billingsley 1968) to a Gaussian process!

Exercises

8 Show that for Z as defined in this section, $K(s,t) = (2 - |s-t|)^+\lambda$.

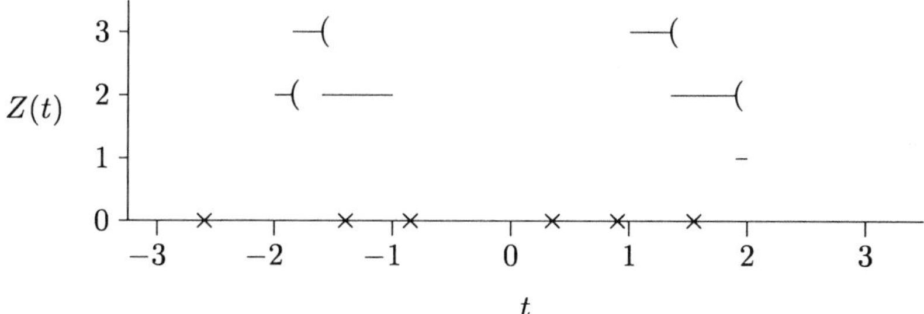

FIGURE 1. A partial realization of the process Z described in 1.4. The $\times$s on the horizontal axis indicate events of the Poisson process N. Values for $Z(t)$ are plotted for $t \in R = [-2,-1] \cup [1,2]$.

9 Verify that (4) holds with probability 1.

10 Verify that (5) gives the BLP of $Z(0)$ by using the characterization for the BLP in the last paragraph of 1.3.

1.5 Best linear unbiased prediction

Suppose we have the following model for a random field Z,

$$Z(\mathbf{x}) = \mathbf{m}(\mathbf{x})^T\boldsymbol{\beta} + \varepsilon(\mathbf{x}), \tag{6}$$

where ε is a mean 0 random field with known covariance structure, $\mathbf{m}$ is a known function with values in $\mathbb{R}^p$ and $\boldsymbol{\beta}$ is a vector of p unknown coefficients. We observe $\mathbf{Z} = (Z(\mathbf{x}_1), \ldots, Z(\mathbf{x}_n))^T$ and wish to predict $Z(\mathbf{x}_0)$. If $\boldsymbol{\beta}$ were known, we could use the BLP

$$\mathbf{m}(\mathbf{x}_0)^T\boldsymbol{\beta} + \mathbf{k}^T\mathbf{K}^{-1}(\mathbf{Z} - \mathbf{M}\boldsymbol{\beta}), \tag{7}$$

where $\mathbf{M} = (\mathbf{m}(\mathbf{x}_1)\cdots\mathbf{m}(\mathbf{x}_n))^T$ and $\mathbf{K}$ and $\mathbf{k}$ are defined as in 1.2. If $\boldsymbol{\beta}$ is unknown but all covariances are known, a natural approach is to replace $\boldsymbol{\beta}$ in (7) by the generalized least squares estimator $\hat{\boldsymbol{\beta}} = (\mathbf{M}^T\mathbf{K}^{-1}\mathbf{M})^{-1}\mathbf{M}^T\mathbf{K}^{-1}\mathbf{Z}$, assuming $\mathbf{K}$ is nonsingular and $\mathbf{M}$ is of full rank. The estimator $\hat{\boldsymbol{\beta}}$ is best linear unbiased for $\boldsymbol{\beta}$ (see Exercise 11).

An alternative approach is to minimize the mse of prediction among all predictors of the form $\lambda_0 + \boldsymbol{\lambda}^T\mathbf{Z}$ subject to the unbiasedness constraint $E(\lambda_0 + \boldsymbol{\lambda}^T\mathbf{Z}) = EZ(\mathbf{x}_0)$ for all $\boldsymbol{\beta}$. The unbiasedness constraint is identical to $\lambda_0 + \boldsymbol{\lambda}^T\mathbf{M}\boldsymbol{\beta} = \mathbf{m}(\mathbf{x}_0)^T\boldsymbol{\beta}$ for all $\boldsymbol{\beta}$, or

$$\lambda_0 = 0 \quad \text{and} \quad \mathbf{M}^T\boldsymbol{\lambda} = \mathbf{m}(\mathbf{x}_0). \tag{8}$$

Our goal then is to minimize $E\{Z(\mathbf{x}_0) - \boldsymbol{\lambda}^T\mathbf{Z}\}^2$ subject to $\boldsymbol{\lambda}$ satisfying (8). If $\boldsymbol{\lambda}$ solves this constrained minimization problem, then $\boldsymbol{\lambda}^T\mathbf{Z}$ is called a best linear unbiased predictor (BLUP) for $Z(\mathbf{x}_0)$. To solve this problem, first note that there exists a LUP (linear unbiased predictor) if and only if

$\mathbf{m}(\mathbf{x}_0) \in C(\mathbf{M}^T)$, so let us suppose so from now on. If $\boldsymbol{\lambda}$ satisfies $\mathbf{M}^T\boldsymbol{\lambda} = \mathbf{m}(\mathbf{x}_0)$, then any LUP can be written as $(\boldsymbol{\lambda} + \boldsymbol{\nu})^T\mathbf{Z}$ where $\mathbf{M}^T\boldsymbol{\nu} = \mathbf{0}$. Considering (1) and (8), $\boldsymbol{\lambda}^T\mathbf{Z}$ is a BLUP if $(\mathbf{K}\boldsymbol{\lambda} - \mathbf{k})^T\boldsymbol{\nu} = 0$ for all $\boldsymbol{\nu}$ satisfying $\mathbf{M}^T\boldsymbol{\nu} = \mathbf{0}$, or equivalently, if there exists a vector $\boldsymbol{\mu}$ such that $\mathbf{K}\boldsymbol{\lambda} - \mathbf{k} = \mathbf{M}\boldsymbol{\mu}$. Thus, $\boldsymbol{\lambda}^T\mathbf{Z}$ is a BLUP if

$$\begin{pmatrix} \mathbf{K} & \mathbf{M} \\ \mathbf{M}^T & \mathbf{O} \end{pmatrix} \begin{pmatrix} \boldsymbol{\lambda} \\ \boldsymbol{\mu} \end{pmatrix} = \begin{pmatrix} \mathbf{k} \\ \mathbf{m}(\mathbf{x}_0) \end{pmatrix},$$

for some $\boldsymbol{\mu}$, where $\mathbf{O}$ is a matrix of zeroes. This set of linear equations has a solution if and only if $\mathbf{m}(\mathbf{x}_0) \in C(\mathbf{M}^T)$ (Exercise 12). If $\mathbf{K}$ and $\mathbf{M}$ are of full rank, then

$$\begin{pmatrix} \boldsymbol{\lambda} \\ \boldsymbol{\mu} \end{pmatrix} = \begin{pmatrix} \mathbf{K} & \mathbf{M} \\ \mathbf{M}^T & \mathbf{O} \end{pmatrix}^{-1} \begin{pmatrix} \mathbf{k} \\ \mathbf{m}(\mathbf{x}_0) \end{pmatrix}. \tag{9}$$

From (9) it can then be shown that (Exercise 13)

$$\begin{aligned} \boldsymbol{\lambda} =& \{\mathbf{K}^{-1} - \mathbf{K}^{-1}\mathbf{M}(\mathbf{M}^T\mathbf{K}^{-1}\mathbf{M})^{-1}\mathbf{M}^T\mathbf{K}^{-1}\}\mathbf{k} \\ &+ \mathbf{K}^{-1}\mathbf{M}(\mathbf{M}^T\mathbf{K}^{-1}\mathbf{M})^{-1}\mathbf{m}(\mathbf{x}_0), \end{aligned} \tag{10}$$

so that the resulting predictor is

$$\boldsymbol{\lambda}^T\mathbf{Z} = \mathbf{k}^T\mathbf{K}^{-1}(\mathbf{Z} - \mathbf{M}\hat{\boldsymbol{\beta}}) + \mathbf{m}(\mathbf{x}_0)^T\hat{\boldsymbol{\beta}},$$

which is identical to what we obtained by replacing $\boldsymbol{\beta}$ in the BLP by the generalized least squares estimator $\hat{\boldsymbol{\beta}}$. The mse of the BLUP is

$$k_0 - \mathbf{k}^T\mathbf{K}^{-1}\mathbf{k} + \boldsymbol{\gamma}^T(\mathbf{M}^T\mathbf{K}^{-1}\mathbf{M})^{-1}\boldsymbol{\gamma}, \tag{11}$$

where $\boldsymbol{\gamma} = \mathbf{m}(\mathbf{x}_0) - \mathbf{M}^T\mathbf{K}^{-1}\mathbf{k}$ and $k_0 = \mathbf{K}(\mathbf{x}_0, \mathbf{x}_0)$ as in 1.2 (Exercise 14).

Best linear unbiased prediction is called kriging in the geostatistical literature, named after the South African mining engineer D. G. Krige (Krige 1951; Journel and Huijbregts 1978). If $\mathbf{m}(\mathbf{x}) \equiv 1$, so that the mean of the process is assumed to be an unknown constant, then best linear unbiased prediction is called ordinary kriging. Best linear unbiased prediction for more general $\mathbf{m}$ is known as universal kriging and best linear prediction with the mean assumed 0 is called simple kriging. Simple kriging is generally called objective analysis in the atmospheric sciences (Thiébaux and Pedder 1987 and Daley 1991, Chapter 4). Goldberger (1962) described best linear unbiased prediction for regression models with correlated errors but did not explicitly consider the spatial setting. Cressie (1989, 1990) provides further discussion on the history of various forms of kriging.

As noted in 1.3, A useful characterization of the BLP is that its error is orthogonal (uncorrelated) to all possible linear predictions. The BLUP has a similar characterization, which is implicit in the derivation of (10). Suppose a random field Z is of the form given in (6). The random variable $\sum_{j=1}^{\ell} \alpha_j Z(\mathbf{y}_j)$ is called a contrast if it has mean 0 for all $\boldsymbol{\beta}$, or equivalently, if $\sum_{j=1}^{\ell} \alpha_j \mathbf{m}(\mathbf{y}_j) = \mathbf{0}$. A BLUP of $Z(\mathbf{x}_0)$ based on some set of observations

$Z(\mathbf{x}_1), \ldots, Z(\mathbf{x}_n)$ is characterized by the following two properties: its error is a contrast and its error is orthogonal to any contrast of the observations (Exercise 15). It follows that to find BLUPs and to evaluate their mses, it is only necessary to know the covariance structure of all contrasts of the random field. This property is of value when using intrinsic random functions (see 2.9), which is a class of models for which variances of linear combinations that are not contrasts are undefined.

The BLUP has a Bayesian interpretation (Kitanidis 1986 and Omre 1987). Let $N(\boldsymbol{\mu}, \boldsymbol{\Sigma})$ be the multivariate normal distribution with mean vector $\boldsymbol{\mu}$ and covariance matrix $\boldsymbol{\Sigma}$. Suppose Z is given by (6), where the random field ε is Gaussian and independent of the random vector $\boldsymbol{\beta}$ which has prior distribution $N(\boldsymbol{\mu}, \sigma^2\mathbf{V})$ for some positive definite $\mathbf{V}$. Define $\mathbf{W}(\sigma^2) = (\mathbf{M}^T\mathbf{K}^{-1}\mathbf{M} + \sigma^{-2}\mathbf{V}^{-1})^{-1}$. Then the posterior distribution of $\boldsymbol{\beta}$ given $\mathbf{Z}$ is (Exercise 16)

$$\boldsymbol{\beta} \mid \mathbf{Z} \sim N\left(\mathbf{W}(\sigma^2)(\mathbf{M}^T\mathbf{K}^{-1}\mathbf{Z} + \sigma^{-2}\mathbf{V}^{-1}), \mathbf{W}(\sigma^2)\right). \tag{12}$$

For making predictions about $Z(\mathbf{x}_0)$, the natural Bayesian solution is to use the conditional distribution of $Z(\mathbf{x}_0)$ given $\mathbf{Z}$ but averaging over the posterior of $\boldsymbol{\beta}$ given $\mathbf{Z}$. This distribution is known as the predictive distribution of $Z(\mathbf{x}_0)$ (Zellner 1971, Section 2.8) and is given by (Exercise 17)

$$Z(\mathbf{x}_0) \mid \mathbf{Z} \sim N\Big(\mathbf{k}^T\mathbf{K}^{-1}\mathbf{Z} + \boldsymbol{\gamma}^T\mathbf{W}(\sigma^2)\left(\mathbf{M}^T\mathbf{K}^{-1}\mathbf{Z} + \sigma^{-2}\mathbf{V}^{-1}\boldsymbol{\mu}\right), \\ k_0 - \mathbf{k}^T\mathbf{K}^{-1}\mathbf{k} + \boldsymbol{\gamma}^T\mathbf{W}(\sigma^2)\boldsymbol{\gamma}\Big). \tag{13}$$

Letting σ^2 grow means letting the prior on $\boldsymbol{\beta}$ get increasingly uninformative, and as $\sigma^2 \to \infty$ (assuming $\mathbf{M}$ is of full rank), the limiting predictive distribution of $Z(\mathbf{x}_0)$ given $\mathbf{Z}$ is Gaussian with the BLUP as its conditional expectation and conditional variance given by (11).

Exercises

11 Show that if $\hat{\boldsymbol{\beta}}$ is the generalized least squares estimator for $\boldsymbol{\beta}$, then for any fixed vector $\mathbf{q} \in \mathbb{R}^p$, $\mathbf{q}^T\hat{\boldsymbol{\beta}}$ is the BLUP for $\mathbf{q}^T\boldsymbol{\beta}$. Since the quantity being predicted here is not random, $\mathbf{q}^T\hat{\boldsymbol{\beta}}$ is more commonly called the best linear unbiased estimator. Thus, we have that best linear unbiased estimation is just a special case of best linear unbiased prediction.

12 Show that if a LUP exists, then the BLUP exists and is unique in the sense that the BLP was shown to be unique in Exercise 1.

13 If $\mathbf{K}$ and $\mathbf{M}$ are of full rank, verify that (9) implies (10).

14 Show that (11) gives the mse of the BLUP.

15 Show that a BLUP based on some set of observations is characterized by the following two properties: its error is a contrast and its error is orthogonal to any contrast of the observations.

16 Verify (12).

17 Verify (13).

18 Suppose as in Exercise 2 that X_0, X_1, X_2 are random variables with variance 1, $\operatorname{cov}(X_0, X_1) = \operatorname{cov}(X_1, X_2) = \rho$ for $|\rho| \leq 2^{-1/2}$ and $\operatorname{cov}(X_0, X_2) = 0$ but now assume that all three random variables have an unknown common mean. Find the BLUP of X_0 based on X_1 and X_2. Find the mse of the BLUP. Compare your results to those for Exercise 2.

19 Suppose as in Exercise 3 that $X_0, X_1, \ldots, X_n$ are random variables with variance 1 and $\operatorname{cov}(X_i, X_j) = \rho$ for $i \neq j$. Find the BLUP of X_0 based on $X_1, \ldots, X_n$ if all X_is have a common unknown mean. Find the mse of this BLUP. Find the BLUP of X_0 based on $X_1, \ldots, X_n$ and its mse if $EX_i = \beta i$ for some unknown constant β. Compare your results with those of Exercise 3.

1.6 Some recurring themes

There are four recurring and interrelated themes that underlie my approach to problems in spatial prediction. In order to provide the reader with some guidance as to what is most important in the upcoming chapters, it is worthwhile to spell out these themes here. I make a number of statements without justification in the present section and I hope that the reader who questions these statements will be thereby motivated to continue on to the rest of the work.

The first of these themes is the contrast between interpolation and extrapolation. Although these words do not have a sharp distinction in the spatial setting, by interpolation I mean predictions at locations that are "surrounded" by available observations or, alternatively, are not near or beyond the boundaries of the region in which there are observations. By extrapolation, I mean predictions at locations beyond the boundaries of the observation region. My main goal in this work is to develop a mathematical framework that is most appropriate for studying interpolation problems. In most problems in which spatial prediction is contemplated, interpolation will be of greater interest than extrapolation, since one would generally take observations in any region in which prediction were to be done unless there were some physical impediment to doing so. To the extent that someone is interested in extrapolation, which is generally the case in time series analysis, the results and approach taken in this work are decidedly less relevant. Sections 3.4–3.6, 6.8 and 6.9 provide some comparisons between interpolation and extrapolation.

This focus on interpolation leads to the second theme, which is that the properties of interpolation schemes depend mainly on the local behavior of the random field under study. In particular, 3.4–3.6 provide theoretical and numerical evidence that the behavior of a random field over longer distances is much less relevant when interpolating than when extrapolating. Accordingly, Chapter 2, which provides background material on second-order properties of random fields, emphasizes their local behavior. Chapter 2 focuses on random fields on $\mathbb{R}^d$ with covariance functions $K(\mathbf{x}, \mathbf{y})$ depending only on $\mathbf{x} - \mathbf{y}$, in which case, I call the function $K(\mathbf{x} - \mathbf{y}) = K(\mathbf{x}, \mathbf{y})$ the autocovariance function of the random field. If the autocovariance function is continuous, then it can be written as the Fourier transform of a positive finite measure. In most cases of practical interest, this measure has a density with respect to Lebesgue measure known as the spectral density. More specifically, the spectral density f satisfies

$$K(\mathbf{x}) = \int_{\mathbb{R}^d} \exp(i\boldsymbol{\omega}^T \mathbf{x}) f(\boldsymbol{\omega}) d\boldsymbol{\omega}$$

for all $\mathbf{x} \in \mathbb{R}^d$. It turns out that the local behavior of a random field is intimately related to how the spectral density f behaves for large values of $|\boldsymbol{\omega}|$. Generally speaking, the more quickly the spectral density decreases as $|\boldsymbol{\omega}|$ increases, the smoother the random field.

As in many areas of statistics, it is not possible to make much progress on the theory of spatial interpolation from finite sample results. Thus, much of the theory in the following chapters is asymptotic. The third theme of this work is that the most appropriate asymptotic framework for problems of spatial interpolation is to consider taking more and more observations in a fixed region, which I call fixed-domain asymptotics. Most existing asymptotic theory concerning inference for stochastic processes and random fields based on discrete observations allows the observation region to grow with the number of observations, which I call increasing-domain asymptotics. Chapter 3, and 3.3 in particular, detail my arguments for preferring fixed-domain asymptotics for studying spatial interpolation. For now, I would point out that if the goal is to develop a theory that shows the relationship between the local behavior of a random field and the properties of interpolation methods, then the fixed-domain approach is quite natural in that the degree of differentiability of the random field, which is a fundamental aspect of its local behavior, plays a central role in any fixed-domain asymptotic results.

The final theme is the connection between what aspects of a random field model are important for purposes of spatial interpolation and what aspects of the model can be well estimated from available data. This issue is particularly crucial when using fixed-domain asymptotics because there will commonly be parameters of models that cannot be consistently estimated as the number of observations in a fixed region increases. However, at least for Gaussian random fields, results in 4.3 demonstrate that models that cannot be distinguished well based on observations in a fixed region yield asymptotically indistiguishable spatial interpolants. These results are an example of what Dawid (1984, p. 285) calls Jeffreys's law: "things we shall never find much out about cannot be very important for prediction." Sections 4.4 and 6.8 provide some quantitative formulations of Jeffreys's law.

The following subsections describe a few implications for the practice of spatial statistics that arise from the consideration of these themes, with a focus on those implications that suggest problems with current common practices and conceptions.

The Matérn model

The second theme states that properties of spatial interpolants depend strongly on the local behavior of the random field. In practice, this local behavior is not known and must be estimated from the same data that will be used to do the interpolation. This state of affairs strongly suggests that it is critical to select models for the covariance structures that include at least one member whose local behavior accurately reflects the actual local behavior of the spatially varying quantity under study. A number of commonly used models for the covariance structure, including the spherical (see 2.10), the exponential and the Gaussian (see 2.7) provide no flexibility with regard to this local behavior and essentially assume it is known a priori. An alternative model that I recommend for general adoption is the Matérn model (see 2.7, 2.10 and 6.5). This model includes a parameter that allows for any degree of differentiability for the random field and includes the exponential model as a special case and the Gaussian model as a limiting case.

BLPs and BLUPs

Best linear unbiased prediction provides an elegant and satisfying solution to the problem of linear prediction when the mean function of the random field is of the form $\mathbf{m}(\mathbf{x})^T\boldsymbol{\beta}$ with $\boldsymbol{\beta}$ unknown. However, when best linear unbiased prediction is used in practice, the components of $\mathbf{m}$ are quite commonly highly regular functions such as monomials and have little impact on the local behavior of the random field. Thus, considering our second theme, it should also be the case that such highly smooth mean functions have little impact on spatial interpolation. In fact, under fixed-domain asymptotics, BLUPs generally do as well asymptotically as BLPs (that is, assuming $\boldsymbol{\beta}$ is known), but one also does as well asymptotically by just setting $\boldsymbol{\beta} = \mathbf{0}$ (see 4.3). It seems to me that many texts in spatial statistics and geostatistics place too great an emphasis on modeling mean functions and on BLUPs, perhaps because it distracts attention from the more important but less well understood problem of modeling and estimation of the local behavior of random fields. In particular, intrinsic random function models (see 2.9), although of some mathematical interest, are not a helpful generalization of stationary random fields for spatial interpolation problems because the mean functions are just polynomials in the coordinates and the local behavior of these models is indistinguishable from the local behavior of stationary models.

In arguing for less emphasis on modeling the mean function when the goal is spatial interpolation, it is important to exclude mean functions that do have a strong effect on the local behavior of a random field. As an example, when interpolating monthly average surface temperatures in a region based on scattered observations, one might use altitude as a component of $\mathbf{m}$. In a mountainous region, variations in altitude may largely explain local variations in average temperatures and hence, including altitude as a component of $\mathbf{m}$ may have a profound effect on the spatial interpolation of average temperatures.

Inference for differentiable random fields

The most commonly used geostatistical tool for making inferences about spatial covariance functions is the empirical semivariogram. Specifically, for a random field Z observed at locations $\mathbf{x}_1, \ldots, \mathbf{x}_n$, the empirical semivariogram at a distance h is the average of $\frac{1}{2}\{Z(\mathbf{x}_i) - Z(\mathbf{x}_j)\}^2$ over pairs of points $(\mathbf{x}_i, \mathbf{x}_j)$ that are very nearly a distance of h apart. Although the empirical semivariogram can be a useful tool for random fields that are not differentiable, it is much less useful and can even be seriously misleading for differentiable random fields. Indeed, Matheron (1971, 1989) states that "statistical inference is impossible" for differentiable random fields. Section 6.2 explains what he means by this statement and shows why it is incorrect. At the heart of the problem is his unstated and erroneous presumption that the empirical semivariogram contains all possible information about the local behavior of a random field. Once one is willing to consider methods for estimating spatial covariance structures that are not based on the empirical semivariogram, inference for differentiable random fields is just as possible as it is for nondifferentiable ones. In particular, in Sections 6.4, 6.9 and 6.10, I advocate the use of likelihood-based or Bayesian methods for estimating the parameters of a random field. These methods are just as appropriate for differentiable as for nondifferentiable random fields.

Nested models are not tenable

It is fairly common practice in the geostatistical literature to model covariance structures as linear combinations of spherical semivariogram functions with different ranges (see 2.7 for definitions). See Journel and Huijbregts

(1978, p. 167), Wackernagel (1995, p. 95) and Goovaerts (1997, p. 159) for examples where such models are advocated or employed. However, because all spherical semivariograms correspond to random fields with the same local behavior, there is little to be gained for purposes of spatial interpolation in employing such models. Furthermore, there is little hope of estimating the parameters of such models with any degree of accuracy for datasets of the size that generally occur in geological applications. I believe such models would not be employed if users had a proper appreciation of the inherent uncertainties in empirical semivariograms.

1.7 Summary of practical suggestions

Use the Matérn model. Calculate and plot likelihood functions for unknown parameters of models for covariance structures. Do not put too much faith in empirical semivariograms.

2

Properties of Random Fields

2.1 Preliminaries

This chapter provides the necessary background on random fields for understanding the subsequent chapters on prediction and inference for random fields. The focus here is on weakly stationary random fields (defined later in this section) and the associated spectral theory. Some previous exposure to Fourier methods is assumed. A knowledge of the theory of characteristic functions at the level of a graduate course in probability (see, for example, Billingsley (1995), Chung (1974), or Feller (1971)) should, for the most part, suffice. When interpolating a random field, the local behavior of the random field turn out to be critical (see Chapter 3). Accordingly, this chapter goes into considerable detail about the local behavior of random fields and its relationship to spectral theory.

For a real random field Z on R with $E\{Z(\mathbf{x})^2\} < \infty$ for all $\mathbf{x} \in R$, the covariance function $K(\mathbf{x}, \mathbf{y}) = \operatorname{cov}\{Z(\mathbf{x}), Z(\mathbf{y})\}$ must satisfy

$$\sum_{j,k=1}^{n} c_j c_k K(\mathbf{x}_j, \mathbf{x}_k) \geq 0 \tag{1}$$

for all finite n, all $\mathbf{x}_1, \ldots, \mathbf{x}_n \in R$ and all real $c_1, \ldots, c_n$, which follows by noting

$$\operatorname{var}\left\{\sum_{j=1}^{n} c_j Z(\mathbf{x}_j)\right\} = \sum_{j,k=1}^{n} c_j c_k K(\mathbf{x}_j, \mathbf{x}_k).$$

A function K satisfying (1) is said to be positive definite on $R \times R$. We have just demonstrated that (1) is necessary for there to exist a random field with covariance function K and mean function $EZ(\mathbf{x}) = m(\mathbf{x})$. It is also sufficient, which follows by showing that for K positive definite, there exists a Gaussian random field with this covariance function and mean function m. Specifically, take the joint distribution of $(Z(\mathbf{x}_1), \ldots, Z(\mathbf{x}_n))$ to be multivariate normal with mean $(m(\mathbf{x}_1), \ldots, m(\mathbf{x}_n))$ and covariance matrix with jkth element $K(\mathbf{x}_j, \mathbf{x}_k)$. This family of finite-dimensional distributions satisfies the consistency conditions of Kolmogorov's theorem (Billingsley 1995, Section 36) and hence there exists a random field (Gaussian by definition) with these finite-dimensional distributions (Gihman and Skorohod 1974, p. 147).

Stationarity

If we do not make any assumptions restricting the class of random fields we wish to consider, making inferences about its probability law from observing a single realization of the field is hopeless. A common simplifying assumption is that the probabilistic structure in some sense looks similar in different parts of R. Supposing $R = \mathbb{R}^d$ for instance, one way to define this concept is through strict stationarity: for all finite n, $\mathbf{x}_1, \ldots, \mathbf{x}_n \in \mathbb{R}^d$, $t_1, \ldots, t_n \in \mathbb{R}$ and $\mathbf{x} \in \mathbb{R}^d$,

$$\begin{aligned} &\Pr\{Z(\mathbf{x}_1 + \mathbf{x}) \le t_1, \ldots, Z(\mathbf{x}_n + \mathbf{x}) \le t_n\} \\ &\qquad = \Pr\{Z(\mathbf{x}_1) \le t_1, \ldots, Z(\mathbf{x}_n) \le t_n\}. \end{aligned}$$

A different type of stationarity is defined in terms of the first two moments of Z. Suppose the covariance function of Z depends on $\mathbf{x}$ and $\mathbf{y}$ only through $\mathbf{x} - \mathbf{y}$. Then there is a function K on $\mathbb{R}^d$, which I call the autocovariance function for Z, such that $\text{cov}\{Z(\mathbf{x}), Z(\mathbf{y})\} = K(\mathbf{x} - \mathbf{y})$ for all $\mathbf{x}$ and $\mathbf{y}$ in $\mathbb{R}^d$. A random field is called weakly stationary if it has finite second moments, its mean function is constant and it possesses an autocovariance function. Note that a strictly stationary random field with finite second moment is also weakly stationary. For describing strength of associations between random variables it is more natural to consider correlations than covariances, so we sometimes make use of the autocorrelation function of a weakly stationary random field, defined as $C(\mathbf{x}) = K(\mathbf{x})/K(\mathbf{0})$ assuming $K(\mathbf{0}) > 0$.

Since weak stationarity is a less restrictive assumption than strict stationarity whenever the second moments are finite, it is tempting to claim in practice that one is only assuming weak stationarity and then make inferences that only depend on specifying the first two moments of the random field. This temptation is perhaps encouraged by results in discrete time series showing that certain asymptotic properties of the periodogram (the squared modulus of the discrete Fourier transform of the observations) do not depend on the time series being Gaussian (Priestley 1981, Section 6.2).

However, as the example in Section 1.4 demonstrates, considering only the first two moments can lead to infinitely suboptimal predictions. A further example illustrating problems that can occur by just considering the first two moments of a random field is given in 2.2.

Isotropy

Stationarity can be thought of as an invariance property under the translation group of transformations of the coordinates. For a random field on $\mathbb{R}^d$, we can also consider invariance under rotations and reflections. I call a random field Z on $\mathbb{R}^d$ strictly isotropic if its finite-dimensional joint distributions are invariant under all rigid motions. That is, for any orthogonal $d \times d$ matrix $\mathbf{H}$ and any $\mathbf{x} \in \mathbb{R}^d$,

$$\begin{aligned}\Pr\{Z(\mathbf{H}\mathbf{x}_1 + \mathbf{x}) \le t_1, \ldots, Z(\mathbf{H}\mathbf{x}_n + \mathbf{x}) \le t_n\} \\ = \Pr\{Z(\mathbf{x}_1) \le t_1, \ldots, Z(\mathbf{x}_n) \le t_n\}\end{aligned}$$

for all finite n, $\mathbf{x}_1, \ldots, \mathbf{x}_n \in \mathbb{R}^d$ and $t_1, \ldots, t_n \in \mathbb{R}$. A random field on $\mathbb{R}^d$ is weakly isotropic if there exists a constant m and a function K on $[0, \infty)$ such that $m(\mathbf{x}) = m$ and $\operatorname{cov}\{Z(\mathbf{x}), Z(\mathbf{y})\} = K(|\mathbf{x} - \mathbf{y}|)$ for all $\mathbf{x}, \mathbf{y} \in \mathbb{R}^d$, where $|\cdot|$ indicates Euclidean distance. I call the function K on $[0, \infty)$ the isotropic autocovariance function for Z. Note that I am implicitly assuming a (strictly/weakly) isotropic random field is always (strictly/weakly) stationary. The isotropy condition amounts to assuming there is no reason to distinguish one direction from another for the random field under consideration. A simple but useful extension of isotropic random fields is to random fields that become isotropic after a linear transformation of coordinates. We say Z exhibits a geometric anisotropy if there exists an invertible matrix $\mathbf{V}$ such that $Z(\mathbf{V}\mathbf{x})$ is isotropic (Journel and Huijbregts 1978, p. 177).

Exercise

1 Show that a Gaussian random field on $\mathbb{R}^d$ is strictly stationary if and only if it is weakly stationary. Show that a Gaussian random field on $\mathbb{R}^d$ is strictly isotropic if and only if it is weakly isotropic.

2.2 The turning bands method

The turning bands method (Matheron 1973) is a procedure for simulating isotropic random fields on $\mathbb{R}^d$ based on simulating processes on $\mathbb{R}$. The method is clever and useful but I am mainly introducing it here as a further example of the problems that can occur by just considering the first two moments of a random field. Define b_d to be the unit ball in $\mathbb{R}^d$ centered at the origin, so that its boundary ∂b_d is the unit sphere. Matheron gives the following procedure for generating a weakly isotropic random field in $\mathbb{R}^d$.

(i) For an autocovariance function B on $\mathbb{R}$, simulate a stochastic process Y on $\mathbb{R}$ with $EY(t) = 0$ and $\operatorname{cov}\{Y(t), Y(s)\} = B(t-s)$.

(ii) Independently of Y, choose a random unit vector $\mathbf{V}$ from the uniform distribution on ∂b_d.

(iii) Let $Z(\mathbf{x}) = Y(\mathbf{x}^T\mathbf{V})$ for $\mathbf{x} \in \mathbb{R}^d$.

The resulting random field Z is weakly isotropic: it has mean 0 since $EZ(\mathbf{x}) = E\left[E\left\{Y(\mathbf{x}^T\mathbf{V}) \mid \mathbf{V}\right\}\right] = 0$ and

$$\begin{aligned}
\operatorname{cov}\{Z(\mathbf{x}), Z(\mathbf{y})\} &= E\left[E\left\{Y(\mathbf{x}^T\mathbf{V})Y(\mathbf{y}^T\mathbf{V}) \mid \mathbf{V}\right\}\right] \\
&= E\left\{B\left((\mathbf{x}-\mathbf{y})^T\mathbf{V}\right)\right\} \\
&= \int_{\partial b_d} B\left((\mathbf{x}-\mathbf{y})^T\mathbf{v}\right) U(d\mathbf{v}), \qquad (2)
\end{aligned}$$

where U is the uniform probability measure on ∂b_d. By symmetry considerations, $\operatorname{cov}\{Z(\mathbf{x}), Z(\mathbf{y})\}$ depends on $\mathbf{x}$ and $\mathbf{y}$ only through $|\mathbf{x}-\mathbf{y}|$, so we can write $\operatorname{cov}\{Z(\mathbf{x}), Z(\mathbf{y})\} = K(|\mathbf{x}-\mathbf{y}|)$. In $\mathbb{R}^3$,

$$\begin{aligned}
K(r) &= \frac{1}{4\pi}\int_0^{2\pi}\int_0^{\pi} B(r\cos\phi)\sin\phi \, d\phi \, d\theta \\
&= \int_0^1 B(rt)\, dt.
\end{aligned}$$

The inverse relationship is given by $B(r) = (d/dr)\{rK(r)\}$ (Exercise 2). Is this a sensible algorithm for simulating isotropic random fields? Clearly not in general, since $Z(\mathbf{x})$ is constant on planes of the form $\mathbf{x}^T\mathbf{V} = c$ for any real c. The problem is that the first two moments of the random field do not adequately describe its properties.

Another subtle point arises by taking Y in step (i) of the algorithm to be Gaussian and then supposing, based on (iii), that Z must also be Gaussian. However, Z cannot be Gaussian with covariance function given by (2), since a Gaussian random field with this covariance function would not be constant along certain planes. The resolution of this apparent paradox is that conditionally on $\mathbf{V} = \mathbf{v}$, Z is Gaussian but not isotropic, and unconditionally, Z is isotropic but not Gaussian.

The conclusion I draw from this example and that in 1.4 is that anyone who claims to be only making assumptions about the first two moments of a random field is being naive at the least. To make sensible predictions, it is unavoidable at least implicitly to make further assumptions about the law of the random field.

Of course, Matheron (1973) did not intend for anyone to use the algorithm described here to simulate isotropic random fields in practice. To use the turning bands method to simulate an approximately isotropic Gaussian random field in $\mathbb{R}^d$, what is done in practice is to simulate a large number of independent realizations of Gaussian processes $Y_1, Y_2, \ldots, Y_n$ on $\mathbb{R}$ with autocovariance function B and then set

$$Z(\mathbf{x}) = \frac{1}{n^{1/2}} \sum_{j=1}^{n} Y_j(\mathbf{x}^T \mathbf{V}_j),$$

where the $\mathbf{V}_j$s are random unit vectors independent of $Y_1, \ldots, Y_n$. If, in addition, $\mathbf{V}_1, \ldots, \mathbf{V}_n$ are independent and uniformly distributed on the unit sphere, then Z has covariance function given by (2). For n large, a central limit effect should make at least the low-order finite-dimensional distributions approximately normal. There may be some advantages in choosing the $\mathbf{V}_j$s more systematically to obtain a more evenly spaced distribution on ∂b_d. For example, for $d = 3$, Journel and Huijbregts (1978, p. 503) suggest taking $n = 15$ and the $\mathbf{V}_j$s to be along the lines joining the midpoints of opposite edges on a regular icosahedron centered at the origin.

Note that the approximate Gaussianity of Z should hold even if the Y_is are not Gaussian due to the central limit theorem effect. Thus, the turning bands method cannot be used directly to simulate a non-Gaussian random field. For a random field Z such that, for example, $\log Z$ is Gaussian, we can of course use turning bands to simulate $\log Z$ and then transform pointwise to obtain Z. See Cressie (1993) for further discussion on simulating random fields.

Exercise

2 In using the turning bands method to simulate an isotropic random field on $\mathbb{R}^3$ with K as its isotropic autocovariance function, show that B in step (i) of the algorithm is given by $(d/dr)\{rK(r)\}$.

2.3 Elementary properties of autocovariance functions

Suppose Z is weakly stationary on $\mathbb{R}^d$ with autocovariance function K. Then K must satisfy

$$K(\mathbf{0}) \geq 0,$$

$$K(\mathbf{x}) = K(-\mathbf{x}) \quad \text{and}$$

$$|K(\mathbf{x})| \leq K(\mathbf{0}).$$

The first two conditions are trivial and the last follows from the Cauchy–Schwarz inequality. We say the real-valued function K is positive definite if

$$\sum_{j,k=1}^{n} c_j c_k K(\mathbf{x}_j - \mathbf{x}_k) \geq 0$$

for all finite n, all real $c_1, \ldots, c_n$ and all $\mathbf{x}_1, \ldots, \mathbf{x}_n \in \mathbb{R}^d$. This condition is necessary and sufficient for there to exist a weakly stationary random field with autocovariance function K by the same argument as given in 2.1 for positive definite functions on $\mathbb{R}^d \times \mathbb{R}^d$.

Some other properties of positive definite (p.d.) functions include the following.

If K_1 and K_2 are p.d., then $a_1 K_1 + a_2 K_2$ is p.d. for all nonnegative a_1 and a_2. (3)

If $K_1, K_2, \ldots$ are p.d. and $\lim_{n\to\infty} K_n(\mathbf{x}) = K(\mathbf{x})$ for all $\mathbf{x} \in \mathbb{R}^d$, then K is p.d. (4)

If K_1 and K_2 are p.d., then $K(\mathbf{x}) = K_1(\mathbf{x})K_2(\mathbf{x})$ is p.d. (5)

The proofs of (3) and (4) are straightforward. The easiest way to prove (5) is to consider independent mean 0 Gaussian random fields Z_1 and Z_2 with autocovariance functions K_1 and K_2, respectively, and to show that K is the autocovariance function of the random field Z defined by $Z(\mathbf{x}) = Z_1(\mathbf{x})Z_2(\mathbf{x})$.

Exercise

3 If K_θ is a p.d. autocovariance function on $\mathbb{R}^d$ for all $\theta \in \mathbb{R}$ and is continuous in θ for all $\mathbf{x}$, show that $\int_{\mathbb{R}} K_\theta(\mathbf{x})\mu(d\theta)$ is p.d. if μ is a positive finite measure on $\mathbb{R}$ and $\int_{\mathbb{R}} K_\theta(\mathbf{0})\mu(d\theta) < \infty$.

2.4 Mean square continuity and differentiability

There is no simple relationship between the autocovariance function of a random field and the smoothness of its realizations. However, it is possible to relate the autocovariance function to what are known as mean square properties of a random field. For a sequence of random variables $X_1, X_2, \ldots$ and a random variable X defined on some common probability space, define $X_n \xrightarrow{L^2} X$ to mean $E(X_n - X)^2 \to 0$ and $EX^2 < \infty$. We say $\{X_n\}$ converges in L^2 if there exists X such that $X_n \xrightarrow{L^2} X$.

Suppose Z is a random field on $\mathbb{R}^d$. Then Z is mean square continuous at $\mathbf{x}$ if

$$\lim_{\mathbf{y}\to\mathbf{x}} E\{Z(\mathbf{y}) - Z(\mathbf{x})\}^2 = 0.$$

For Z weakly stationary with autocovariance function K, $E\{Z(\mathbf{y}) - Z(\mathbf{x})\}^2 = 2\{K(\mathbf{0}) - K(\mathbf{x}-\mathbf{y})\}$, so that Z is mean square continuous at $\mathbf{x}$ if and only if K is continuous at the origin. Since a weakly stationary random field is either mean square continuous everywhere or nowhere, we can say Z is mean square continuous if and only if K is continuous at the origin.

The mean square continuity of Z does not imply that its realizations are continuous. The process Z in Section 1.4 is mean square continuous but $\Pr(Z \text{ is continuous on } \mathbb{R}) = 0$.

If K is continuous at $\mathbf{0}$, then it is continuous everywhere, which follows by noting

$$\begin{aligned}|K(\mathbf{x}) - K(\mathbf{y})| &= |\mathrm{cov}\,\{Z(\mathbf{x}) - Z(\mathbf{y}), Z(\mathbf{0})\}| \\ &\leq [\mathrm{var}\,\{Z(\mathbf{x}) - Z(\mathbf{y})\}\,\mathrm{var}\,\{Z(\mathbf{0})\}]^{1/2} \\ &= [2\,\{K(\mathbf{0}) - K(\mathbf{x}-\mathbf{y})\}\,K(\mathbf{0})]^{1/2} \\ &\to 0\end{aligned}$$

as $\mathbf{y} \to \mathbf{x}$. On the other hand, for a weakly stationary process on $\mathbb{R}$, if K is not continuous at the origin, it may have other discontinuities (Exercises 4 and 5).

Mean square differentiability has a similar definition as an L^2 limit. A process Z on $\mathbb{R}$ with finite second moments is called mean square differentiable at t if $\{Z(t+h_n) - Z(t)\}/h_n$ converges in L^2 for all sequences $\{h_n\}$ converging to 0 as $n \to \infty$ with limit independent of $\{h_n\}$. If such a limit exists, we call it $Z'(t)$. A weakly stationary process Z on $\mathbb{R}$ is either mean square differentiable everywhere or nowhere. For Z weakly stationary with autocovariance function K, define the process

$$Z_h(t) = \frac{Z(t+h) - Z(t)}{h},$$

which has autocovariance function

$$K_h(t) = \frac{1}{h^2}\,\{2K(t) - K(t+h) - K(t-h)\}\,.$$

If K is twice differentiable, then

$$\lim_{h \to 0} K_h(t) = -K''(t),$$

so that $-K''$ is positive definite by (4). In Section 2.6 I prove that Z is mean square differentiable if and only if $K''(0)$ exists and is finite, and that if Z is mean square differentiable, then Z' has autocovariance function $-K''$.

To define higher-order mean square derivatives, we say Z is m-times mean square differentiable if it is $(m-1)$-times mean square differentiable and $Z^{(m-1)}$ is mean square differentiable. By repeated application of the stated results in the preceding paragraph on the mean square differentiability of a process, it follows that Z is m-times mean square differentiable if and only if $K^{(2m)}(0)$ exists and is finite and, if so, the autocovariance function of $Z^{(m)}$ is $(-1)^m K^{(2m)}$.

The following example shows that Z can have analytic realizations and not be mean square differentiable. Let $Z(t) = \cos(Xt + Y)$ where X and Y are independent random variables with X following a standard Cauchy distribution (i.e., has density $1/\{\pi(1+x^2)\}$ for $x \in \mathbb{R}$) and Y following a uniform distribution on $[0, 2\pi]$ (i.e., has density $1/(2\pi)$ on $[0, 2\pi]$). Then $EZ(t) = 0$ and

$$\begin{aligned}\operatorname{cov}\{Z(s), Z(t)\} &= E\{\cos(Xs+Y)\cos(Xt+Y)\} \\ &= \tfrac{1}{2}E\cos\{X(s-t)\} + \tfrac{1}{2}E\cos\{X(s+t)+2Y\} \\ &= \tfrac{1}{2}e^{-|s-t|},\end{aligned}$$

so Z is weakly stationary (it is also strictly stationary) and mean square continuous but not mean square differentiable, even though all realizations of Z are analytic.

Exercises

4 Find a p.d. autocovariance function on $\mathbb{R}$ that is discontinuous at $t = -1, 0$ and 1 and continuous elsewhere.

5 Find a p.d. autocovariance function on $\mathbb{R}$ that is discontinuous at all $t \in \mathbb{R}$.

6 For $Z(t) = \cos(Xt+Y)$, where X and Y are independent random variables with distributions as given in the last paragraph of 2.4, consider predicting $Z(t)$ for $t > 0$ based on observing $Z(s)$ for all $s \leq 0$. Find the conditional expectation of $Z(t)$ and the BLP of $Z(t)$ (see 3.4) and compare. Why is the conditional expectation not a linear predictor?

2.5 Spectral methods

Spectral methods are a powerful tool for studying random fields. In Fourier analysis, it is somewhat more natural to consider complex-valued functions rather than restricting to real-valued functions. We say Z is a complex random field if $Z(\mathbf{x}) = U(\mathbf{x}) + iV(\mathbf{x})$, where U and V are real random fields. If U and V are jointly weakly stationary, by which we mean U and V are weakly stationary and that $\operatorname{cov}\{U(\mathbf{x}), V(\mathbf{y})\}$ depends only on $\mathbf{x} - \mathbf{y}$, then Z is weakly stationary and we define

$$\begin{aligned}K(\mathbf{y}) &= \operatorname{cov}\left\{Z(\mathbf{x}+\mathbf{y}), \overline{Z(\mathbf{x})}\right\} \\ &= \operatorname{cov}\{U(\mathbf{x}+\mathbf{y}), U(\mathbf{x})\} + \operatorname{cov}\{V(\mathbf{x}+\mathbf{y}), V(\mathbf{x})\} \\ &\quad + i\left[\operatorname{cov}\{U(\mathbf{x}), V(\mathbf{x}+\mathbf{y})\} - \operatorname{cov}\{U(\mathbf{x}+\mathbf{y}), V(\mathbf{x})\}\right]\end{aligned}$$

as the autocovariance function of Z. Then $K(-\mathbf{x}) = \overline{K(\mathbf{x})}$ and for $c_1, \ldots, c_n$ complex

$$\sum_{j,k=1}^{n} c_j \bar{c}_k K(\mathbf{x}_j - \mathbf{x}_k) \geq 0,$$

since the left side equals $E\left|\sum_{j=1}^{n} c_j Z(\mathbf{x}_j) - E\sum_{j=1}^{n} c_j Z(\mathbf{x}_j)\right|^2$. A function K satisfying this condition for all finite n, all $\mathbf{x}_1, \ldots, \mathbf{x}_n \in \mathbb{R}^d$ and all complex $c_1, \ldots, c_n$ is said to be a positive definite complex function on $\mathbb{R}^d$.

Spectral representation of a random field

As an example of a complex random field, suppose $\boldsymbol{\omega}_1, \ldots, \boldsymbol{\omega}_n \in \mathbb{R}^d$ and let $Z_1, \ldots, Z_n$ be mean 0 complex random variables with $E(Z_i \overline{Z}_j) = 0$ if $i \neq j$ and $E|Z_i|^2 = f_i$. Consider

$$Z(\mathbf{x}) = \sum_{k=1}^{n} Z_k \exp(i\boldsymbol{\omega}_k^T \mathbf{x}), \tag{6}$$

so that Z_k is the complex random amplitude for Z at frequency $\boldsymbol{\omega}_k$. Then Z is a weakly stationary complex random field in $\mathbb{R}^d$ with autocovariance function $K(\mathbf{x}) = \sum_{k=1}^{n} f_k \exp(i\boldsymbol{\omega}_k^T \mathbf{x})$.

Equation (6) is an example of a spectral representation of a random field. By taking L^2 limits of sums like those in (6) in an appropriate manner, spectral representations of all mean square continuous weakly stationary random fields can be obtained (Yaglom 1987a). That is, all mean square continuous weakly stationary random fields are, in an appropriate sense, L^2 limits of linear combinations of complex exponentials with uncorrelated random amplitudes. To make this concept more precise, we need to consider complex random measures, which map Borel sets on $\mathbb{R}^d$ into complex-valued random variables. Suppose M is a complex random measure on $\mathbb{R}^d$. Since it is a measure, $M(\Delta_1 \cup \Delta_2) = M(\Delta_1) + M(\Delta_2)$ for disjoint Borel sets Δ_1 and Δ_2. In addition, suppose that for some positive finite measure F and all Borel sets,

$$EM(\Delta) = 0,$$
$$E|M(\Delta)|^2 = F(\Delta)$$

and for all disjoint Borel sets Δ_1 and Δ_2,

$$E\left\{M(\Delta_1)\overline{M(\Delta_2)}\right\} = 0.$$

I assume such a random measure exists; see, for example, Gihman and Skorohod (1974) for mathematical details. Next consider how to interpret the integral

$$Z(\mathbf{x}) = \int_{\mathbb{R}^d} \exp(i\boldsymbol{\omega}^T \mathbf{x}) M(\mathbf{d}\boldsymbol{\omega}). \tag{7}$$

The idea is to think of the integral as a limit in L^2 of sums of the form (6). For simplicity, suppose that with probability 1, M is identically 0 outside $[-1, 1]^d$. Now set $\mathbf{j} = (j_1, \ldots, j_d)^T$, $\Delta_n(\mathbf{j}) = \times_{k=1}^{d}(n^{-1}(j_k - 1), n^{-1}j_k]$ (so that $\Delta_n(\mathbf{j})$ is the cube with edges of length n^{-1} and "upper right corner" at $n^{-1}\mathbf{j}$), $S_n = \{-n, -n+1, \dots, n\}^d$ and define

$$Z_n(\mathbf{x}) = \sum_{\mathbf{j}\in S_n} \exp(in^{-1}\mathbf{j}^T\mathbf{x})M(\Delta_n(\mathbf{j})), \tag{8}$$

which is a sum of the form given in (6). It is possible to show that for any $\mathbf{x}$, $Z_n(\mathbf{x})$ converges in L^2 (Exercise 7) and the integral on the right side of (7) is defined to be this limit. More specifically, by defining $F_n(\Delta) = \sum_{\mathbf{j}\in S_n} F(\Delta_n(\mathbf{j}))1\{n^{-1}\mathbf{j} \in \Delta\}$ we get

$$\begin{aligned} E\{Z_n(\mathbf{x})\overline{Z_n(\mathbf{y})}\} &= \sum_{\mathbf{j}\in S_n} \exp\left\{in^{-1}\mathbf{j}^T(\mathbf{x}-\mathbf{y})\right\}F(\Delta_n(\mathbf{j})) \\ &= \int_{\mathbb{R}^d} \exp\left\{i\boldsymbol{\omega}^T(\mathbf{x}-\mathbf{y})\right\}F_n(\mathbf{d}\boldsymbol{\omega}). \end{aligned}$$

Since F_n converges weakly to F (Exercise 8) and $\exp\left\{i\boldsymbol{\omega}^T(\mathbf{x}-\mathbf{y})\right\}$ is bounded and uniformly continuous for $\boldsymbol{\omega} \in [-1,1]^d$,

$$E\{Z_n(\mathbf{x})\overline{Z_n(\mathbf{y})}\} \to \int_{\mathbb{R}^d} \exp\left\{i\boldsymbol{\omega}^T(\mathbf{x}-\mathbf{y})\right\}F(\mathbf{d}\boldsymbol{\omega})$$

as $n \to \infty$ for any fixed $\mathbf{x}$ and $\mathbf{y}$ (see Chapter 1 of Billingsley (1968) for definitions and results on weak convergence of measures on metric spaces). In conjunction with the L^2 convergence of $Z_n(\mathbf{x})$ for all $\mathbf{x}$, this implies that the autocovariance function of Z is (Exercise 8)

$$K(\mathbf{x}) = \int_{\mathbb{R}^d} \exp(i\boldsymbol{\omega}^T\mathbf{x})F(\mathbf{d}\boldsymbol{\omega}). \tag{9}$$

The function F is called the spectral measure or spectrum for Z.

Bochner's Theorem

It is easy to see that for any finite positive measure F, the function K given in (9) is p.d., since

$$\begin{aligned} \sum_{j,k=1}^{n} c_j\bar{c}_k K(\mathbf{x}_j - \mathbf{x}_k) &= \sum_{j,k=1}^{n} c_j\bar{c}_k \int_{\mathbb{R}^d} \exp\{i\boldsymbol{\omega}^T(\mathbf{x}_j - \mathbf{x}_k)\}F(\mathbf{d}\boldsymbol{\omega}) \\ &= \int_{\mathbb{R}^d} \left|\sum_{j=1}^{n} c_j \exp(i\boldsymbol{\omega}^T\mathbf{x}_j)\right|^2 F(\mathbf{d}\boldsymbol{\omega}) \\ &\geq 0. \end{aligned}$$

Furthermore, all continuous positive definite complex functions are of the form (9) with F a positive measure of finite mass.

Theorem 1 (Bochner's Theorem). *A complex-valued function K on $\mathbb{R}^d$ is the autocovariance function for a weakly stationary mean square continuous complex-valued random field on $\mathbb{R}^d$ if and only if it can be represented as in (9) where F is a positive finite measure.*

A proof of this fundamental result is given in Gihman and Skorohod (1974, p. 208). Although the following argument does not constitute a proof, it is instructive to use the existence of spectral representations to indicate why the Fourier transform of a measure assigning negative mass to any measurable set cannot be a positive definite function. For simplicity, suppose Z is a complex process on $\mathbb{R}$ with spectral measure F. Suppose $\omega_1 < \omega_2$ are not mass points of F and set $I = (\omega_1, \omega_2)$. Proceeding formally, consider the process Y defined by

$$Y(t) = \frac{1}{2\pi} \int_{-\infty}^{\infty} \frac{e^{i\omega_2 u} - e^{i\omega_1 u}}{iu} Z(t-u)\, du. \tag{10}$$

Using the spectral representation of Z and again proceeding formally,

$$\begin{aligned} Y(t) &= \frac{1}{2\pi} \int_{\mathbb{R}} \frac{e^{i\omega_2 u} - e^{i\omega_1 u}}{iu} \left\{ \int_{\mathbb{R}} e^{i\omega(t-u)} M(d\omega) \right\} du \\ &= \frac{1}{2\pi} \int_{\mathbb{R}} e^{i\omega t} \left\{ \int_{\mathbb{R}} \frac{e^{i(\omega_2-\omega)u} - e^{i(\omega_1-\omega)u}}{iu}\, du \right\} M(d\omega) \\ &= \int_I e^{i\omega t} M(d\omega) \end{aligned}$$

(see Exercise 9 for the last step). The autocovariance function of Y is $\int_I e^{i\omega t} F(d\omega)$ and in particular $E|Y(0)|^2 = F(I)$, which must be nonnegative, so that F must be a positive measure. We are a long way from proving Bochner's Theorem even in $\mathbb{R}$, but the physical intuition should be clear. Given any process Z, we can define another process Y that is made up of only those frequencies in the spectral representation contained in the interval I. The transformation from Z to Y defined by (10) is called a band-pass filter. Since $F(I) = E|Y(0)|^2 \geq 0$ for any interval I, F must be a positive measure.

If F has a density with respect to Lebesgue measure, I call this density the spectral density and generally denote it by f. When the spectral density exists, we have the inversion formula (Yaglom 1987a, p. 332)

$$f(\boldsymbol{\omega}) = \frac{1}{(2\pi)^d} \int_{\mathbb{R}^d} \exp(-i\boldsymbol{\omega}^T \mathbf{x}) K(\mathbf{x})\, d\mathbf{x}. \tag{11}$$

Exercises

7 Show that $Z_n(\mathbf{x})$ as defined in (8) converges in L^2. If you have trouble proving this, show that the subsequence $Z_{2^n}(\mathbf{x})$ converges in L^2.

8 For Z_n as defined in (8), show that F_n converges weakly to F. Show that (9) gives the autocovariance function for Z as defined in (7).

9 For $\omega_1 < \omega_2$, evaluate

$$\int_{-\infty}^{\infty} \frac{e^{i(\omega_2-\omega)u} - e^{i(\omega_1-\omega)u}}{iu}\, du$$

for all real ω.

10 Use the inversion formula (11) to determine which of these functions on $\mathbb{R}$ are positive definite:

(i) $e^{-|t|}\cos t$.

(ii) $e^{-|t|}(1-|t|)$.

(iii) $(1-t^2)^+$.

2.6 Two corresponding Hilbert spaces

A technique that is used on several occasions in this work is to turn questions about Hilbert spaces of random variables into questions about Hilbert spaces of sums of complex exponentials and their limits. Ibragimov and Rozanov (1978) make extensive use of this idea. Suppose Z is a mean 0 weakly stationary real random field on $\mathbb{R}^d$ with autocovariance function K and corresponding spectrum F. For a subset R of $\mathbb{R}^d$, let $\mathcal{H}_R(F) = \mathcal{H}_R(0, K)$ be the closed linear manifold of $Z(\mathbf{x})$ for $\mathbf{x} \in R$, where the 0 in $\mathcal{H}_R(0, K)$ refers to the mean of Z. Similarly, define $\mathcal{L}_R(F)$ to be the closed linear manifold of functions of $\boldsymbol{\omega}$ of the form $\exp(i\boldsymbol{\omega}^T\mathbf{x})$ for $\mathbf{x} \in R$ under the inner product $(\phi, \mu)_F = \int_{\mathbb{R}} \phi(\omega)\overline{\mu(\omega)}\, F(d\omega)$. If we identify $\sum_{j=1}^n a_j Z(\mathbf{x}_j)$ with $\sum_{j=1}^n a_j \exp(i\boldsymbol{\omega}^T\mathbf{x}_j)$ and extend this correspondence to respective limits of such sums, $\mathcal{H}_R(F)$ and $\mathcal{L}_R(F)$ are essentially two ways of describing the same Hilbert space since

$$\operatorname{cov}\left\{\sum_{j=1}^n a_j Z(\mathbf{x}_j), \sum_{k=1}^m b_k Z(\mathbf{y}_k)\right\} = \left(\sum_{j=1}^n a_j \exp(i\boldsymbol{\omega}^T\mathbf{x}_j), \sum_{k=1}^m b_k \exp(i\boldsymbol{\omega}^T\mathbf{y}_k)\right)_F .$$

Indeed, if $Z(\mathbf{x}) = \int_{\mathbb{R}^d} \exp(i\boldsymbol{\omega}^T\mathbf{x})M(\mathbf{d}\boldsymbol{\omega})$ is the spectral representation for Z, then for $V \in \mathcal{L}_R(F)$, the corresponding random variable in $\mathcal{H}_R(F)$ is given by $\int_R V(\boldsymbol{\omega})M(\mathbf{d}\boldsymbol{\omega})$ (Gihman and Skorohod 1974, p. 244).

An application to mean square differentiability

Let us make use of this correspondence to prove two results stated in 2.4 for a weakly stationary process Z on $\mathbb{R}$: first, we show that Z is mean square differentiable if and only if $K''(0)$ exists and is finite and, second, that if Z is mean square differentiable, then Z' has autocovariance function $-K''$. Since a constant mean obviously does not affect the mean square differentiability of a process, assume $EZ(t) = 0$. Because of the correspondence between $\mathcal{L}_{\mathbb{R}}(F)$ and $\mathcal{H}_{\mathbb{R}}(F)$, to study the convergence of $Z_h(t) = \{Z(t+h) - Z(t)\}/h$

as $h \to 0$ in $\mathcal{H}_{\mathbb{R}}(F)$ it is completely equivalent to study the convergence of $\tau_h(\omega) = (e^{i\omega(t+h)} - e^{i\omega t})/h$ as $h \to 0$ in $\mathcal{L}_{\mathbb{R}}(F)$. But $\lim_{h\to 0} \tau_h(\omega) = i\omega e^{i\omega t}$ for every ω, so that τ_h converges in $\mathcal{L}_{\mathbb{R}}(F)$ if and only if it converges to $i\omega e^{i\omega t}$ in $\mathcal{L}_{\mathbb{R}}(F)$ (Exercise 11). Obviously, $\int_{\mathbb{R}} \omega^2 F(d\omega) < \infty$ is necessary for this convergence, since otherwise $i\omega e^{i\omega t} \notin \mathcal{L}_{\mathbb{R}}(F)$. But $\int_{\mathbb{R}} \omega^2 F(d\omega) < \infty$ is also sufficient, since it implies

$$\lim_{h\to 0} \int_{\mathbb{R}} |\tau_h(\omega) - i\omega e^{i\omega t}|^2 F(d\omega) = 0$$

by a simple application of the Dominated Convergence Theorem. It is well known (Chung 1974, Theorem 6.4.1; Lukacs 1970, Section 2.3) that the finiteness of the second moment of a finite positive measure is equivalent to its Fourier transform possessing a second derivative at the origin, which proves that Z is mean square differentiable if and only if K is twice differentiable at the origin. Furthermore, $\int_{\mathbb{R}} \omega^2 F(d\omega) < \infty$ implies that K is twice differentiable with $-K''(t) = \int_{\mathbb{R}} \omega^2 e^{i\omega t} F(d\omega)$, so that $K''(0)$ exists implies K is twice differentiable. The claim that when K is twice differentiable, Z' has autocovariance function $-K''$ follows by showing that

$$\lim_{h\to 0, k\to 0} \operatorname{cov}\{Z_h(s), Z_k(t)\} = -K''(s-t),$$

where $Z_h = \{Z(t+h) - Z(t)\}/h$ as in 2.4.

Exercises

11 For a sequence of complex-valued functions $\tau_1, \tau_2, \ldots$ on $\mathbb{R}$ converging pointwise to the function τ, prove that τ_n converges in $\mathcal{L}_{\mathbb{R}}(F)$ if and only if it converges to τ in $\mathcal{L}_{\mathbb{R}}(F)$. Suggestion: use a subsequence argument similar to the one in the proof of Theorem 19.1 (the completeness of L^p spaces) of Billingsley (1995).

12 For $R = [0,1]$ and $K(t) = e^{-|t|}$, show that every element of $\mathcal{L}_R(F)$ can be written in the form $a + (1+i\omega)\int_0^1 e^{i\omega t} c(t)dt$ for some real constant a and real-valued function c that is square-integrable on $[0,1]$. This result is a special case of (1.3) of Ibragimov and Rozanov (1978, p. 30).

2.7 Examples of spectral densities on $\mathbb{R}$

This section describes some commonly used classes of spectral densities and a class of spectral densities that should be commonly used. I consider only real processes here, in which case, we can always take the spectral density to be an even function.

Rational spectral densities

Rational functions that are even, nonnegative and integrable have corresponding autocovariance functions that can be expressed in terms of elementary functions (Yaglom 1987a, pp. 133–136). For example, for positive constants ϕ and α, if $f(\omega) = \phi(\alpha^2 + \omega^2)^{-1}$, then $K(t) = \pi\phi\alpha^{-1}e^{-\alpha|t|}$, which can be obtained by contour integration (Carrier, Krook and Pearson 1966, p. 80). Since K does not even have a first derivative at 0, we have that the corresponding process Z is not mean square differentiable. Alternatively, we reach the same conclusion by noting $\int_{-\infty}^{\infty} \omega^2 f(\omega)d\omega = \infty$.

As a second example, suppose $f(\omega) = \phi(\alpha^2 + \omega^2)^{-2}$, which implies $K(t) = \frac{1}{2}\pi\phi\alpha^{-3}e^{-\alpha|t|}(1 + \alpha|t|)$. In this case, $\int_{-\infty}^{\infty} \omega^2 f(\omega)d\omega < \infty$ and $\int_{-\infty}^{\infty} \omega^4 f(\omega)d\omega = \infty$, so the corresponding process Z is once but not twice mean square differentiable. This result is not so easy to see via the autocovariance function. However, if care is taken, it is possible to calculate directly $-K''(t) = \frac{1}{2}\pi\phi\alpha^{-1}e^{-\alpha|t|}(1 - \alpha|t|)$ for all t, including $t = 0$. Alternatively, one can get this result from $-K''(t) = \phi\int_{-\infty}^{\infty} \omega^2(\alpha^2 + \omega^2)^{-2}e^{i\omega t}d\omega$ by using contour integration.

The general form for a rational spectral density for a real process on $\mathbb{R}$ is given by

$$f(\omega) = |P_n(i\omega)|^2/|Q_m(i\omega)|^2, \tag{12}$$

where P_n and Q_m are polynomials with real coefficients of order n and m, respectively, $m > n$, and Q_m has no zeroes on the imaginary axis (Exercise 13). These last two conditions ensure the integrability of f. Processes on $\mathbb{R}$ with rational spectral densities can be thought of as continuous time analogues of the familiar autoregressive moving-average models for discrete time series (Priestley 1981, Chapter 3). A process Z with spectral density given by (12) has exactly $m-n-1$ mean square derivatives. Thus, the class of processes with rational spectral densities includes processes with exactly p mean square derivatives for any nonnegative integer p. However, later in this section I describe a class of processes with even greater flexibility in their local behavior.

Principal irregular term

Before giving any further examples of spectral densities, it is worthwhile to consider more generally the behavior of autocovariance functions in a neighborhood of 0. A natural way to describe this behavior of an autocovariance function $K(t)$ is to take a series expansion in $|t|$ about 0. For $K(t) = \pi\phi\alpha^{-1}e^{-\alpha|t|}$, we have $K(t) = \pi\phi\alpha^{-1} - \pi\phi|t| + O(|t|^2)$ as $|t| \downarrow 0$. It follows that K is not differentiable at 0 due to the nonzero coefficient for $|t|$. For $K(t) = \frac{1}{2}\pi\phi\alpha^{-3}e^{-\alpha|t|}(1 + \alpha|t|)$,

$$K(t) = \tfrac{1}{2}\pi\phi\alpha^{-3} - \tfrac{1}{4}\pi\phi\alpha^{-1}|t|^2 + \tfrac{1}{6}\pi\phi|t|^3 + O(|t|^4) \tag{13}$$

as $|t| \downarrow 0$. The nonzero coefficient multiplying $|t|^3$ implies that K is not three times differentiable as a function of t at $t = 0$. Note that a function of the form $b_0 + b_1 t^2 + b_2 |t|^3 + O(t^4)$ as $t \to 0$ need not be even twice differentiable at 0 (see Exercise 14). However, we also know K is p.d. The following result is an easy consequence of Theorem 2.3.1 of Lukacs (1970) (Exercise 15).

Theorem 2. *If K is p.d. on $\mathbb{R}$ and $K(t) = \sum_{j=0}^{n} c_j t^{2j} + o(t^{2n})$ as $t \to 0$, then K has $2n$ derivatives.*

Theorem 2 in conjunction with (13) does imply that $K(t) = \frac{1}{2}\pi\phi\alpha^{-3}e^{-\alpha|t|} \times (1 + \alpha|t|)$ is twice differentiable.

For an autocovariance function K, let us informally define its principal irregular term as the first term in the series expansion about 0 for K as a function of $|t|$ that is not proportional to $|t|$ raised to an even power (Matheron 1971, p. 58). For $K(t) = \pi\phi\alpha^{-1}e^{-\alpha|t|}$, the principal irregular term is $-\pi\phi|t|$, and for $K(t) = \frac{1}{2}\pi\phi\alpha^{-3}e^{-\alpha|t|}(1 + \alpha|t|)$ it is $\frac{1}{6}\pi\phi|t|^3$. In both cases, the coefficient of the principal irregular term does not depend on α. This fact suggests that for either class of models, the local behavior of the corresponding process is not much affected by α. Note that $f(\omega) = \phi(\alpha^2 + \omega^2)^{-1} \sim \phi\omega^{-2}$ and $f(\omega) = \phi(\alpha^2 + \omega^2)^{-2} \sim \phi\omega^{-4}$ as $\omega \to \infty$, so that the high frequency behavior of the spectral densities also does not depend on α. Section 2.8 explores this close connection between the high frequency behavior of the spectral density and the coefficient of the principal irregular term of the autocovariance function more generally.

It is not so easy to give a formal definition of the principal irregular term, since as we show in (16), it need not be of the form $\alpha|t|^\beta$. One possible definition is to call g a principal irregular term for K if $g(t)t^{-2n} \to 0$ and $|g(t)|t^{-2n-2} \to \infty$ as $t \to 0$ and K is of the form $K(t) = \sum_{j=0}^{n} c_j t^{2j} + g(t) + o(|g(t)|)$ as $t \to 0$. It follows from Theorem 2 that the corresponding process is exactly n times mean square differentiable.

A problem with this definition for a principal irregular term g is that any function h such that $h(t)/g(t) \to 1$ as $t \downarrow 0$ is also a principal irregular term. If $g(t) = \alpha|t|^\beta$ is a principal irregular term for K, I call β the power and α the coefficient of the principal irregular term. Note that if such α and β exist they must be unique, so there is no ambiguity in their definition. For models used in practice, if there is a principal irregular term, it can usually be taken to be of the form $g(t) = \alpha|t|^\beta$ for β positive and not an even integer or $g(t) = \alpha t^{2k} \log|t|$ for some positive integer k.

Gaussian model

A somewhat commonly used form for the autocovariance function of a smooth process on $\mathbb{R}$ is $K(t) = ce^{-\alpha t^2}$, for which the corresponding spectral density is $f(\omega) = \frac{1}{2}c(\pi\alpha)^{-1/2}e^{-\omega^2/(4\alpha)}$. Because of its functional form, it is sometimes called the Gaussian model (Journel and Huijbregts 1978,

p. 165). This name is unfortunate, as it apparently suggests that this model is of central importance in the same way as Gaussian probability distributions. Nothing could be farther from the truth. Note that K is infinitely differentiable, and correspondingly, all moments of the spectral density are finite, so that the corresponding Z has mean square derivatives of all orders. In fact, a much stronger result holds: for any $t > 0$, as $n \to \infty$, $\sum_{j=0}^{n} Z^{(j)}(0)t^j/j! \to Z(t)$ in L^2 (Exercise 16). That is, it is possible to predict $Z(t)$ perfectly for all $t > 0$ based on observing $Z(s)$ for all $s \in (-\epsilon, 0]$ for any $\epsilon > 0$. Such behavior would normally be considered unrealistic for a physical process. One might argue that a process cannot practically be observed continuously in time, but we show in 3.5 that even with discrete observations, the use of this autocovariance yields unreasonable predictors. Figure 1 plots $e^{-t^2/2}$ and $e^{-|t|}(1 + |t|)$. Both functions are of the form $1 - \frac{1}{2}t^2 + O(|t|^3)$ as $t \to 0$. It is difficult from looking at these plots to see that the first function is analytic whereas the second only has two derivatives at the origin. One important practical conclusion we can draw is that plots of empirical autocovariance functions are likely to be a poor way to distinguish between possible models for the autocovariance function of a smooth process.

Triangular autocovariance functions

A class of autocovariance functions that we have seen before and will see again is $K(t) = c(a - |t|)^+$ for c and a positive. Such autocovariance functions are sometimes called triangular due to the shape of the graph of K. Although these autocovariance functions are not commonly used in applications, some BLPs under this model have unusual behavior (see 3.5) and it is important to explore the reasons for this behavior in order to develop a good understanding of the properties of BLPs. Using the inversion formula (11)

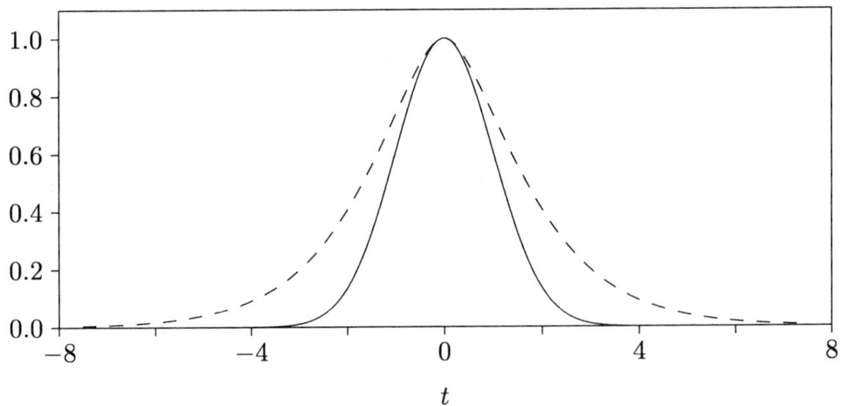

FIGURE 1. Plots of $e^{-t^2/2}$ (solid line) and $e^{-|t|}(1 + |t|)$ (dashed line).

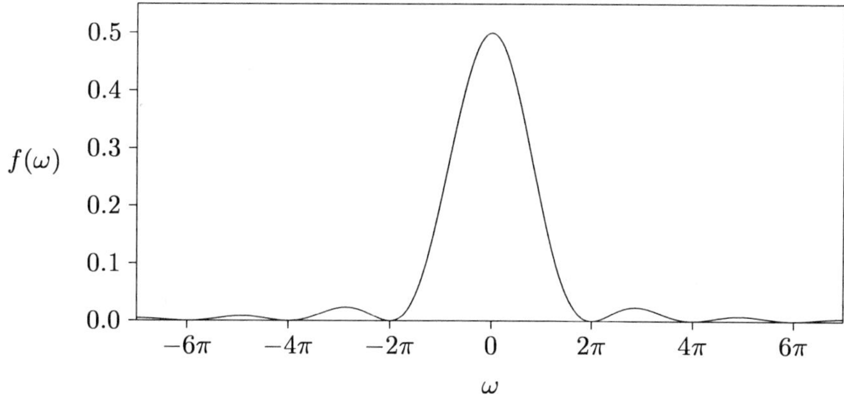

FIGURE 2. Plot of the spectral density $f(\omega) = (1 - \cos\omega)/\omega^2$ for the triangular autocovariance function $K(t) = \pi^{-1}(1 - |t|)^+$.

(Exercise 17), the corresponding spectral density is $c\pi^{-1}\{1 - \cos(a\omega)\}/\omega^2$, plotted in Figure 2. The oscillating behavior of the spectral density would be unrealistic for many physical processes. More specifically, there would usually be no reason for assuming the spectrum has much more mass near the frequency $(2n + 1)\pi$ than near $2n\pi$ for n large, which is the case for the spectral density $(1 - \cos\omega)/\omega^2$. We show in 3.5 that the fact that some BLPs under this model have strange properties is a direct consequence of the oscillations of the spectral density at high frequencies.

Matérn class

A class of autocovariance functions that I believe has considerable practical value is obtained from spectral densities of the form $f(\omega) = \phi(\alpha^2 + \omega^2)^{-\nu-1/2}$ for $\nu > 0$, $\phi > 0$ and $\alpha > 0$. The corresponding autocovariance function is

$$K(t) = \frac{\pi^{1/2}\phi}{2^{\nu-1}\Gamma(\nu + 1/2)\alpha^{2\nu}} (\alpha|t|)^\nu \mathcal{K}_\nu(\alpha|t|), \tag{14}$$

where $\mathcal{K}_\nu$ is a modified Bessel function (Abramowitz and Stegun 1965, pp. 374–379). I call this class of autocovariance functions the Matérn class after Bertil Matérn (Matérn 1960, 1986). The critical parameter here is ν: the larger ν is, the smoother Z is. In particular, Z will be m times mean square differentiable if and only if $\nu > m$, since $\int_{-\infty}^{\infty} \omega^{2m} f(\omega) d\omega < \infty$ if and only if $\nu > m$. When ν is of the form $m + \frac{1}{2}$ with m a nonnegative integer, the spectral density is rational and the autocovariance function is of the form $e^{-\alpha|t|}$ times a polynomial in $|t|$ of degree m (Abramowitz and Stegun 1965, 10.2.15). For example, as we have already seen, when $\nu = \frac{1}{2}$, $K(t) = \pi\phi\alpha^{-1}e^{-\alpha|t|}$ and when $\nu = \frac{3}{2}$, $K(t) = \frac{1}{2}\pi\phi\alpha^{-3}e^{-\alpha|t|}(1 + \alpha|t|)$.

We can also determine the mean square smoothness of a process in the Matérn class through its behavior at the origin. Using results on Bessel and gamma functions from Abramowitz and Stegun (1965, Section 9.6 and 6.1.18), for K as in (14) with ν not an integer and $m < \nu < m+1$,

$$K(t) = \sum_{j=0}^{m} b_j t^{2j} - \frac{\pi\phi}{\Gamma(2\nu+1)\sin(\nu\pi)} |t|^{2\nu} + O(|t|^{2m+2}) \text{ as } t \to 0 \qquad (15)$$

for appropriate real constants $b_0, \ldots, b_m$ depending on ϕ, ν and α (Exercise 18). For K as in (14) and $\nu = m+1$ a positive integer,

$$K(t) = \sum_{j=0}^{m} b_j t^{2j} + \frac{2(-1)^m\phi}{(2m+2)!} t^{2m+2} \log|t| + O(t^{2m+2}) \text{ as } t \to 0 \qquad (16)$$

for appropriate constants $b_0, \ldots, b_m$ depending on ϕ, m and α. Note that in both (15) and (16) the coefficient multiplying the principal irregular term does not depend on α and that $f(\omega) \sim \phi|\omega|^{-2\nu-1}$ as $|\omega| \to \infty$, so that the high frequency behavior of f also does not depend on α. Theorem 2 together with (15) and (16) implies that $(\alpha|t|)^\nu \mathcal{K}_\nu(\alpha|t|)$ is $2m$ times differentiable if and only if $\nu > m$. Thus, we recover the fact that the corresponding process Z is m times mean square differentiable if and only if $\nu > m$.

We can obtain a more precise result on the local behavior of a process with autocovariance function given by (14). Specifically, termwise differentiation of either (15) or (16) can be justified, from which it follows that for $m < \nu < m+1$,

$$\operatorname{var}\left\{Z^{(m)}(h) - Z^{(m)}(0)\right\} \sim \frac{2\phi}{\Gamma(2\nu - m + 1)\sin(\nu\pi)} h^{2(\nu-m)} \qquad (17)$$

as $h \downarrow 0$ and for $\nu = m+1$,

$$\operatorname{var}\left\{Z^{(m)}(h) - Z^{(m)}(0)\right\} \sim 2\phi h^2 \log h \qquad (18)$$

as $h \downarrow 0$ (Exercise 18). Thus, the continuous parameter ν has a direct interpretation in the time domain as a measure of smoothness of the process, with larger values of ν corresponding to smoother processes.

In comparison, if Z has a rational spectral density and exactly m mean square derivatives, it is possible to show that for some $c > 0$,

$$\operatorname{var}\left\{Z^{(m)}(h) - Z^{(m)}(0)\right\} \sim ch$$

as $h \downarrow 0$. Therefore, in terms of the local behavior of Z, rational spectral densities only cover the Matérn models with $\nu = m + \frac{1}{2}$. Of course, this conclusion is transparent in the spectral domain, since a rational spectral density f for a process with exactly m mean square derivatives must satisfy $f(\omega) \sim c\omega^{-2m-2}$ as $\omega \to \infty$ for some $c > 0$.

We can use the fact that functions in the Matérn class are positive definite to show that $e^{-|t|^\delta}$ is positive definite for $0 < \delta < 2$ (we already know it is positive definite for $\delta = 2$). Specifically, for $0 < \nu < 1$,

$$\alpha|t|^{\nu}\mathcal{K}_{\nu}(\alpha|t|) = \gamma - \beta|\alpha t|^{2\nu} + O(t^2)$$

in a neighborhood of 0, where

$$\gamma = \frac{\pi 2^{\nu-1}}{\sin(\nu\pi)\Gamma(1-\nu)} \quad \text{and} \quad \beta = \frac{\pi}{2^{\nu+1}\sin(\nu\pi)\Gamma(1+\nu)}$$

(see Exercise 18). Then $B_n(t) = c_n \{\alpha_n|t|^{\nu}\mathcal{K}_{\nu}(\alpha_n|t|)\}^n$ is p.d. for α_n and c_n positive and taking $c_n = \gamma^{-n}$, $\alpha_n = n^{-1/(2\nu)}$ yields

$$B_n(t) \to e^{-(\beta/\gamma)|t|^{2\nu}}$$

for every fixed t. By (4), $e^{-(\beta/\gamma)|t|^{\delta}}$ is p.d. for $0 < \delta < 2$ and hence so is $e^{-|t|^{\delta}}$. Finally, $e^{-|t|^{\delta}}$ is not p.d. for $\delta > 2$, which can be seen by noting that the second derivative of this function is 0 for $t = 0$, which would imply $\text{var}\{Z'(0)\} = 0$. Yaglom (1987b, p. 48) provides some historical notes on determining the positive definiteness of the function $e^{-|t|^{\delta}}$.

Exercises

13 Show that if f is the spectral density of a real process on $\mathbb{R}$ and is rational, then f can be written as in (12).

14 Show that the function $t^4\cos(t^{-3})$ (defined by continuity at $t = 0$) is of the form $b_0 + b_1 t^2 + b_2|t|^3 + O(t^4)$ as $t \to 0$ but is not twice differentiable at 0.

15 Look up Theorem 2.3.1 of Lukacs (1970) and show that Theorem 2 follows.

16 Suppose Z is a weakly stationary process on $\mathbb{R}$ with analytic autocovariance K. Show that $\sum_{j=0}^{n} Z^{(j)}(0)t^j/j! \to Z(t)$ in L^2 as $n \to \infty$ for any $t > 0$.

17 Using the inversion formula (11), show that the spectral density corresponding to $K(t) = c(a - |t|)^+$ is $c\pi^{-1}\{1 - \cos(\omega a)\}/\omega^2$.

18 Verify (15) and (16) by looking up the relevant series expansions for modified Bessel functions in, for example, Abramowitz and Stegun (1965). Give explicit expressions for $b_0, \dots, b_m$ in both cases. Verify (17) and (18).

2.8 Abelian and Tauberian theorems

We have now seen a number of examples in which the tail behavior of the spectrum is closely related to the smoothness at the origin of the autocovariance function. General results on properties of the transform of a measure in terms of properties of the measure are known as Abelian theorems; results on the converse problem of finding properties of a measure in terms of those of its transform are known as Tauberian theorems (Feller 1971; Bingham, Goldie and Teugels 1987).

Pitman (1968) proved both Abelian and Tauberian theorems useful for our purposes. We first need to define the notion of a regularly varying function. A function $h: (0,\infty) \to [0,\infty)$ is said to have index ρ at ∞, or to be regularly varying at ∞ with index ρ, if for every $\lambda > 0$,

$$\frac{h(\lambda t)}{h(t)} \to \lambda^{\rho} \quad \text{as} \quad t \to \infty.$$

Obviously, $h(t) = t^{\rho}$ or any function for which $h(t)t^{-\rho}$ converges to a positive finite constant as $t \to \infty$ has index ρ, but so does $t^{\rho}(\log t)^{q}$ for any real q (Exercise 19). A function $h: (0,\infty) \to [0,\infty)$ is said to have index ρ at 0 if $h(1/t)$ has index $-\rho$ at ∞. See Bingham, Goldie and Teugels (1987) for a comprehensive account of regularly varying functions. For a positive finite measure F on $\mathbb{R}$, define the tail area of the spectrum

$$H(x) = F(\mathbb{R}) - F((-\infty, x]) + F((-\infty, -x])$$

and the cosine transform $U(t) = \int_{\mathbb{R}} \cos tx F(dx)$. Pitman (1968) also gives results for the sine transform of F, but this is 0 when F is the spectrum of a real process.

Theorem 3 (Abelian Theorem). *Suppose, for an integer* $n \geq 0$, $\mu_{2n} = \int_{\mathbb{R}} x^{2n} F(dx) < \infty$. *Define*

$$U_{2n}(t) = \sum_{r=0}^{n} (-1)^r \frac{\mu_{2r} t^{2r}}{(2r)!} - U(t).$$

If H has index $-\tau$ at ∞ with $2n < \tau < 2n+2$ then $U_{2n}(t) \sim S(\tau)H(1/t)$ as $t \downarrow 0$, where $S(\tau) = \pi/\{2\Gamma(\tau)\sin\left(\frac{1}{2}\pi\tau\right)\}$ for $\tau > 0$. If $\tau = 2n+2$, then

$$U_{2n}(t) \sim (-1)^n t^{2n+2} \int_0^{1/t} \frac{x^{2n+1}}{(2n+1)!} H(x)dx. \tag{19}$$

Pitman (1968) gives the proof for this general result. I only consider the special case for which $H(x) \sim cx^{-\tau}$ as $x \to \infty$ for some $c > 0$ and $0 < \tau < 2$. Note that the function U_{2n} is generally a principal irregular part of U.

Using integration by parts,

$$\frac{\mu_0 - U(t)}{t} = \int_0^{\infty} H(x) \sin tx \, dx,$$

so that

$$\frac{\mu_0 - U(t)}{H(1/t)} = \int_0^{\infty} \frac{H(x/t)}{H(1/t)} \sin x \, dx.$$

Since H is bounded and $H(x) \sim cx^{-\tau}$ as $x \to \infty$, there exists a finite constant A such that $H(x) \leq A(1+x)^{-\tau}$ for all $x > 0$ and $H(x) \geq \frac{1}{2}cx^{-\tau}$ for all x sufficiently large. Hence, for all $x > 0$ and all t sufficiently small,

$$\left|\frac{H(x/t)}{H(1/t)}\sin x\right| \leq \frac{2A}{c(t+x)^{\tau}},$$

which is integrable on $[0,p]$ for any finite p. Furthermore, for any $x > 0$,

$$\lim_{t\downarrow 0}\frac{H(x/t)}{H(1/t)}\sin x = x^{-\tau}\sin x,$$

so that

$$\lim_{t\downarrow 0}\int_0^p \frac{H(x/t)}{H(1/t)}\sin x\,dx = \int_0^p x^{-\tau}\sin x\,dx$$

for any finite p. Let q be the smallest integer such that $2\pi q \geq p$. Then

$$\begin{aligned}
&\left|\int_p^\infty \frac{H(x/t)}{H(1/t)}\sin x\,dx\right| \\
&\quad\leq 2\pi\frac{H(p/t)}{H(1/t)} + \frac{1}{H(1/t)}\sum_{j=q}^{\infty}\left|\int_{2\pi j}^{2\pi(j+1)} H\left(\frac{x}{t}\right)\sin x\,dx\right| \\
&\quad\leq 2\pi\frac{H(p/t)}{H(1/t)} + \frac{2\pi}{H(1/t)}\sum_{j=q}^{\infty}\left|H\left(\frac{2\pi j}{t}\right) - H\left(\frac{2\pi(j+1)}{t}\right)\right| \\
&\quad\leq \frac{4\pi H(p/t)}{H(1/t)},
\end{aligned}$$

which can be made arbitrarily small by taking t small and p large. It follows that

$$\lim_{t\downarrow 0}\int_0^\infty \frac{H(x/t)}{H(1/t)}\sin x\,dx = \int_0^\infty x^{-\tau}\sin x\,dx = S(\tau),$$

where the last equality is by 3.761.4 of Gradshteyn and Ryzhik (1994). Thus, $\mu_0 - U(t) \sim S(\tau)H(1/t)$, proving the theorem when $H(x) \sim cx^{-\tau}$ for $0 < \tau < 2$.

Theorem 4 (Tauberian Theorem). *If $\mu_0 - U(t)$ is of index τ as $t \downarrow 0$ and $0 < \tau < 2$, then*

$$H(x) \sim \frac{1 - U(1/x)}{S(\tau)} \quad \text{as} \quad x \to \infty.$$

This is part of Theorem 5 of Pitman (1968). Even if we restrict to $\mu_0 - U(t) \sim ct^{\tau}$ as $t \downarrow 0$, the proof is quite a bit more delicate than for Theorem 3.

Exercises

19 Show that $t^{\rho}(\log t)^{q}$ has index ρ as $t \to \infty$ for any real q.

20 In the proof of the special case of Theorem 3, use the Second Mean Value Theorem for Integrals (see Spivak (1980, p. 367) or Phillips (1984, p. 302) for a more general version) to provide an alternative argument to justify that $\int_p^\infty \{H(x/t)/H(1/t)\} \sin x \, dx$ can be made arbitrarily small by taking t small and p large. This is the argument Pitman (1968) uses. The Second Mean Value Theorem for Integrals was unfamiliar to me prior to my reading Pitman's paper but is a result well worth knowing.

21 If $H(x) \sim x^{-2} \log x$ as $x \to \infty$, find an explicit expression in terms of elementary functions for a function U_2 satisfying (19).

22 For the autocovariance function $e^{-|t|^\delta}$ on $\mathbb{R}$ with $0 < \delta < 2$, show that the corresponding spectral density is asymptotic to $c\omega^{-\delta-1}$ as $\omega \to \infty$ and find c as a function of δ.

2.9 Random fields with nonintegrable spectral densities

This section explores what one might mean by a random field with nonintegrable spectral density. Bochner's Theorem tells us that the corresponding random field cannot be both weakly stationary and mean square continuous.

Intrinsic random functions

If f is nonintegrable in a neighborhood of the origin, then the corresponding random field is nonstationary and corresponds to what Matheron (1973) calls an intrinsic random function. Intuitively, there is then so much variation at low frequencies that the random field cannot have some constant level of variation about its mean. For example, consider the function on $\mathbb{R}$ $f(\omega) = |\omega|^{-\alpha}$ for some $\alpha \in (1, 3)$. If this were the spectral density of a weakly stationary process Z, then we would have

$$\operatorname{var}\left\{\sum_{j=1}^n c_j Z(t_j)\right\} = \int_{-\infty}^\infty \left|\sum_{j=1}^n c_j \exp(i\omega t_j)\right|^2 |\omega|^{-\alpha} d\omega. \tag{20}$$

Formally evaluating this expression for $n = 1$, $c_1 = 1$ and any t_1 gives $\operatorname{var}\{Z(t_1)\} = \int_{-\infty}^\infty |\omega|^{-\alpha} d\omega = \infty$. However, if $\sum_{j=1}^n c_j = 0$, then

$$\left|\sum_{j=1}^n c_j \exp(i\omega t_j)\right|^2 = O(\omega^2) \tag{21}$$

in a neighborhood of the origin (Exercise 23), so the integral on the right side of (20) is then finite for $1 < \alpha < 3$. Furthermore, using $\sum_{j=1}^n c_j = 0$

to justify the following second equality,

$$\begin{aligned}
&\int_{-\infty}^{\infty} \left| \sum_{j=1}^{n} c_j \exp(i\omega t_j) \right|^2 |\omega|^{-\alpha} d\omega \\
&= \int_{-\infty}^{\infty} \sum_{j,k=1}^{n} c_j c_k \cos\{\omega(t_j - t_k)\} |\omega|^{-\alpha} d\omega \\
&= \int_{-\infty}^{\infty} \sum_{j,k=1}^{n} c_j c_k [\cos\{\omega(t_j - t_k)\} - 1] |\omega|^{-\alpha} d\omega \\
&= \sum_{j,k=1}^{n} c_j c_k \int_{-\infty}^{\infty} [\cos\{\omega(t_j - t_k)\} - 1] |\omega|^{-\alpha} d\omega \\
&= -\frac{\pi}{\Gamma(\alpha)\sin\left\{\frac{1}{2}\pi(\alpha-1)\right\}} \sum_{j,k=1}^{n} c_j c_k |t_j - t_k|^{\alpha-1}
\end{aligned}$$

by 3.823 of Gradshteyn and Ryzhik (1994). We see that the function

$$G(t) = -\frac{\pi|t|^{\alpha-1}}{\Gamma(\alpha)\sin\left\{\frac{1}{2}\pi(\alpha-1)\right\}}$$

behaves like an autocovariance function in that $\sum_{j,k=1}^{n} c_j c_k G(t_j - t_k) \geq 0$ whenever $\sum_{j=1}^{n} c_j = 0$. It is possible to show that for $1 < \alpha < 3$, there exist processes (nonstationary, of course) for which

$$\operatorname{var}\left\{\sum_{j=1}^{n} c_j Z(t_j)\right\} = -\phi \sum_{j,k=1}^{n} c_j c_k |t_j - t_k|^{\alpha-1}$$

for some $\phi > 0$ whenever $\sum_{j=1}^{n} c_j = 0$. Brownian motion is an example of such a process with $\alpha = 2$. A Gaussian process with $1 < \alpha < 3$ but $\alpha \neq 2$ is known as a fractional Brownian motion (Mandelbrot and Van Ness 1968); see Voss (1988) for an elementary introduction to such processes.

Let us consider extending these ideas to positive symmetric measures F on $\mathbb{R}^d$ satisfying

$$\int_{\mathbb{R}^d} \frac{|\boldsymbol{\omega}|^{2r+2}}{1 + |\boldsymbol{\omega}|^{2r+2}} F(d\boldsymbol{\omega}) < \infty \tag{22}$$

for a nonnegative integer r. If we restrict attention to $c_1, \ldots, c_n \in \mathbb{R}$ and $\mathbf{x}_1, \ldots, \mathbf{x}_n \in \mathbb{R}^d$ such that $\sum_{j=1}^{n} c_j \exp(i\boldsymbol{\omega}^T \mathbf{x}_j) = O(|\boldsymbol{\omega}|^{r+1})$ in $\boldsymbol{\omega}$, then (22) implies $\int_{\mathbb{R}^d} \left|\sum_{j=1}^{n} c_j \exp(i\boldsymbol{\omega}^T \mathbf{x}_j)\right|^2 F(d\boldsymbol{\omega}) < \infty$ (Exercise 24). For $\mathbf{x} = (x_1, \ldots, x_d)$ and $\boldsymbol{\alpha} = (\alpha_1, \ldots, \alpha_d)$, define $\mathbf{x}^{\boldsymbol{\alpha}} = \prod_{i=1}^{d} x_i^{\alpha_i}$ and let D_r be the set of all d-tuples whose components are nonnegative integers summing to at most r. Then if $\sum_{j=1}^{n} c_j \mathbf{x}_j^{\boldsymbol{\alpha}} = 0$ for all $\boldsymbol{\alpha} \in D_r$, $\int_{\mathbb{R}^d} \left|\sum_{j=1}^{n} c_j \exp(i\boldsymbol{\omega}^T \mathbf{x}_j)\right|^2 F(d\boldsymbol{\omega}) < \infty$, as required. Since the Fourier transform of F will not be defined in the ordinary sense when F has infinite mass in any neighborhood of the origin, we need to modify our definition of its transform. Set $Q_r(t) = \sum_{j=0}^{r}(-t^2)^j/(2j)!$, which is just the first $r+1$ nonzero terms in the Taylor series for $\cos t$. Define

$$G(\mathbf{x}) = \int_{b_d} \left\{\cos(\boldsymbol{\omega}^T\mathbf{x}) - Q_r(\boldsymbol{\omega}^T\mathbf{x})\right\} F(d\boldsymbol{\omega}) + \int_{b_d^c} \cos(\boldsymbol{\omega}^T\mathbf{x}) F(d\boldsymbol{\omega}), \quad (23)$$

where b_d is the ball of radius 1 centered at the origin and the superscript c indicates complement. Since $|\cos(\boldsymbol{\omega}^T\mathbf{x}) - Q_r(\boldsymbol{\omega}^T\mathbf{x})| = O\left(|\boldsymbol{\omega}|^{2r+2}\right)$ for any fixed $\mathbf{x}$, the first integral on the right side of (23) is well defined for F satisfying (22). Furthermore, if $\sum_{j=1}^{n} c_j \mathbf{x}_j^{\boldsymbol{\alpha}} = 0$ for all $\boldsymbol{\alpha} \in D_r$, then

$$\sum_{j,k=1}^{n} c_j c_k G(\mathbf{x}_j - \mathbf{x}_k) = \int_{\mathbb{R}^d} \left|\sum_{j=1}^{n} c_j \exp(i\boldsymbol{\omega}^T\mathbf{x}_j)\right|^2 F(d\boldsymbol{\omega}). \quad (24)$$

The choice of b_d in (23) is arbitrary; we could just as well use any bounded region containing a neighborhood of the origin and (24) would still hold. Matheron (1973) shows that for any positive, symmetric measure F satisfying (22) there is a real random field Z for which $\operatorname{var}\left\{\sum_{j=1}^{n} c_j Z(\mathbf{x}_j)\right\}$ is given by $\int_{\mathbb{R}^d} \left|\sum_{j=1}^{n} c_j \exp(i\boldsymbol{\omega}^T\mathbf{x}_j)\right|^2 F(d\boldsymbol{\omega})$ whenever $\sum_{j=1}^{n} c_j \mathbf{x}_j^{\boldsymbol{\alpha}} = 0$ for all $\boldsymbol{\alpha} \in D_r$. Matheron calls such a random field an intrinsic random function of order r, or r-IRF. In addition, he calls G a generalized covariance function for Z if it is symmetric and

$$\operatorname{var}\left\{\sum_{j=1}^{n} c_j Z(\mathbf{x}_j)\right\} = \sum_{j,k=1}^{n} c_j c_k G(\mathbf{x}_j - \mathbf{x}_k) \quad (25)$$

whenever $\sum_{j=1}^{n} c_j \mathbf{x}_j^{\boldsymbol{\alpha}} = 0$ for all $\boldsymbol{\alpha} \in D_r$. To be consistent with the terminology used here, I call G a generalized autocovariance function. A symmetric real-valued function G on $\mathbb{R}^d$ is said to be conditionally positive definite of order r if $\sum_{j,k=1}^{n} c_j c_k G(\mathbf{x}_k - \mathbf{x}_j) \geq 0$ whenever $\sum_{j=1}^{n} c_j \mathbf{x}_j^{\boldsymbol{\alpha}} = 0$ for all $\boldsymbol{\alpha} \in D_r$, so that any generalized autocovariance function for an r-IRF is conditionally positive definite of order r. A minor adaptation of Theorem 2.1 of Matheron (1973) gives that a continuous symmetric G on $\mathbb{R}^d$ is conditionally positive definite of order r if and only if it can be written in the form

$$G(\mathbf{x}) = \int_{\mathbb{R}^d} \left[\cos(\boldsymbol{\omega}^T\mathbf{x}) - Q_r(\boldsymbol{\omega}^T\mathbf{x}) 1\left\{|\boldsymbol{\omega}| \leq 1\right\}\right] F(d\boldsymbol{\omega}) + P(\mathbf{x}),$$

where F is a positive symmetric measure satisfying (22) and P is an even polynomial of order at most $2r+2$ that is conditionally positive definite of order r. It is trivially true that every even polynomial of order at most $2r$ is conditionally positive definite of order r, since for any such polynomial P, $\sum_{j,k=1}^{n} c_j c_k P(\mathbf{x}_j - \mathbf{x}_k) = 0$ whenever $\sum_{j=1}^{n} c_j \mathbf{x}_j^{\boldsymbol{\alpha}} = 0$ for all $\boldsymbol{\alpha} \in D_r$. It follows that if G is a generalized autocovariance function for the r-IRF Z, then so is G plus any even polynomial of order at most $2r$.

An even polynomial of order $2r+2$ may or may not be conditionally positive definite of order r. For example, for $\sum_{j=1}^n c_j = 0$ and $t_1, \dots, t_n \in \mathbb{R}$, $\sum_{j,k=1}^n c_j c_k (t_j - t_k)^2 = -2\left(\sum_{j=1}^n c_j t_j\right)^2$, so that $P(t) = -at^2$ is conditionally positive definite of order 0 if and only if $a \geq 0$. Micchelli (1986) gives useful conditions under which functions of $\mathbf{x}$ that depend only on $|\mathbf{x}|$ are conditionally positive definite.

IRFs can be written as a sum of a very smooth IRF and a stationary random field. Specifically, suppose F has infinite mass in any neighborhood of $\mathbf{0}$ but has finite mass on any set bounded away from $\mathbf{0}$. The form of G in (23) implies that the corresponding r-IRF Z can be written as $Z_1 + Z_2$ where Z_1 and Z_2 are uncorrelated random fields, Z_1 has spectral measure $F(\mathbf{d}\boldsymbol{\omega})1\{|\boldsymbol{\omega}| \leq 1\}$ and hence is very smooth but nonstationary and Z_2 has spectral measure $F(\mathbf{d}\boldsymbol{\omega})1\{|\boldsymbol{\omega}| > 1\}$ and hence is stationary. In particular, in one dimension $Z_1^{(r+1)}$ will be a stationary analytic process with spectral measure $\omega^{2r+2}F(d\omega)1\{|\omega| \leq 1\}$. Matheron (1973, Theorem 1.5) gives a different decomposition of an r-IRF into a very smooth r-IRF and a stationary random field. These decompositions imply that in terms of the local behavior of a random field, r-IRFs do not provide any additional flexibility over stationary random fields. In spectral terms, the spectrum of an r-IRF must have finite mass on any set bounded away from $\mathbf{0}$, so that r-IRFs are no more general than stationary random fields in their high frequency behavior.

There is a nice mathematical connection between r-IRFs and best linear unbiased prediction. If we suppose the mean function of an r-IRF Z is of the form $EZ(\mathbf{x}) = \sum_{\boldsymbol{\alpha} \in D_r} \beta_{\boldsymbol{\alpha}} \mathbf{x}^{\boldsymbol{\alpha}}$, where the $\beta_{\boldsymbol{\alpha}}s$ are unknown, then the variance of the error of any linear unbiased predictor depends on the covariance structure only through a generalized autocovariance function G for Z and is independent of the equivalent form of the generalized autocovariance function that is selected (Exercise 26). Hence, if $Z(\mathbf{x})$ possesses a linear unbiased predictor, then we can find its BLUP and the mse of the BLUP from just knowing G. In other words, the r-IRF model only defines the covariance structure for contrasts of the random field Z, but in order to determine BLUPs and their mses, that is all we need to know (see 1.5).

Semivariograms

In practice, the most frequently used class of IRFs is the 0-IRFs. For a 0-IRF Z with generalized autocovariance function G, $\operatorname{var}\{Z(\mathbf{x}) - Z(\mathbf{y})\} = 2G(\mathbf{0}) - 2G(\mathbf{x} - \mathbf{y})$. Define the semivariogram γ of a 0-IRF by $\gamma(\mathbf{x}) = \frac{1}{2}\operatorname{var}\{Z(\mathbf{x}) - Z(\mathbf{0})\}$. Then $-\gamma$ is a generalized autocovariance function for Z. The semivariogram is commonly used for modeling random fields in the geostatistical literature (Journel and Huijbregts 1978; Isaaks and Srivastava 1989; Cressie 1993). See Cressie (1988) for some historical notes on semivariograms. One reason for its popularity is that there is a convenient way to estimate $\gamma(\mathbf{x})$. For simplicity, suppose the observations $\mathbf{x}_1, \ldots, \mathbf{x}_n$ form some repeating pattern so that there are vectors $\mathbf{x}$ for which $\mathbf{x}_i - \mathbf{x}_j = \mathbf{x}$ for many pairs of observations $\mathbf{x}_i$ and $\mathbf{x}_j$. For such a vector $\mathbf{x}$, an unbiased estimator of $\gamma(\mathbf{x})$ is

$$\hat{\gamma}(\mathbf{x}) = \frac{1}{2n(\mathbf{x})} \sum_{\mathbf{x}_i - \mathbf{x}_j = \mathbf{x}} \{Z(\mathbf{x}_i) - Z(\mathbf{x}_j)\}^2,$$

where $n(\mathbf{x})$ is the number of pairs of observations whose difference equals $\mathbf{x}$. Note that $E\{Z(\mathbf{x}_i) - Z(\mathbf{x}_j)\}^2$ can be determined from γ because $Z(\mathbf{x}_i) - Z(\mathbf{x}_j)$ satisfies (25) with $r = 0$. If the observations are irregularly located, it is usually necessary to average over pairs of points whose difference is nearly $\mathbf{x}$, although how one defines "nearly" is a nontrivial problem (Cressie 1993, pp. 69–70). Although plots of these estimates versus $\mathbf{x}$ or, more commonly, $|\mathbf{x}|$, can be helpful in identifying structures in spatial data, they do not directly provide a method for estimating the function γ at all $\mathbf{x}$ up to some magnitude, which is what we need for prediction. Chapter 6 provides further discussion on estimating semivariograms.

Generalized random fields

By considering spectral densities that are not integrable at infinity, we get what are known as generalized random fields. The random field is generalized in the sense that its pointwise evaluation is not defined, but only certain linear functionals of it. In terms of the spectral representation of the random field given in (7), there is so much variation at high frequencies that the Fourier transform of the random measure does not converge pointwise. The best-known generalized process is white noise, which can be thought of as a continuous time analogue to a sequence of independent and identically distributed observations. White noise has constant spectral density over all of $\mathbb{R}$. The name derives from the fact that white light is approximately an equal mixture of all visible frequencies of light, which was demonstrated by Isaac Newton. For $d = 1$, the autocovariance function corresponding to the density $2\pi c$ should then be $K(t) = 2\pi c \int_{-\infty}^{\infty} e^{i\omega t} d\omega$. However, this integral is ∞ for $t = 0$ and is undefined in the ordinary sense for $t \neq 0$. Using the theory of generalized functions (Gel'fand and Vilenkin 1964), it is possible to show that a reasonable definition for $K(t)$ is $c\delta_t$, where δ_t is the Dirac delta-function, a generalized function satisfying $\int_{-\infty}^{\infty} \delta_t g(t)\, dt = g(0)$ for all sufficiently smooth g. To see this relationship between the Dirac delta-function and the uniform spectral density on $\mathbb{R}$, consider evaluating $\operatorname{var}\left\{\int_{-\infty}^{\infty} h(t) Z(t)\, dt\right\}$ when Z is white noise and h is square integrable. From the time domain, proceeding formally,

$$\begin{aligned}\operatorname{var}\left\{\int_{-\infty}^{\infty} h(t)Z(t)\,dt\right\} &= \int_{-\infty}^{\infty}\int_{-\infty}^{\infty} h(s)h(t)\operatorname{cov}\{Z(s),Z(t)\}\,ds\,dt \\ &= c\int_{-\infty}^{\infty}\int_{-\infty}^{\infty} h(s)h(t)\delta_{s-t}ds\,dt \\ &= c\int_{-\infty}^{\infty} h(t)^2dt.\end{aligned}$$

From the spectral domain, letting $H(\omega) = \int_{-\infty}^{\infty} h(t)e^{i\omega t}dt$ and again proceeding formally,

$$\begin{aligned}&\operatorname{var}\left\{\int_{-\infty}^{\infty} h(t)Z(t)\,dt\right\} \\ &= \int_{-\infty}^{\infty}\int_{-\infty}^{\infty} h(s)h(t)\left\{\int_{-\infty}^{\infty} 2\pi c e^{i\omega(s-t)}d\omega\right\} ds\,dt \\ &= 2\pi c\int_{-\infty}^{\infty}\left\{\int_{-\infty}^{\infty} h(s)e^{i\omega s}ds\right\}\left\{\int_{-\infty}^{\infty} h(t)e^{-i\omega t}dt\right\} d\omega \\ &= 2\pi c\int_{-\infty}^{\infty} |H(\omega)|^2\,d\omega \\ &= c\int_{-\infty}^{\infty} h(t)^2dt,\end{aligned}$$

where the last equality is by Parseval's relation. If $h(t) = 1\{a < t < b\}/(b-a)$, then $\operatorname{var}\left\{\int_{-\infty}^{\infty} h(t)Z(t)\,dt\right\} = c/(b-a)$, so that the variance of an average of Z over an interval tends to ∞ as the length of the interval tends to 0. This result is in line with my previous statement that white noise is not defined pointwise.

One way to think about white noise is as a generalized derivative of Brownian motion, so that the spectral density of white noise should be ω^2 times the spectral density of Brownian motion, which is the case since, as we noted earlier, the spectral density for Brownian motion is proportional to ω^{-2}. Equivalently, Brownian motion can be interpreted as an integral of white noise. This result is the continuous time analogue to a random walk being a sum of independent and identically distributed random variables. Gel'fand and Vilenkin (1964) provide a rigorous development of random fields that includes nonintegrability of the spectral density at both the origin and infinity as special cases. Yaglom (1987a, Section 24) provides a more accessible treatment of these topics.

Exercises

23 For $\sum_{j=1}^{n} c_j = 0$, show that (21) holds for ω in a neighborhood of the origin.

24 If $c_1, \ldots, c_n \in \mathbb{R}$ and $\mathbf{x}_1, \ldots, \mathbf{x}_n \in \mathbb{R}^d$ satisfy $\sum_{j=1}^n c_j \exp(i\boldsymbol{\omega}^T \mathbf{x}_j) = O(|\boldsymbol{\omega}|^{r+1})$ in $\boldsymbol{\omega}$, show that (22) implies

$$\int_{\mathbb{R}^d} \left| \sum_{j=1}^n c_j \exp(i\boldsymbol{\omega}^T \mathbf{x}_j) \right|^2 F(\mathbf{d}\boldsymbol{\omega}) < \infty.$$

25 Show that the function $K(s,t) = |s|^\alpha + |t|^\alpha - |s-t|^\alpha$ is positive definite on $\mathbb{R} \times \mathbb{R}$ for $0 < \alpha < 2$.

26 For $\alpha > 0$ and $\lfloor \cdot \rfloor$ the greatest integer function, show that $G(t) = (-1)^{1+\lfloor \alpha/2 \rfloor} |t|^\alpha$ is a generalized autocovariance function for an $\lfloor \alpha/2 \rfloor$-IRF on $\mathbb{R}$. Find the corresponding spectral measure.

27 For a positive integer m, show that the function $G(t) = (-1)^{m-1}|t|^{2m} \times \log|t|$ is conditionally positive definite of order m. Find the corresponding spectral measure.

28 Extend the preceding two exercises to isotropic random fields on $\mathbb{R}^d$.

29 Suppose the mean function of an r-IRF Z is of the form $EZ(\mathbf{x}) = \sum_{\alpha \in D_r} \beta_\alpha \mathbf{x}^\alpha$, where the β_αs are unknown. Show that the variance of the error of any linear unbiased predictor depends on the covariance structure only through a generalized autocovariance function G for Z and is independent of the equivalent form of the generalized autocovariance function that is selected.

2.10 Isotropic autocovariance functions

The class of all continuous autocovariance functions on $\mathbb{R}^d$ can be characterized as the Fourier transforms of all finite positive measures on $\mathbb{R}^d$. Adding the requirement that the random field be weakly isotropic, we now seek an analogous characterization of isotropic autocovariance functions for random fields on $\mathbb{R}^d$. In addition, we consider smoothness properties of isotropic autocovariance functions at positive distances. A number of the topics in this section follow the development in Section 22.1 of Yaglom (1987a).

Characterization

Suppose $K(r)$, $r \geq 0$, is an isotropic autocovariance function in $\mathbb{R}^d$; that is, there exists a weakly isotropic complex-valued random field Z on $\mathbb{R}^d$ such that $\operatorname{cov}\{Z(\mathbf{x}), \overline{Z(\mathbf{y})}\} = K(|\mathbf{x} - \mathbf{y}|)$ for all $\mathbf{x}, \mathbf{y} \in \mathbb{R}^d$. For $\mathbf{x} \in \mathbb{R}^d$, we have $K(|\mathbf{x}|) = K(|-\mathbf{x}|) = \overline{K(|\mathbf{x}|)}$, so that K must be real. So, if $Z(\mathbf{x}) = V(\mathbf{x}) + iW(\mathbf{x})$ with V and W real, then $\operatorname{cov}\{V(\mathbf{x}), W(\mathbf{y})\} = \operatorname{cov}\{V(\mathbf{y}), W(\mathbf{x})\}$.

By Bochner's Theorem, there exists a positive finite measure F such that for all $\mathbf{x} \in \mathbb{R}^d$

$$K(|\mathbf{x}|) = \int_{\mathbb{R}^d} \exp(i\boldsymbol{\omega}^T\mathbf{x}) F(\mathbf{d}\boldsymbol{\omega}).$$

Since $K(\mathbf{x}) = K(-\mathbf{x})$, without loss of generality, we can take F to be symmetric about the origin. Furthermore, by isotropy,

$$K(r) = \int_{\partial b_d} K(r|\mathbf{x}|) U(\mathbf{dx}),$$

where U is the uniform probability measure on ∂b_d, the d-dimensional unit sphere. Thus,

$$\begin{aligned} K(r) &= \int_{\partial b_d} \left\{ \int_{\mathbb{R}^d} \exp(ir\boldsymbol{\omega}^T\mathbf{x}) F(\mathbf{d}\boldsymbol{\omega}) \right\} U(\mathbf{dx}) \\ &= \int_{\mathbb{R}^d} \left\{ \int_{\partial b_d} \exp(ir\boldsymbol{\omega}^T\mathbf{x}) U(\mathbf{dx}) \right\} F(\mathbf{d}\boldsymbol{\omega}) \\ &= \int_{\mathbb{R}^d} \left\{ \int_{\partial b_d} \cos(r\boldsymbol{\omega}^T\mathbf{x}) U(\mathbf{dx}) \right\} F(\mathbf{d}\boldsymbol{\omega}), \end{aligned}$$

since the imaginary part of the integral drops out due to the symmetry of F. It is clear that the inner integral over $\mathbf{x}$ depends on $\boldsymbol{\omega}$ only through its length $|\boldsymbol{\omega}|$, so take $\boldsymbol{\omega}$ to point in the direction of the "north pole" and switch to spherical coordinates with ϕ measuring the angle from the pole. For given ϕ, the region of integration over the other $d-2$ coordinates of ∂b_d is a $(d-1)$-dimensional sphere of radius $\sin\phi$, so using the fact that the surface area of a unit sphere in d dimensions is $A_d = 2\pi^{d/2}/\Gamma(d/2)$,

$$\begin{aligned} &\int_{\partial b_d} \cos(r\boldsymbol{\omega}^T\mathbf{x})\, U(\mathbf{dx}) \\ &\qquad = \frac{1}{A_d} \int_0^\pi \cos\left(r|\boldsymbol{\omega}|\cos\phi\right) A_{d-1} (\sin\phi)^{d-2} d\phi \\ &\qquad = \Gamma(d/2) \left(\frac{2}{r|\boldsymbol{\omega}|}\right)^{(d-2)/2} J_{(d-2)/2}(r|\boldsymbol{\omega}|) \end{aligned}$$

using a standard integral representation for the ordinary Bessel function J_ν (see 9.1.20 of Abramowitz and Stegun 1965). Thus,

$$K(r) = \Gamma(d/2) \int_{\mathbb{R}^d} \left(\frac{2}{r|\boldsymbol{\omega}|}\right)^{(d-2)/2} J_{(d-2)/2}\left(r|\boldsymbol{\omega}|\right) F(\mathbf{d}\boldsymbol{\omega}).$$

Letting $G(u) = \int_{|\omega|<u} F(\mathbf{d}\boldsymbol{\omega})$, then for $r > 0$,

$$K(r) = 2^{(d-2)/2}\Gamma(d/2) \int_0^\infty (ru)^{-(d-2)/2} J_{(d-2)/2}(ru)\, dG(u), \qquad (26)$$

where G is nondecreasing, bounded on $[0, \infty)$ and $G(0) = 0$. The right side of (26) is known as the Hankel transform of order $\frac{1}{2}(d-2)$ of G. We have the following.

Theorem 5. *For $d \geq 2$, a function K is a continuous isotropic autocovariance function for a random field on $\mathbb{R}^d$ if and only if it can be represented as in (26) with G nondecreasing, bounded on $[0, \infty)$ and $G(0) = 0$.*

For d odd, (26) can be expressed in terms of elementary functions. For example, for $d = 3$, $K(r) = \int_0^\infty (ru)^{-1} \sin(ru) dG(u)$. It is often difficult to determine whether a given function can be written as in (26). Christakos (1984) and Pasenchenko (1996) give sufficient conditions for a function to be an isotropic autocovariance function that can be easily verified in some circumstances.

Let $\mathcal{D}_d$ be the class of continuous isotropic autocovariance functions in d dimensions and $\mathcal{D}_\infty = \bigcap_{d=1}^\infty \mathcal{D}_d$ the class of functions that are isotropic continuous autocovariance functions in all dimensions. By considering a d-dimensional weakly isotropic random field along m coordinates, $m < d$, we obtain an m-dimensional weakly isotropic random field, so that $\mathcal{D}_d \subset \mathcal{D}_m$. Thus, $\mathcal{D}_1 \supset \mathcal{D}_2 \supset \cdots \supset \mathcal{D}_\infty$.

To characterize the elements of $\mathcal{D}_\infty$, define

$$\begin{aligned}\Lambda_d(t) &= 2^{(d-2)/2}\Gamma(d/2)t^{-(d-2)/2}J_{(d-2)/2}(t) \\ &= \Gamma(d/2)\sum_{j=0}^\infty \frac{\left(-\frac{1}{4}t^2\right)^j}{j!\Gamma\left(\frac{d}{2}+j\right)} \\ &= 1 - \frac{t^2}{2d} + \frac{t^4}{8d(d+2)} \cdots\end{aligned}$$

so that for fixed t,

$$\Lambda_d\left((2d)^{1/2}t\right) \to e^{-t^2} \quad \text{as } d \to \infty. \tag{27}$$

This suggests that a function is in $\mathcal{D}_\infty$ if and only if it has the representation

$$K(r) = \int_0^\infty e^{-r^2u^2} dG(u) \tag{28}$$

for G bounded nondecreasing on $[0, \infty)$. To prove that all functions of this form are in $\mathcal{D}_\infty$, note that the density $(2\pi)^{-d/2} \exp\{-|\mathbf{x}|^2/(2\sigma^2)\}$ on $\mathbb{R}^d$ has joint characteristic function $\exp(-\sigma^2|\boldsymbol{\omega}|^2/2)$, so that $e^{-r^2u^2} \in \mathcal{D}_d$ for all d and all $r \geq 0$, hence $e^{-r^2u^2} \in \mathcal{D}_\infty$. Then $\int_0^\infty e^{-r^2u^2} dG(u)$ can be expressed as a pointwise limit of positive sums of functions of the form $e^{-r^2u^2}$, so a function of the form (28) with G bounded nondecreasing on $[0, \infty)$ is in $\mathcal{D}_\infty$.

Schoenberg (1938) proved the converse result. His argument was to first show the convergence in (27) is uniform in t; I refer the reader to Schoenberg's paper for this part of the proof. Write G in (26) as G_d now to indicate its dependence on d and set, without loss of generality, $G_d(0) = 0$. Define $\tilde{G}_d$ by $\tilde{G}_d(u) = G_d\{(2d)^{1/2}u\}$. If $K \in \mathcal{D}_\infty$, then for all d,

$$K(r) = \int_0^\infty \Lambda_d\{(2d)^{1/2}ru\}\, d\tilde{G}_d(u),$$

where the $\tilde{G}_d$s are uniformly bounded since $K(0) = \tilde{G}_d(\infty)$. By Helly's selection principle, there is a subsequence $\{\tilde{G}_{d_j}\}_{j=1}^\infty$ converging vaguely to monotone G with $G(0) = 0$ (Chung 1974, p. 83). Now, for given $r \geq 0$ and all j,

$$\begin{aligned}
&\left|K(r) - \int_0^\infty e^{-r^2u^2} dG(u)\right| \\
&\quad \leq \left|\int_0^\infty \Lambda_{d_j}\{(2d_j)^{1/2}ru\}\, d\tilde{G}_{d_j}(u) - \int_0^\infty e^{-r^2u^2} d\tilde{G}_{d_j}(u)\right| \\
&\qquad + \left|\int_0^\infty e^{-r^2u^2} d\tilde{G}_{d_j}(u) - \int_0^\infty e^{-r^2u^2} dG(u)\right|.
\end{aligned}$$

The first term on the right side tends to 0 as $j \to \infty$ because of the uniform convergence in (27) and $\{\tilde{G}_{d_j}\}$ uniformly bounded. The second tends to 0 for any given $r > 0$ by the vague convergence of $\{\tilde{G}_{d_j}\}$ and $e^{-r^2u^2} \to 0$ as $u \to \infty$ (Chung 1974, Theorem 4.4.1). Since neither $K(r)$ nor $\int_0^\infty e^{-r^2u^2} dG(u)$ depend on j, the two functions must be identical.

Lower bound on isotropic autocorrelation functions

Define the isotropic autocorrelation function $C(r) = K(r)/K(0)$. A function C is an isotropic autocorrelation function for a random field on $\mathbb{R}^d$ if and only if it is of the form $C(r) = \int_0^\infty \Lambda_d(ru)dG(u)$, where $\int_0^\infty dG(u) = 1$ and G is nondecreasing. Thus, for all r,

$$C(r) \geq \inf_{s \geq 0} \Lambda_d(s).$$

For $d = 2$,

$$C(r) \geq \inf_{s \geq 0} J_0(s) \approx -0.403,$$

for $d = 3$,

$$C(r) \geq \inf_{s \geq 0} \frac{\sin s}{s} \approx -0.218$$

and for $d = \infty$, $C(r) \geq 0$. For all finite d, the infimum of $\Lambda_d(s)$ is attained at a finite value of s, so the lower bounds on C are achievable (Exercise 30). For $d = \infty$, the bound cannot be achieved since $e^{-t^2} > 0$ for all t, so that $K(r) > 0$ for all r if $K \in \mathcal{D}_\infty$ (unless $K(0) = 0$).

Inversion formula

The relationship (26) has a simple inversion if $\int_0^\infty r^{d-1}|K(r)|\,dr < \infty$. In this case, there exists a nonnegative function f with $\int_0^\infty u^{d-1}f(u)\,du < \infty$ such that

$$f(u) = (2\pi)^{-d/2}\int_0^\infty (ur)^{-(d-2)/2}J_{(d-2)/2}(ur)r^{d-1}K(r)\,dr$$

and

$$K(r) = (2\pi)^{d/2}\int_0^\infty (ru)^{-(d-2)/2}J_{(d-2)/2}(ru)u^{d-1}f(u)\,du \qquad (29)$$

(Yaglom 1987a). Note that if $\tilde{K}$ and $\tilde{f}$ are the functions on $\mathbb{R}^d$ such that $\tilde{K}(\mathbf{x}) = K(|\mathbf{x}|)$ and $\tilde{f}(\boldsymbol{\omega}) = f(|\boldsymbol{\omega}|)$, then the conditions $\int_0^\infty r^{d-1}|K(r)|\,dr < \infty$ and $\int_0^\infty u^{d-1}f(u)\,du < \infty$ are equivalent to the absolute integrability of $\tilde{K}$ and $\tilde{f}$, respectively.

Smoothness properties

We have shown in 2.4 that an autocovariance function on $\mathbb{R}$ that is continuous at 0 is continuous. For continuous isotropic autocovariance functions in more than one dimension, we can make stronger statements about the smoothness of $K(r)$ for $r \geq 0$. From standard properties of Bessel functions, $|J_\nu(t)| < C_\nu(1+|t|)^{-1/2}$ for some constant C_ν and all t, $t^{-\nu}J_\nu(t)$ is bounded and $(d/dt)\{t^{-\nu}J_\nu(t)\} = -t^{-\nu}J_{\nu+1}(t)$. From (26), for $d \geq 3$ and $r > 0$,

$$\begin{aligned} K'(r) &= 2^{(d-2)/2}\Gamma(d/2)\,\frac{d}{dr}\left\{\int_0^\infty (ru)^{-(d-2)/2}J_{(d-2)/2}(ru)\,dG(u)\right\} \\ &= -2^{(d-2)/2}\Gamma(d/2)\int_0^\infty u(ru)^{-(d-2)/2}J_{d/2}(ru)\,dG(u), \end{aligned}$$

where differentiating inside the integral can be justified by G bounded and the preceding properties of J_ν (note that for $d \geq 3$, the last integrand is bounded). More generally, for $K \in \mathcal{D}_d$, K is $\lfloor \frac{1}{2}(d-1)\rfloor$ times differentiable on $(0,\infty)$ (see Trebels (1976) or Exercise 31).

Although $K \in \mathcal{D}_2$ may not be differentiable on $(0,\infty)$ (see Exercise 32 for an example), we can draw a stronger conclusion than that K is continuous, which is automatically true since $K \in \mathcal{D}_1$ is continuous. For an interval I, let us say that a function f is Lipschitz with parameter α or $f \in \mathrm{Lip}(\alpha)$, if there exists C finite such that $|f(s)-f(t)| \leq C|s-t|^\alpha$ for all $s,t \in I$. Then I claim $K \in \mathcal{D}_2$ implies $K \in \mathrm{Lip}(\frac{1}{2})$ on $I = [a,\infty)$ for any $a > 0$. To prove this, note that for $s > t \geq a$,

$$
\begin{aligned}
&|K(s) - K(t)| \\
&\leq \int_0^\infty |J_0(su) - J_0(tu)|\, dG(u) \\
&\leq \int_0^{u_0^+} u(s-t) \sup_{tu \leq \xi \leq su} |J_1(\xi)|\, dG(u) + \int_{u_0^+}^\infty \frac{2C_0}{(1+tu)^{1/2}}\, dG(u) \\
&\leq C_1(s-t) \int_0^{u_0^+} \frac{u}{(1+tu)^{1/2}}\, dG(u) + \frac{2C_0}{(1+tu_0)^{1/2}} \{G(\infty) - G(0)\} \\
&\leq \left\{ \frac{C_1(s-t)u_0 + 2C_0}{(1+au_0)^{1/2}} \right\} \{G(\infty) - G(0)\}.
\end{aligned}
$$

Choosing $u_0 = (s-t)^{-1}$ yields the desired result. Similarly, for $d \geq 4$ and even, it is possible to show that $K^{((d-2)/2)}$ is $\mathrm{Lip}(\frac{1}{2})$ on any interval $[a, \infty)$ with $a > 0$ (Exercise 31). Since a continuous function is not necessarily $\mathrm{Lip}(\alpha)$ for any $\alpha > 0$ and a function that is Lip(1) is absolutely continuous and hence nearly differentiable, it is not unreasonable to characterize a function that is $\mathrm{Lip}(\frac{1}{2})$ as being $\frac{1}{2}$ times differentiable. Using this loose interpretation, we might now say that $K \in \mathcal{D}_d$ implies K is $\frac{1}{2}(d-1)$ times differentiable on $(0, \infty)$.

Additional smoothness beyond continuity at the origin for $K \in \mathcal{D}_d$ should imply additional smoothness away from the origin. For example, consider whether $K(r) = (1-r)^+$ is in $\mathcal{D}_2$. We have already seen that K is in D_1 and since it is not differentiable at $r = 1$, it is not in $\mathcal{D}_3$. Using the inversion formula (29), we can show that $K \notin \mathcal{D}_2$ by showing that $\int_0^1 J_0(ur)r(1-r)\, dr$ is negative for some $u \geq 0$. More specifically, by applying asymptotic expansions for J_0 and J_1 (Abramowitz and Stegun 1965, p. 364), as $u \to \infty$,

$$
\begin{aligned}
&\int_0^1 J_0(ur)r(1-r)\, dr \\
&\quad = \int_0^1 J_0(ur)r\, dr - \int_0^1 J_0(ur)r^2 dr \\
&\quad = u^{-1}J_1(u) - u^{-3} \int_1^u r^2 J_0(r)\, dr + O(u^{-3}) \\
&\quad = \left(\frac{2}{\pi}\right)^{1/2} \left\{ u^{-3/2} \cos\left(u - \frac{3\pi}{4}\right) - \frac{3}{8}\, u^{-5/2} \sin\left(u - \frac{3\pi}{4}\right) \right\} \\
&\qquad - \left(\frac{2}{\pi}\right)^{3/2} u^{-3} \int_1^u r^2 \left\{ \cos\left(r - \frac{\pi}{4}\right) + \frac{1}{8r} \sin\left(r - \frac{\pi}{4}\right) \right. \\
&\qquad \left. - \frac{9}{128r^2} \cos\left(r - \frac{\pi}{4}\right) + O(r^{-3}) \right\} dr + O(u^{-3})
\end{aligned}
$$

$$= \left(\frac{2}{\pi}\right)^{1/2} u^{-5/2} \sin\left(u - \frac{3\pi}{4}\right) + O(u^{-3}), \tag{30}$$

which can be obtained by repeated integration by parts (Exercise 33). Thus, for some u sufficiently large f is negative, so that $K(r) = (1-r)^+$ is not in $\mathcal{D}_2$.

More generally, for some $K \in \mathcal{D}_2$, suppose $K(r) = 1 - \gamma r + o(r)$ as $r \downarrow 0$. Then I claim that K is differentiable on $(0, \infty)$ and that $K'(r)$ is in $\mathrm{Lip}(\frac{1}{2})$ on $[a, \infty)$ for any $a > 0$. To prove this claim, first apply Theorem 1 of Bingham (1972), which is a Tauberian theorem for Hankel transforms. This theorem shows that $K(r) = 1 - \gamma r + o(r)$ as $r \downarrow 0$ implies $G(\infty) - G(u) \sim C/u$ as $u \to \infty$ for an appropriate positive constant C. Using the properties of Bessel functions described earlier, it follows that (Exercise 35)

$$K'(r) = -\int_0^\infty u J_1(ur)\, dG(u) \tag{31}$$

for $r > 0$. An argument similar to the one proving the $\mathrm{Lip}(\frac{1}{2})$ property of elements of $\mathcal{D}_2$ yields $K'(r)$ is in $\mathrm{Lip}(\frac{1}{2})$ on $[a, \infty)$ for any $a > 0$ as claimed (Exercise 36). I am unaware of a general result on what some degree of smoothness at 0 for an element of $\mathcal{D}_d$ implies about its smoothness elsewhere. Thinking of an element of $\mathcal{D}_d$ as being a function on $\mathbb{R}$ by setting $K(-r) = K(r)$, then for any $\epsilon > 0$, I would expect K to have at least $\frac{1}{2}(d-1) - \epsilon$ more "derivatives" away from the origin than it does at the origin, where, for a positive noninteger t, a function is said to have t derivatives at a point if in some neighborhood of this point it has $\lfloor t \rfloor$ derivatives and this $\lfloor t \rfloor$th derivative is $\mathrm{Lip}(t - \lfloor t \rfloor)$.

To see why I have included the $-\epsilon$ term in this conjecture, consider the following example. Pasenchenko (1996) shows that $K(r) = \{1 - r^{1/2}\}^+$ is in $\mathcal{D}_2$. This function is in $\mathrm{Lip}(\frac{1}{2})$ in a neighborhood of 0 but K is not differentiable at 1. Hence, the proposition that $K \in \mathcal{D}_d$ has $\frac{1}{2}(d-1)$ more derivatives away from the origin than at the origin is false using the definition of fractional derivatives given here. Of course, such a proposition may be true under a different definition of fractional differentiation.

Matérn class

The Matérn class of functions

$$K(r) = \phi(\alpha r)^\nu \mathcal{K}_\nu(\alpha r), \qquad \phi > 0,\ \alpha > 0,\ \nu > 0,$$

which we saw in (14) to be positive definite on $\mathbb{R}$, are in fact all in $\mathcal{D}_\infty$. This can be verified by using the inversion formula and (6.576.7) of Gradshteyn and Ryzhik (1994) to obtain the corresponding isotropic spectral density

$$f(u) = \frac{2^{\nu-1}\phi\Gamma\left(\nu+\frac{d}{2}\right)\alpha^{2\nu}}{\pi^{d/2}(\alpha^2+u^2)^{\nu+d/2}}, \tag{32}$$

which is nonnegative and satisfies $\int_0^\infty f(u)u^{d-1}du < \infty$. This model is used by Handcock (1989), Handcock and Stein (1993) and Handcock and Wallis (1994). Matérn (1960) appears to be the first author to have recommended these functions as a sensible class of models for isotropic random fields in any number of dimensions. I believe this class of models has much to recommend it. As we show in Chapter 3, the smoothness of a random field plays a critical role in interpolation problems. Furthermore, there is often no basis for knowing a priori the degree of smoothness of some physical process modeled as a random field. Thus, it is prudent to use classes of models that allow for the degree of smoothness to be estimated from the data rather than restricted a priori. The Matérn model does allow for great flexibility in the smoothness of the random field while still keeping the number of parameters manageable.

It is often convenient to describe the smoothness of an isotropic random field through the principal irregular term of the isotropic autocovariance function. Extending the definition I gave in 2.7 for processes on $\mathbb{R}$ in the obvious way, I call g a principal irregular term for the isotropic autocovariance function K if $g(r)r^{-2n} \to 0$ and $|g(r)|r^{-2n-2} \to \infty$ as $r \downarrow 0$ and K is of the form $K(r) = \sum_{j=0}^{n} c_j r^{2j} + g(r) + o(|g(r)|)$ as $r \downarrow 0$. As in the one-dimensional setting, if $g(r) = \alpha|r|^\beta$ is a principal irregular term for K, I call β the power and α the coefficient of the principal irregular term. It follows from (15) and (16) that for the Matérn model with ν a noninteger, 2ν is the power of the principal irregular term and when ν is a positive integer, there is a principal irregular term proportional to $r^{2\nu}\log r$.

For statistical purposes, the parameterization in (32) may not be best. In particular, for fixed α, f becomes more and more concentrated around the origin as ν increases. In particular, suppose $C_{\alpha,\nu}$ is the isotropic autocorrelation function corresponding to $f(\boldsymbol{\omega}) = \phi(\alpha^2 + |\boldsymbol{\omega}|^2)^{-\nu-d/2}$. Then $\lim_{\nu\to\infty} C_{\alpha,\nu}(r) = 1$ for all $r \geq 0$ (Exercise 39). One way to solve this problem is to use the class of models $f(\boldsymbol{\omega}) = \phi\left\{\alpha^2\left(\nu + \frac{1}{2}d\right) + |\boldsymbol{\omega}|^2\right\}^{-\nu-d/2}$ (see Exercise 39 for the limiting behavior of the corresponding autocorrelation functions as $\nu \to \infty$). I implicitly use this parameterization in the numerical studies in 3.5 and elsewhere. Handcock and Wallis (1994) recommend the alternative parameterization

$$g_\eta(u) = \frac{\sigma c(\nu,\rho)}{\left(\frac{4\nu}{\rho^2}+u^2\right)^{\nu+d/2}}, \tag{33}$$

where $\boldsymbol{\eta} = (\sigma, \nu, \rho)$ and

$$c(\nu, \rho) = \frac{\Gamma\left(\nu + \frac{d}{2}\right)(4\nu)^\nu}{\pi^{d/2}\Gamma(\nu)\rho^{2\nu}}.$$

The corresponding isotropic autocovariance function is

$$K_\eta(r) = \frac{\sigma}{2^{\nu-1}\Gamma(\nu)} \left(\frac{2\nu^{1/2}r}{\rho}\right)^\nu \mathcal{K}_\nu\left(\frac{2\nu^{1/2}r}{\rho}\right),$$

which has the nice property that it does not depend on d. In either parameterization, ν has the same interpretation as a measure of the differentiability of the random field. The parameter σ in (33) is just $\text{var}\{Z(\mathbf{x})\}$. Finally, ρ measures how quickly the correlations of the random field decay with distance. It is thus closely related to what is known as the practical range of an isotropic autocovariance function in the geostatistical literature, which is informally defined as the distance at which the correlations are nearly 0, say 0.05, for example (Journel and Huijbregts 1978, p. 164). Figure 3 plots the autocovariance functions corresponding to g_η for $\sigma = 1$, $\nu = 1$ and several values of ρ and shows how the correlations decay more slowly as ρ increases. Although α^{-1} has a similar interpretation in (32), ρ has the attractive feature that its interpretation is largely independent of ν, which is not the case for α. To illustrate this point, Figure 4 plots the autocorrelation functions corresponding to $\rho = 1$ for $\nu = 1, 2$ or 3 under (33) and Figure 5 plots the autocorrelation functions corresponding to $\alpha = 1$ for $\nu = 1, 2$ or 3 under (32). The autocorrelation functions in Figure 4 are much more similar at longer distances than those in Figure 5. Another way to see that the interpretation of ρ is only weakly dependent on ν is to consider the limit of (33) as $\nu \to \infty$. Specifically, for fixed σ, ρ and u,

$$\lim_{\nu \to \infty} g_\eta(u) = \sigma \left(\frac{\rho^2}{4\pi}\right)^{d/2} \exp\left(-\tfrac{1}{4}\rho^2 u^2\right) \tag{34}$$

and the corresponding isotropic autocovariance function is $K(r) = \sigma e^{-r^2/\rho^2}$. This calculation shows that for fixed ρ and σ and two different but large values of ν, the corresponding covariance functions are nearly the same.

An alternative to the Matérn class of models that is sometimes used is $K(r) = Ce^{-ar^\gamma}$ (Diggle, Tawn and Moyeed 1998; De Oliveira, Kedem and Short 1997). These functions are also in $\mathcal{D}_\infty$ for all C and a positive and all $\gamma \in (0, 2]$, which follows by the same reasoning as held in the one-dimensional setting treated in 2.7. For $\gamma > 2$, we noted that $K \notin \mathcal{D}_1$ so it is not in $\mathcal{D}_d$ for any d. The power of the principal irregular part of K is γ, so in terms of local behavior of the random field, γ corresponds to 2ν in the Matérn model when $\gamma < 2$ and, roughly speaking, $\gamma = 2$ corresponds to $\nu = \infty$. We see that $K(r) = Ce^{-ar^\gamma}$ has no elements providing similar local behavior as the Matérn class for $1 \le \nu < \infty$. Thus, although the

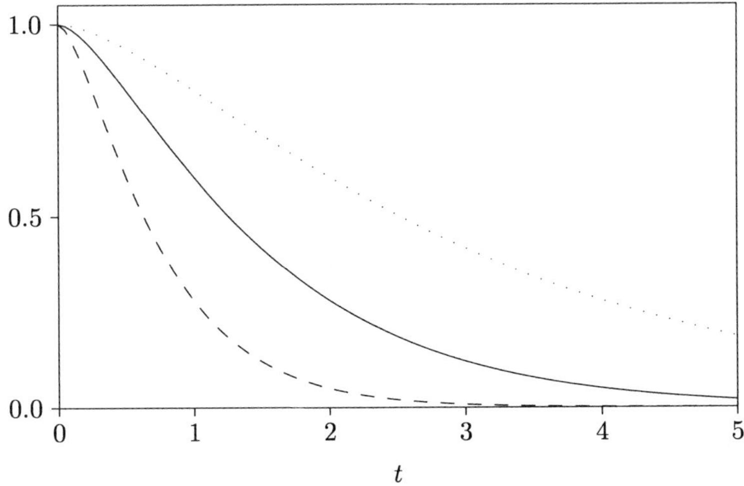

FIGURE 3. Plots of Matérn autocovariance functions under the parameterization given in (33) with $\sigma = 1$, $\nu = 1$ and several values of ρ. Solid line corresponds to $\rho = 2$, dashed line to $\rho = 1$ and dotted line to $\rho = 4$.

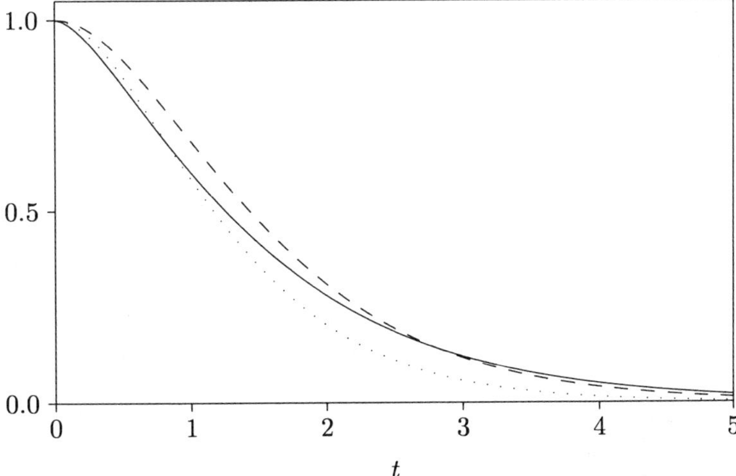

FIGURE 4. Plots of Matérn autocovariance functions under the parameterization given in (33) with $\sigma = 1$, $\rho = 2$ and several values of ν. Solid line corresponds to $\nu = 1$, dashed line to $\nu = 2$ and dotted line to $\nu = 3$.

Matérn class has no more parameters than this model, it provides much greater range for the possible local behavior of the random field. The fact that its use requires the calculation of a Bessel function does not create a serious obstacle to its adoption as programs that calculate all manners of Bessel functions are readily available (Cody 1987). Section 6.5 provides further discussion on the use of the Matérn model.

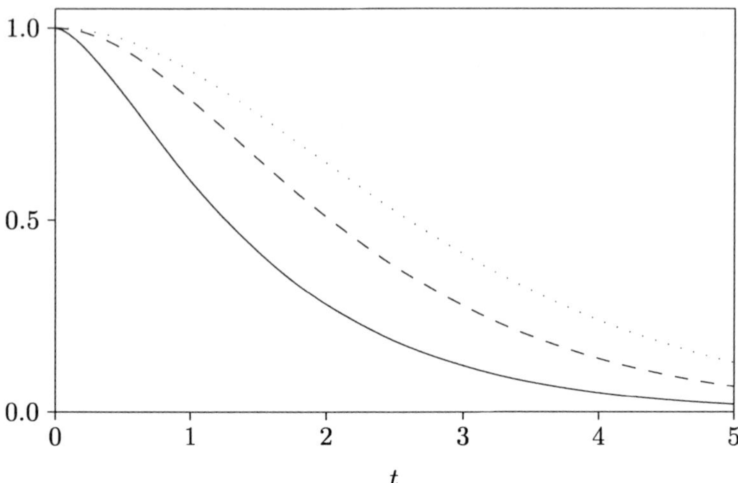

FIGURE 5. Plots of Matérn autocorrelation functions under the parameterization given in (32) with $\alpha = 1$ and several values of ν. Solid line corresponds to $\nu = 1$, dashed line to $\nu = 2$ and dotted line to $\nu = 3$.

Spherical model

Perhaps the most commonly used model for isotropic autocovariance functions in geological and hydrological applications is the spherical: for positive constants c and ρ,

$$K(r) = \begin{cases} c\left(1 - \dfrac{3}{2\rho}r + \dfrac{1}{2\rho^3}r^3\right), & r \le \rho \\ 0, & r > \rho. \end{cases} \tag{35}$$

This function is in $\mathcal{D}_3$ but is not in $\mathcal{D}_4$ (Exercise 40). The parameter ρ is called the range parameter and is the distance at which correlations become exactly 0. Its popularity in the geostatistical literature (Journel and Huijbregts 1978, p. 116; Isaaks and Srivastava 1989, p. 374; Bras and Rodríguez-Iturbe 1985, p. 418; Christakos 1992, p. 71; Wackernagel 1995, p. 42; Kitanidis 1997, p. 56; and Goovaerts 1997, p. 88) is a bit of a mystery to me. Perhaps its superficially simple functional form is attractive or perhaps there is a mistaken belief that there is some statistical advantage in having the autocorrelation function being exactly 0 beyond some finite distance. However, the fact that this function is only once differentiable at $r = \rho$ can lead to problems when using likelihood methods for estimating the parameters of this model (see 6.4). Furthermore, in three dimensions, the corresponding isotropic spectral density (Exercise 40) has oscillations at high frequencies similar to the spectral density corresponding to the triangular autocovariance function in one dimension (Figure 2 in 2.7). As I argued in 2.7, such oscillations would generally not make much physical sense. Stein and Handcock (1989) show that when using the spherical model in three dimensions, certain prediction problems have rather pathological behavior. I consider the spherical model to be a poor substitute for the exponential model, since both have linear behavior at the origin, but the exponential model has none of the pathologies of the spherical.

There can, in some circumstances, be a computational advantage in using a model such as the spherical for which the covariance is identically 0 beyond some distance. Specifically, if ρ in (35) is much smaller than the dimensions of the observation region, then most of the covariances between the observations will be 0, which saves storage space and makes it possible to use sparse matrix methods to more efficiently calculate kriging predictors (Cohn et al. 1998). For this reason, Gaspari and Cohn (1999) give a large number of isotropic autocovariance functions that are identically 0 beyond a fixed distance. However, I suspect that all of the examples of autocovariance functions derived in Gaspari and Cohn (1999) will, like the spherical model, lead to problems when using likelihood-based methods for estimating unknown parameters. If the computational advantage of having the autocovariance function identically 0 beyond some distance is needed in a particular application, I would suggest at least using an autocovariance function that has two more derivatives away from the origin than it does at the origin. For example, consider $K(r)^2$, where $K(r)$ is a spherical autovariance function as in (35). Equation (5) in 2.3 implies that this function is in $\mathcal{D}_3$. Furthermore, like the spherical model, it behaves linearly at the origin and is identically 0 beyond a certain distance, but unlike the spherical model, it has two derivatives on $(0, \infty)$.

Exercises

30 Using standard properties of Bessel functions given in, say, Chapter 9 of Abramowitz and Stegun (1965), show that for any $d \geq 2$, the infimum of $\Lambda_d(s)$ for $s \geq 0$ is attained at a finite value of s. Use this to show that if $r > 0$, there is an isotropic autocorrelation function $C \in \mathcal{D}_d$ such that $C(r) = \inf_{s \geq 0} \Lambda_d(s)$.

31 Show that $K \in \mathcal{D}_d$ implies K is $\lfloor \frac{1}{2}(d-1) \rfloor$ times differentiable on $(0, \infty)$. For $d \geq 4$ and even, show that $K \in \mathcal{D}_d$ implies $K^{((d-2)/2)}$ is $\mathrm{Lip}(\frac{1}{2})$ on any interval $[a, \infty)$ with $a > 0$.

32 Show that the isotropic autocovariance function for a random field on $\mathbb{R}^2$ corresponding to the isotropic spectral density $f(u) = u^{-5/2}(1 - \cos u)$ is not differentiable at $u = 1$ by the following steps.

(i) Find the isotropic autocovariance function by using (29), integration by parts and formulas 6.669.1, 6.669.2 and 6.561.14 of Gradshteyn and Ryzhik (1994).

(ii) Show that the resulting function is continuous but not differentiable at 1 by using properties of hypergeometric functions given, for example, in Chapter 15 of Abramowitz and Stegun (1965).

33 Provide the details for (30).

34 By considering the correlation of $Z(\epsilon) - Z(0)$ and $Z(1+\epsilon) - Z(1)$ as $\epsilon \downarrow 0$, show that $\{(1-r)^+\}^\gamma \notin \mathcal{D}_1$ if $0 < \gamma < 1$.

35 Verify (31).

36 Using (31), show that if $K \in \mathcal{D}_2$ and $K'(0^+)$ exists and is finite, then $K'(r)$ is in $\mathrm{Lip}(\frac{1}{2})$ on $[a, \infty)$ for any $a > 0$.

37 (Pólya's criteria). Prove that if K is even and is continuous, nonnegative, nonincreasing and convex on $[0, \infty)$ then it is in $\mathcal{D}_1$ by using (3), (4) and the fact that $(1 - |t|)^+$ is in $\mathcal{D}_1$.

38 For a weakly isotropic random field in $\mathbb{R}^d$, $d \geq 2$, the results given in this section do not resolve whether such functions must be continuous away from the origin. Read Crum (1956) and determine to what extent the results in this paper resolve this issue.

39 Show that if $C_{\alpha,\nu}$ is the isotropic autocorrelation function corresponding to $f(\boldsymbol{\omega}) = \phi(\alpha^2 + |\boldsymbol{\omega}|^2)^{-\nu-d/2}$, $\lim_{\nu\to\infty} C_{\alpha,\nu}(t) = 1$ for all t. For $f(\boldsymbol{\omega}) = \phi\{\alpha^2(\nu + \frac{1}{2}d) + |\boldsymbol{\omega}|^2\}^{-\nu-d/2}$, find the limiting behavior of the corresponding isotropic autocorrelation function as $\nu \to \infty$ for fixed α.

40 Consider a Poisson process N on $\mathbb{R}^3$ with constant intensity λ, so that for A a Borel subset of $\mathbb{R}^3$, $N(A)$ is the number of events of the process in A. Let $Z(\mathbf{x}) = N(b_3(r) + \mathbf{x})$, the number of events in the ball in $\mathbb{R}^3$ of radius r centered at $\mathbf{x}$. Show that the isotropic autocovariance function of Z is of the form given in (35). Find the corresponding spectral density. Show that K as given in (35) is not in $\mathcal{D}_4$.

2.11 Tensor product autocovariances

An easy way to generate autocovariance functions on $\mathbb{R}^d$ is to take products of one dimensional autocovariance functions. Specifically, for autocovariance functions $K_1, \ldots, K_d$ on $\mathbb{R}$ and $\mathbf{x} = (x_1, \ldots, x_d)$, $K(\mathbf{x}) = K_1(x_1) \cdots K_d(x_d)$ is an autocovariance function on $\mathbb{R}^d$, which can be proven using (5). For some mathematical problems, they are easier to study than isotropic autocovariance functions, which has led to their rather widespread use in mathematical works (Ylvisaker 1975; Papageorgiou and Wasilkowski 1990; Woźniakowski 1991; Ying 1993; Ritter 1995; and Müller-Gronbach 1998). However, the extreme dependence of these models on the choice of axes would appear to make them untenable for most physical processes defined on continuous space.

As an example of the physically unrealistic behavior such models imply, consider $K(u,v) = e^{-|u|-|v|}$ for u and v real. Suppose Z has mean 0 and this autocovariance function and we wish to predict $Z(0,0)$ based on observing $Z(t,0)$ and $Z(0,t)$. Then straightforward calculations yield that the BLP of $Z(0,0)$ is $\lambda Z(0,t) + \lambda Z(t,0)$, where $\lambda = 1/(e^t + e^{-t}) = 1/(2\sinh t)$ and the mse of the BLP is $\tanh t$ (Exercise 41). As $t \downarrow 0$, $\lambda = \frac{1}{2} - \frac{1}{4}t^2 + O(t^3)$ and the mse is $t + O(t^3)$. Note that the simple predictor $\frac{1}{2}Z(t,0) + \frac{1}{2}Z(0,t)$ also has mse $t+O(t^3)$ as $t \downarrow 0$, so that the BLP is very nearly the average of the two observations for t small. Next, consider adding a third observation at (t,t). Now the BLP of $Z(0,0)$ is $e^{-t}Z(0,t) + e^{-t}Z(t,0) - e^{-2t}Z(t,t)$ with mse $(1-e^{-2t})^2$ (Exercise 41). As $t \downarrow 0$, its mse is $4t^2 + O(t^3)$, so by adding a third observation that is further away from $(0,0)$ than the other two observations, the mse decreases from $O(t)$ to $O(t^2)$.

The reason for this huge drop in mse is that the random field is locally almost additive. For a function $f(u,v) = f_1(u) + f_2(v)$, note that $f(0,0) = f(u,0) + f(0,v) - f(u,v)$, which is very nearly the form of the BLP for $Z(0,0)$ in terms of $Z(0,t)$, $Z(t,0)$ and $Z(t,t)$ when t is small. Indeed, $\operatorname{var}\{Z(0,0) - Z(0,t) - Z(t,0) + Z(t,t)\} = 4t^2 + O(t^3)$ as $t \downarrow 0$, so this additive approximation does as well asymptotically as the BLP.

If instead of having the first two observations along the axes, we predict $Z(0,0)$ based on observations at $(2^{-1/2}t, 2^{-1/2}t)$ and $(-2^{-1/2}t, 2^{-1/2}t)$ and then add a third observation at $(0, 2^{1/2}t)$, so that the observations have been rotated 45° from the previous setting, the mse of the BLP is $O(t)$ and not $o(t)$ in both cases, so that there is no order of magnitude decrease in the mse when the third observation is added. Thus, predictions based on this model are highly sensitive to the choice of axes and should not be used unless there is some very good reason for thinking the observed random field possesses the required axis dependence. Ripley (1995) has also criticized work in the numerical analysis of deterministic functions that makes use of assumptions depending strongly on the choice of axes.

Finally, note that the only real functions on $\mathbb{R}^d$ that are isotropic and factor into functions of each coordinate are of the form $c\exp(-a|\mathbf{x}|^2)$ (Exercise 42). This fact was used by Maxwell in his famous work on the kinetic theory of gases to argue that the velocity distribution in an ideal gas must be spherical Gaussian (Ruhla 1992). The function ce^{-ar^2} is in $\mathcal{D}_\infty$ for all c and a nonnegative, so in any number of dimensions $c\exp(-a|\mathbf{x}|^2)$ is positive definite for a and c nonnegative. However, as I previously argued in the one-dimensional setting, random fields possessing these autocovariance functions are unrealistically smooth for physical phenomena.

Exercises

41 Suppose Z is a weakly stationary mean 0 random field on $\mathbb{R}^2$ with autocovariance function $K(u,v) = e^{-|u|-|v|}$ for u and v real. Show that the BLP of $Z(0,0)$ based on observing $Z(t,0)$ and $Z(0,t)$ is $\lambda Z(0,t) +$

$\lambda Z(t,0)$, where $\lambda = 1/(e^t + e^{-t}) = 1/(2\sinh t)$ and the mse of the BLP is $\tanh t$. Next, consider adding a third observation at (t,t). Show that the BLP of $Z(0,0)$ is $e^{-t}Z(0,t) + e^{-t}Z(t,0) - e^{-2t}Z(t,t)$ with mse $(1-e^{-2t})^2$.

42 Show that the only real functions on $\mathbb{R}^d$ that are isotropic and factor into functions of each coordinate are of the form $c\exp(-a|\mathbf{x}|^2)$ for a and c real.

3
Asymptotic Properties of Linear Predictors

3.1 Introduction

Suppose we observe a Gaussian random field Z with mean function m and covariance function K at some set of locations. Call the pair (m, K) the second-order structure of the random field. If (m, K) is known, then as noted in 1.2, the prediction of Z at unobserved locations is just a matter of calculation. To review, the conditional distribution of Z at an unobserved location is normal with conditional mean that is a linear function of the observations and constant conditional variance. In practice, (m, K) is at least partially unknown and it is usually necessary to estimate (m, K) from the same data we use to do the prediction. Thus, it might be natural to proceed immediately to methods for estimating second-order structures of Gaussian random fields. However, until we know something about the relationship between the second-order structure and linear predictors, it will be difficult to judge what is meant by a good estimate of the second-order structure. In particular, it will turn out that it is possible to get (m, K) nonnegligibly wrong and yet still get nearly optimal linear predictors. More specifically, for a random field possessing an autocovariance function, if the observations are tightly packed in a region in which we wish to predict the random field, then the low frequency behavior of the spectrum has little impact on the behavior of the optimal linear predictions.

One way to study the behavior of linear predictors when the second-order structure is not perfectly known is to consider the behavior of linear predictors that are optimal under some incorrect second-order structure.

This approach has been used in the atmospheric sciences (Daley 1991, Section 4.9) and in the geostatistical literature (Diamond and Armstrong 1984) as well as in my own work (Stein 1988, 1990a, 1990b, 1993, 1997, 1999 and Stein and Handcock 1989). I generally denote the actual second-order structure for a random field Z by (m_0, K_0) and the incorrect second-order structure used to generate and evaluate linear predictors by (m_1, K_1). I call linear predictors that are best under an incorrect model pseudo-BLPs.

To be more specific, suppose we observe Z for all $\mathbf{x}$ in some set $Q \subset R$ and wish to predict $Z(\mathbf{x}_0)$, $\mathbf{x}_0 \in R \backslash Q$. Define $e_j\left(Z(\mathbf{x}_0), Q\right)$ to be the error of the best linear predictor if (m_j, K_j) is the correct second-order structure and let E_j indicate expected value under (m_j, K_j). One measure as to how well predictions based on K_1 do when K_0 is the correct covariance function is $E_0\{e_1\left(Z(\mathbf{x}_0), Q\right)^2\}/E_0\{e_0\left(Z(\mathbf{x}_0), Q\right)^2\}$, the ratio of the mse of the suboptimal pseudo-BLP to that of the BLP. This ratio is necessarily at least 1. More specifically,

$$\begin{aligned}\frac{E_0 e_1(Z(\mathbf{x}_0), Q)^2}{E_0 e_0(Z(\mathbf{x}_0), Q)^2} &= \frac{E_0[\{e_1(Z(\mathbf{x}_0), Q) - e_0(Z(\mathbf{x}_0), Q)\} + e_0(Z(\mathbf{x}_0), Q)]^2}{E_0 e_0(Z(\mathbf{x}_0), Q)^2} \\ &= 1 + \frac{E_0\{e_1(Z(\mathbf{x}_0), Q) - e_0(Z(\mathbf{x}_0), Q)\}^2}{E_0 e_0(Z(\mathbf{x}_0), Q^2}, \qquad (1)\end{aligned}$$

which follows from the orthogonality of the error of a BLP with all linear combinations of the observations. In addition to the quality of point predictions, another concern is the accuracy of assessments of mse. If we not only compute our prediction under (m_1, K_1) but also assess its mse under this model, this amounts to presuming the mse is $E_1 e_1(Z(\mathbf{x}_0), Q)^2$. The quantity

$$\frac{E_1 e_1(Z(\mathbf{x}_0), Q)^2}{E_0 e_1(Z(\mathbf{x}_0), Q)^2} \qquad (2)$$

is then the ratio of the presumed mse of the pseudo-BLP to its actual mse. If both (1) and (2) are near 1, then little is lost by using (m_1, K_1) instead of the correct (m_0, K_0), at least as far as predicting $Z(\mathbf{x}_0)$ goes.

This chapter considers two ways of investigating the relationship between second-order structures and linear prediction. One way is to study (1) and (2) for various pairs of second-order structures, observations and predictands. The second is to study the spectral characteristics of prediction errors by making use of the correspondence between linear combinations of values of a random field and linear combinations of complex exponentials described in 2.6. It turns out that this second approach is helpful in studying the first.

There are two basic themes to this chapter. One is the differences in the behavior of pseudo-BLPs when interpolating (predicting at locations "surrounded" by observations) and extrapolating (predicting at locations outside the range of observations). The second is the lack of sensitivity of predictions to misspecifications in the spectrum at low frequencies when neighboring observations are highly correlated. In addition, there is an important interaction between these themes: the low frequency behavior of the spectrum matters much less when interpolating than extrapolating. These findings imply that when interpolation is the goal, the focus in model selection and estimation of spectra should be on high frequency behavior.

3.2 Finite sample results

There are some theoretical results we can give on the ratios in (1) and (2) that do not require asymptotic arguments nor specific considerations of the observation locations. Let us first review some relevant definitions on Hilbert spaces and random fields. Suppose Z is a real-valued random field on a closed set $R \subset \mathbb{R}^d$ possessing second-order structure (m, K) with m continuous on R and K continuous on $R \times R$. Let $\mathcal{H}_R^0$ be the real linear manifold of random variables $Z(\mathbf{x})$ for $\mathbf{x} \in R$ and $\mathcal{H}_R(m, K)$ the closure of $\mathcal{H}_R^0$ with respect to the inner product defined by $E(h_1 h_2)$ for h_1 and h_2 in $\mathcal{H}_R^0$. For $h_1, h_2 \in \mathcal{H}_R(m, K)$, define mean and covariance operators m and K such that $Eh_1 = m(h_1)$ and $\operatorname{cov}(h_1, h_2) = K(h_1, h_2)$. Thus, we use K (and m) to indicate both a function and an operator, the meaning being apparent from context. For example, $K(Z(\mathbf{x}), Z(\mathbf{y})) = K(\mathbf{x}, \mathbf{y})$, where K is an operator on $\mathcal{H}_R(m, K) \times \mathcal{H}_R(m, K)$ on the left side of the expression and K is a function on $R \times R$ on the right side.

Suppose

$$0 < a = \inf_{h \in \mathcal{H}_R^0} \frac{E_1 h^2}{E_0 h^2} \leq \sup_{h \in \mathcal{H}_R^0} \frac{E_1 h^2}{E_0 h^2} = b < \infty. \tag{3}$$

Then, as sets, $\mathcal{H}_R(m_0, K_0) = \mathcal{H}_R(m_1, K_1)$, so call this set $\mathcal{H}_R$. The condition (3) simplifies matters because now there is no need to worry about, say, a BLP under $\mathcal{H}_R(m_0, K_0)$ not being an element of $\mathcal{H}_R(m_1, K_1)$. One situation where (3) holds is if $R = \mathbb{R}^d$, $m_0 = m_1$ and K_0 and K_1 are autocovariance functions with corresponding spectral densities f_0 and f_1 such that f_0/f_1 is bounded away from 0 and ∞.

Under (3), we can give some simple bounds on the effects of using the wrong second-order structure. Define $e_j(h, Q)$ to be the error of the BLP of $h \in \mathcal{H}_R$ based on observing Z on Q and let $\mathcal{H}_{-Q}$ be those elements h of $\mathcal{H}_R$ for which $E_0 e_0(h, Q)^2 > 0$. Equation (3) implies a simple bound for the ratio in (2) on assessing mses of pseudo-BLPs:

$$a \leq \inf_{h \in \mathcal{H}_{-Q}} \frac{E_1 e_1(h, Q)^2}{E_0 e_1(h, Q)^2} \leq \sup_{h \in \mathcal{H}_{-Q}} \frac{E_1 e_1(h, Q)^2}{E_0 e_1(h, Q)^2} \leq b. \tag{4}$$

It is not possible to sharpen these bounds without further assumption. Under (3), it is possible to show that $E_0 e_0(h, Q)^2 = 0$ if and only if $E_0 e_0(h, Q)^2 = E_1 e_0(h, Q)^2 = E_1 e_1(h, Q)^2 = E_0 e_1(h, Q)^2 = 0$ (Exercise

1). Thus, the restriction to $\mathcal{H}_{-Q}$ is not a significant one. Cleveland (1971) obtains a more interesting bound for the ratio in (1) on the efficiency of pseudo-BLPs.

Theorem 1. *If $m_0 = m_1 = 0$ and (3) holds, then for all $h \in \mathcal{H}_{-Q}$,*

$$\frac{E_0 e_1(h,Q)^2}{E_0 e_0(h,Q)^2} \leq 1 + \frac{(b-a)^2}{4ab}. \tag{5}$$

PROOF. Following Cleveland (1971), let $\mathcal{P}_j$, $j = 0, 1$, be the operator that maps h into its BLP under K_j based on $\mathcal{H}_Q$. We can then call $\mathcal{P}_j$ an orthogonal projection operator, since $\mathcal{P}_j h$ is the unique element of $\mathcal{H}_Q$ such that $h - \mathcal{P}_j h$ is orthogonal to all elements of $\mathcal{H}_Q$ under the inner product defined by K_j. For $h \in \mathcal{H}_{-Q}$, if $\mathcal{P}_0 h = \mathcal{P}_1 h$, (5) trivially holds, so assume from now on that $\mathcal{P}_0 h \neq \mathcal{P}_1 h$. In this proof and subsequently, I use e_j to denote $e_j(h,Q)$ when it is clear what is being predicted and what are the observations. Define $z = \mathcal{P}_1 e_0$, which is not 0 when $\mathcal{P}_0 h \neq \mathcal{P}_1 h$. Let $\mathcal{R}_j$ be the orthogonal projection operator onto the space spanned by z under the inner product K_j. Then $\mathcal{R}_1 e_0 = \{K_1(\mathcal{P}_1 e_0, e_0)/K_1(\mathcal{P}_1 e_0, \mathcal{P}_1 e_0)\} \mathcal{P}_1 e_0 = \mathcal{P}_1 e_0$, so $e_0 - \mathcal{R}_1 e_0 = h - \mathcal{P}_1 h$. Furthermore, $K_0(e_0, z) = 0$, so $\mathcal{R}_0 e_0 = 0$ and hence $e_0 - \mathcal{R}_0 e_0 = h - \mathcal{P}_0 h$. It follows that there is no loss in generality in taking the prediction space to be one-dimensional, h orthogonal to the prediction space under K_0 and $K_0(h,h) = 1$. Let g be a basis for the prediction space with $K_0(g,g) = 1$ and $\mathcal{S}$ be the space spanned by g and h. By (3), we can choose a basis s_1, s_2 for $\mathcal{S}$ so that for $v = v_1 s_1 + v_2 s_2$, where v_1, v_2 are scalars, $K_0(v,v) = v_1^2 + v_2^2$, $K_1(v,v) = \beta_1 v_1^2 + \beta_2 v_2^2$ and $a \leq \beta_1 \leq \beta_2 \leq b$. Let $h = h_1 s_1 + h_2 s_2$, $g = g_1 s_1 + g_2 s_2$, where g_1, g_2, h_1 and h_2 are scalars, and now take $\mathcal{P}_j$ to be orthogonal projection onto the space spanned by g under inner product K_j, so $\mathcal{P}_1 h = \{K_1(h,g)/K_1(g,g)\} g$. Then $\mathcal{P}_0 h = 0$, $E_0(h - \mathcal{P}_0 h)^2 = 1$ and by (1)

$$\begin{aligned}
&\frac{E_0(h - \mathcal{P}_1 h)^2}{E_0(h - \mathcal{P}_0 h)^2} \\
&\quad = 1 + E_0(\mathcal{P}_1 h)^2 \\
&\quad = 1 + \left\{\frac{K_1(h,g)}{K_1(g,g)}\right\}^2 \\
&\quad = 1 + \left\{\frac{\beta_1 h_1 g_1 + \beta_2 h_2 g_2}{\beta_1 g_1^2 + \beta_2 g_2^2}\right\}^2 \\
&\quad = \frac{\beta_1^2 g_1^2 (g_1^2 + h_1^2) + \beta_2^2 g_2^2 (g_2^2 + h_2^2) + 2\beta_1\beta_2 g_1 g_2 (h_1 h_2 + g_1 g_2)}{(\beta_1 g_1^2 + \beta_2 g_2^2)^2}.
\end{aligned}$$

From $K_0(g,g) = K_0(h,h) = 1$ and $K_0(g,h) = 0$, we get $g_1^2 + h_1^2 = 1$, $g_2^2 + h_2^2 = 1$ and $g_1 g_2 + h_1 h_2 = 0$ (draw a picture of two orthogonal unit vectors in $\mathbb{R}^2$ to see this), so

$$\frac{E_0(h-\mathcal{P}_1h)^2}{E_0(h-\mathcal{P}_0h)^2}=\frac{\beta_1^2g_1^2+\beta_2^2g_2^2}{(\beta_1g_1^2+\beta_2g_2^2)^2}\le\frac{(\beta_1+\beta_2)^2}{4\beta_1\beta_2}=1+\frac{(\beta_2-\beta_1)^2}{4\beta_1\beta_2},$$

which can be verified by calculus, or as a consequence of Kantorovich's inequality (Rao 1973, p. 74). Then Theorem 1 follows from $a\le\beta_1\le\beta_2\le b$ and the fact that $(\beta_2-\beta_1)^2/(4\beta_1\beta_2)$ is increasing as β_2 increases or as β_1 decreases on the region $0<\beta_1<\beta_2$. Cleveland (1971) further shows that this bound cannot be sharpened without adding restrictions on the space of predictors. □

The fact that $b-a$ is squared on the right side of (5) is worth noting. Specifically, suppose $a=1+\epsilon_1$ and $b=1+\epsilon_2$, where both ϵ_1 and ϵ_2 are small, so that all variances are only slightly misspecified under $(0,K_1)$. Then the right side of (5) is approximately $1+\frac{1}{4}(\epsilon_2-\epsilon_1)^2$. For example, if $-\epsilon_1=\epsilon_2=\epsilon$, then $1+\frac{1}{4}(\epsilon_2-\epsilon_1)^2=1+\epsilon^2$, which is much nearer to 1 than either $a=1-\epsilon$ or $b=1+\epsilon$, the bounds in (4). Thus, we see that slight misspecifications of the model can potentially have a much larger effect on the evaluation of mses of pseudo-BLPs than on the efficiency of the pseudo-BLPs.

One other simple finite sample result we can give is that if $E_0h^2\ge E_1h^2$ for all $h\in\mathcal{H}_R$, then $E_1e_1^2\le E_1e_0^2\le E_0e_0^2\le E_0e_1^2$. Consequently,

$$\frac{E_0e_1^2}{E_1e_1^2}\ge\frac{E_0e_1^2}{E_0e_0^2}\ge1, \tag{6}$$

so that if variances under K_1 are always smaller than under the correct K_0, the effect of this misspecification is greater on the evaluation of the mse than on the efficiency of the prediction. There does not appear to be any comparable result when K_1 always gives larger variances than under K_0.

Both (6) and the comparison of the inequalities in (4) and (5) provide some support for the general notion that misspecifying the covariance structure of a random field has a greater impact on evaluating mses than on efficiency of point predictions, which has been noted as an empirical finding by Starks and Sparks (1987). Many of the examples and much of the asymptotic theory in the rest of this chapter also support this finding.

Exercise

1 Assuming (3) holds, show that $E_0e_0(h,Q)^2=0$ if and only if $E_0e_0(h,Q)^2=E_1e_0(h,Q)^2=E_1e_1(h,Q)^2=E_0e_1(h,Q)^2=0$.

3.3 The role of asymptotics

Asymptotic methods provide powerful tools for obtaining approximate results in mathematics and statistics. The most common way to employ asymptotics in statistics is to consider what happens as the number of observations increases. A key question is exactly how this should be done when studying prediction problems for random fields. For someone with a background in statistics, especially in time series analysis, a natural approach would be to let the observation region grow with the number of observations so that the distance between neighboring observations remains at least roughly constant. This approach was taken for predicting area averages of random fields by Quenouille (1949), Matérn (1960) and Dalenius, Hájek and Zubrzycki (1961). A numerical analyst, on the other hand, would more likely consider what happens as the number of observations within a fixed and bounded observation region increases so that the distance between neighboring observations tends to 0 (Novak 1988; Traub, Wasilkowski and Woźniakowski 1988). Although studying a problem from more than one perspective is generally a good idea, I believe that the numerical analyst's asymptotic approach is by far more informative for interpolation problems and is the approach I take here. I call asymptotics based on a growing observation region increasing-domain asymptotics and that based on increasingly dense observations in a fixed and bounded region fixed-domain asymptotics. Cressie (1993) uses the term "infill asymptotics" for this second concept.

A natural argument for using different asymptotics in spatial settings than in time series is the directional nature of time. That is, since it is not possible to go back in time, it does not make sense to think about taking more and more observations in a fixed interval of time. On the other hand, in a fixed region of space, it is possible, at least in principle, to take more and more observations in that region of space as long as the process does not vary over time. However, I feel this argument is slightly off the mark. For any given problem we have a given sample size from which we wish to make predictions. We use asymptotics not because we actually plan to take more and more observations but because we hope the approximations we obtain will be useful for the specific problem at hand. Thus, the fact that we could conceivably take more observations in our fixed region of space is irrelevant to drawing inferences from our given set of observations.

The directionality of time is related to the differences in appropriate asymptotics for temporal and spatial problems, but not, I believe, through the observation sequences that are physically possible. Rather, the difference is due to the types of predictions we are likely to want to make in the two settings. In time series, we usually want to predict the future, or extrapolate. In spatial settings, we usually want to interpolate: predict the process at a location that is, roughly speaking, surrounded by observations. After all, if we wanted to predict a spatial process in some region, we would take observations in that region and not some nearby region unless there were some physical impediment to doing so.

One reasonable expectation about the behavior of a good interpolant of a process Z at $\mathbf{x}_0$ is that it should depend mainly on observations near $\mathbf{x}_0$.

It follows that for a weakly stationary process, the behavior near the origin of the autocovariance function is critical for interpolation. It turns out that it is slightly more accurate to focus on the high frequency behavior of the spectrum, although, as described in 2.6 and 2.8, the two are closely related. A corollary of sorts of this consideration is that the low frequency behavior of the spectrum should have little effect on interpolation. The results in the rest of this chapter provide support for this notion. Furthermore, the results show that focusing on the high frequency behavior works much better when interpolating than extrapolating.

3.4 Behavior of prediction errors in the frequency domain

Suppose Z is a mean 0 weakly stationary random field with spectrum F and spectral representation $Z(\mathbf{x}) = \int_{\mathbb{R}^d} \exp(i\boldsymbol{\omega}^T\mathbf{x})M(d\boldsymbol{\omega})$. We can gain some insight into prediction problems by studying the spectral representation of prediction errors. Recall from 2.6 that $\mathcal{L}_{\mathbb{R}^d}(F)$ is the closed linear manifold of the functions $\exp(i\boldsymbol{\omega}^T\mathbf{x})$ for $\mathbf{x} \in \mathbb{R}^d$ with respect to the inner product defined by F. For a random variable $h \in \mathcal{H}_{\mathbb{R}^d}(F)$, let H be the corresponding function in $\mathcal{L}_{\mathbb{R}^d}(F)$, so that $h = \int_{\mathbb{R}^d} H(\boldsymbol{\omega})M(d\boldsymbol{\omega})$ and $\text{var}(h) = \int_{\mathbb{R}^d} |H(\boldsymbol{\omega})|^2F(d\boldsymbol{\omega})$ (see 2.5). Next, for a symmetric Borel set B, define $Z_B(\mathbf{x}) = \int_B \exp(i\boldsymbol{\omega}^T\mathbf{x})M(d\boldsymbol{\omega})$, the random field obtained by filtering out frequencies not in B. Taking B symmetric makes Z_B real whenever Z is real. Defining $h_B = \int_B H(\boldsymbol{\omega})M(d\boldsymbol{\omega})$, we get $\text{var}(h_B) = \int_B |H(\boldsymbol{\omega})|^2F(d\boldsymbol{\omega})$. Then we may reasonably call

$$\frac{\text{var}(h_B)}{\text{var}(h)} = \frac{\int_B |H(\boldsymbol{\omega})|^2F(d\boldsymbol{\omega})}{\int_{\mathbb{R}^d} |H(\boldsymbol{\omega})|^2F(d\boldsymbol{\omega})} \tag{7}$$

the fraction of the variance of h attributable to the set of frequencies B. This section examines how (7) behaves for prediction errors in some simple interpolation and extrapolation problems on $\mathbb{R}$.

Some examples

As a first example, suppose $f(\omega) = (1+\omega^2)^{-1}$ so that $K(t) = \pi e^{-|t|}$. Consider the extrapolation problem of predicting $Z(0)$ based on $Z(-\delta)$ for some $\delta > 0$. The BLP of $Z(0)$ is $e^{-\delta}Z(-\delta)$ with mse $\pi(1-e^{-2\delta})$ and the BLP is unchanged if further observations are added at locations less than $-\delta$, which follows by noting $\text{cov}\{Z(t), Z(0) - e^{-\delta}Z(-\delta)\} = 0$ for all $t < -\delta$. The prediction error is $Z(0) - e^{-\delta}Z(-\delta)$ and the corresponding function in $\mathcal{L}_{\mathbb{R}^d}(F)$ is $V_\delta(\omega) = 1 - e^{-(1+i\omega)\delta}$, so that

$$\left|V_\delta(\omega)\right|^2 = 2e^{-\delta}(\cosh\delta - \cos\omega\delta).$$

If T_δ is a positive function of δ and $\delta T_\delta \to 0$ as $\delta \downarrow 0$, then

$$\frac{\int_{-T_\delta}^{T_\delta} \left|V_\delta(\omega)\right|^2 (1+\omega^2)^{-1} d\omega}{\int_{-\infty}^{\infty} \left|V_\delta(\omega)\right|^2 (1+\omega^2)^{-1} d\omega} \sim \frac{\delta T_\delta}{\pi} \tag{8}$$

as $\delta \downarrow 0$ (Exercise 2). Thus, even if T_δ is large, as long as δT_δ is small, only a small fraction of the variance of the prediction error is attributable to the frequencies $[-T_\delta, T_\delta]$.

For the same spectral density, now consider the interpolation problem of predicting $Z(0)$ based on $Z(\delta)$ and $Z(-\delta)$. The BLP is $\frac{1}{2}\operatorname{sech}(\delta)\{Z(\delta) + Z(-\delta)\}$ and the mse is $\pi \tanh(\delta)$ (Exercise 3). Moreover, these results are unaffected by taking additional observations outside $[-\delta, \delta]$ (Exercise 3). In this case, the function corresponding to the prediction error is $V_\delta(\omega) = 1 - \operatorname{sech}(\delta)\cos(\omega\delta)$. Thus, $|V_\delta(\omega)|^2 = 4\operatorname{sech}^2(\delta)\left\{\sin^2(\frac{1}{2}\delta\omega) + \sinh^2(\frac{1}{2}\delta)\right\}^2$ and if T_δ is positive and $\delta T_\delta \to 0$ as $\delta \downarrow 0$, then

$$\frac{\int_{-T_\delta}^{T_\delta} |V_\delta(\omega)|^2 (1+\omega^2)^{-1} d\omega}{\int_{-\infty}^{\infty} |V_\delta(\omega)|^2 (1+\omega^2)^{-1} d\omega} \sim \frac{\delta^3}{\pi}\int_{-T_\delta}^{T_\delta} (1+\omega^2)\, d\omega = \frac{2\delta^3}{\pi}(T_\delta + \tfrac{1}{3}T_\delta^3) \tag{9}$$

as $\delta \downarrow 0$ (Exercise 4). For δT_δ small, the fraction of the variance of the prediction error attributable to $[-T_\delta, T_\delta]$ is much smaller than for the extrapolation problem.

As a second example, let us consider a smoother process: $f(\omega) = (2 + \omega^2)^{-2}$ so that $K(t) = 2^{-3/2}\pi \exp(-2^{1/2}|t|)\left(1 + 2^{1/2}|t|\right)$. For an extrapolation problem, consider predicting $Z(0)$ based on $Z(-\delta j)$ for $j = 1, \ldots, 10$. Taking $B = [-T, T]$, Table 1 gives values of (7) for various values of T and δ. It appears that for δT not too large, (7) is very nearly proportional to δT, which is also what happened for the previous spectral density. For the corresponding interpolation problem, predict $Z(0)$ based on $Z(\delta j)$ for $j = \pm 1, \ldots, \pm 10$. For fixed T, it now appears that (7) is proportional to δ^5 for δ sufficiently small, whereas (7) was proportional to δ^3 when $f(\omega) = (1 + \omega^2)^{-1}$.

For both spectral densities, whether extrapolating or interpolating, the fraction of variance of the prediction error attributable to the frequency band $[-T, T]$ is small when δ and δT are small. However, when extrapolating, the rate of convergence appears to be linear in δ regardless of the smoothness of the process, and when interpolating, it is of order δ^3 for the rougher process and appears to be of order δ^5 for the smoother process as long as T is not too large. These results suggest that optimal interpolations are only weakly affected by the low frequency behavior of the spectrum, particularly for smoother processes. We return to this problem in greater generality in 3.5.

Relationship to filtering theory

Christakos (1992), following up on some discussion in Carr (1990), notes that the prediction error process can be viewed as a high-pass filter of sorts. More specifically, he considers best linear unbiased prediction of $Z(\mathbf{x}_0)$ based on observations $Z(\mathbf{x}_1), \ldots, Z(\mathbf{x}_n)$ when $EZ(\mathbf{x}) = \boldsymbol{\beta}^T\mathbf{m}(\mathbf{x})$ and $\mathbf{m}$ contains a constant function as a component. In this case, if $\sum_{i=1}^n \lambda_i Z(\mathbf{x}_i)$ is the BLUP of $Z(\mathbf{x}_0)$, then setting $\lambda_0 = -1$, the prediction error is $\sum_{i=0}^n \lambda_i Z(\mathbf{x}_i)$ with $\sum_{i=0}^n \lambda_i = 0$. Since a BLUP is generally expected to depend mainly on those observations near the predictand $Z(\mathbf{x}_0)$, the error of the BLUP is, roughly speaking, a local difference operator and hence behaves like a high-pass filter (Schowengerdt 1983). Neither Christakos (1992) nor Carr (1990) provide any quantitative theory supporting this viewpoint. The results in this chapter and in Stein (1999) show that it is possible to provide such a quantitative theory. Note that we are considering simple kriging prediction here, for which $\sum_{i=0}^n \lambda_i = 0$ generally does not hold. Nevertheless, under appropriate conditions, the prediction error process still has most of its variation attributable to the high frequency components of Z.

Exercises

2 Verify (8).

3 Show that if $f(\omega) = (1+\omega^2)^{-1}$, then the BLP of $Z(0)$ based on $Z(\delta)$ and $Z(-\delta)$ is $\frac{1}{2}\operatorname{sech}(\delta)\{Z(\delta)+Z(-\delta)\}$ with mse $\pi\tanh(\delta)$. In addition, show that these results are unaffected by taking additional observations outside $[-\delta, \delta]$.

4 Verify (9).

5 Produce results similar to those in Table 1 for $f(\omega) = (3+\omega^2)^{-3}$.

TABLE 1. Values of (7) for $B = [-T, T]$ with $f(\omega) = (2+\omega^2)^{-2}$ when predicting $Z(0)$ based on $Z(-\delta j)$ for $j = 1, \ldots, 10$ (extrapolation) and based on $Z(\delta j)$ for $j = \pm 1, \ldots, \pm 10$ (interpolation).

	Extrapolation		Interpolation	
$\frac{1}{2\pi}T$	$\delta = 0.05$	$\delta = 0.1$	$\delta = 0.05$	$\delta = 0.1$
0.1	0.0100	0.0200	1.99×10^{-8}	6.31×10^{-7}
0.2	0.0200	0.0400	5.75×10^{-8}	1.83×10^{-6}
0.5	0.0500	0.100	7.95×10^{-7}	2.53×10^{-5}
1	0.100	0.200	1.60×10^{-5}	5.10×10^{-4}
2	0.200	0.400	4.52×10^{-4}	1.43×10^{-2}
5	0.499	0.913	4.20×10^{-2}	7.68×10^{-1}

TABLE 2. Interpolating with an incorrect model. Observations at $\pm\delta j$ for $j = 1, \ldots, 2/\delta$ and predict at 0.

Case 1: $m_0 = m_1 = 0, \quad K_0(t) = e^{-|t|}, \quad K_1(t) = \frac{1}{2}e^{-2|t|}$

δ	$E_0 e_0^2$	$E_0 e_1^2 / E_0 e_0^2 - 1$	$E_1 e_1^2 / E_0 e_1^2 - 1$
0.2	0.19738	0.0129	−0.0498
0.1	0.09967	0.0020	−0.0118
0.05	0.04996	0.00026	−0.0028

Case 2: $m_0 = m_1 = 0, \quad K_0(t) = e^{-|t|}(1 + |t|), \quad K_1(t) = \frac{1}{8}e^{-2|t|}(1 + 2|t|)$

δ	$E_0 e_0^2$	$E_0 e_1^2 / E_0 e_0^2 - 1$	$E_1 e_1^2 / E_0 e_1^2 - 1$
0.2	2.1989×10^{-3}	2.9×10^{-3}	−0.0432
0.1	2.7771×10^{-4}	1.5×10^{-4}	−0.0104
0.05	3.4804×10^{-5}	8.3×10^{-6}	−0.0026

3.5 Prediction with the wrong spectral density

If the low frequencies of the spectrum contribute little to the variance of prediction errors, we might then expect that misspecifying the spectrum at low frequencies would have little impact on the predictions. This is in fact the case as the rest of this chapter shows.

Examples of interpolation

Suppose we observe a stationary process Z at δj for $j = \pm 1, \ldots, \pm n$ and wish to predict $Z(0)$. Furthermore, let (m_0, K_0) be the correct second-order structure and (m_1, K_1) the presumed second-order structure. Table 2 gives results for $\delta = 0.2, 0.1$ and 0.05, $n = 2/\delta$ and two different pairs of autocovariance functions.

The first case compares $K_0(t) = e^{-|t|}$ with $K_1(t) = \frac{1}{2}e^{-2|t|}$, for which it is possible to give analytic answers (Exercise 6). For now, just consider the numerical results in Table 2. Note that the values of $E_0 e_1^2 / E_0 e_0^2$ and $E_1 e_1^2 / E_0 e_1^2$ are both near 1, especially for smaller δ, despite the fact that a superficial look at K_0 and K_1 suggests the two autocovariance functions are rather different; after all, $K_0(0) = 1$ and $K_1(0) = \frac{1}{2}$. It is helpful to consider series expansions in powers of $|t|$ of the two functions about the origin: $K_0(t) = 1 - |t| + \frac{1}{2}t^2 + O\left(|t|^3\right)$ and $K_1(t) = \frac{1}{2} - |t| + t^2 + O\left(|t|^3\right)$. We see that $-|t|$ is a principal irregular term for both functions. The fact that the power of the principal irregular term is 1 for both K_0 and K_1 is the key factor in making $E_0 e_1^2 / E_0 e_0^2$ near 1 for δ small. The additional fact that the coefficient of the principal irregular term is -1 for both models is what makes $E_1e_1^2/E_0e_1^2$ near 1 for δ small. So, for example, if we had $K_0(t) = e^{-|t|}$ and $K_2(t) = e^{-2|t|}$, then $E_0e_2^2/E_0e_0^2 = E_0e_1^2/E_0e_0^2$ but $E_2e_2^2/E_0e_2^2 = 2E_1e_1^2/E_0e_1^2$ so that the values of $E_2e_2^2/E_0e_2^2$ are near 2 for δ small, despite the fact that we now have $K_0(0) = K_2(0) = 1$. We can also see the similarities of the autocovariance functions $K_i(t) = \alpha_i^{-1}e^{-\alpha_i|t|}$ with $\alpha_0 = 1$ and $\alpha_1 = 2$ through the spectral densities: the spectral density f_i corresponding to K_i is $f_i(\omega) = 1/\{\pi(\alpha_i^2 + \omega^2)\} = \pi^{-1}\omega^{-2}\{1 - \alpha_i^2\omega^{-2} + O(\omega^{-4})\}$ as $\omega \to \infty$. Thus, for $i = 0, 1$, $f_i(\omega) \sim \pi^{-1}\omega^{-2}$ as $\omega \to \infty$.

The second pair of second-order structures presented in Table 2 is $m_0 = m_1 = 0$, $K_0(t) = e^{-|t|}(1 + |t|)$ and $K_1(t) = \frac{1}{8}e^{-2|t|}(1 + 2|t|)$. Here we see that $E_0e_1^2/E_0e_0^2$ is extremely close to 1, especially for smaller δ and that $E_1e_1^2/E_0e_1^2$ is quite close to 1, although not nearly as close as $E_0e_1^2/E_0e_0^2$. Again, the results support the notion that misspecifying the autocovariance function mainly affects the evaluation of mses. To see why these two autocovariance functions should yield similar linear predictions and assessments of mses, note that $K_0(t) = 1 - \frac{1}{2}t^2 + \frac{1}{3}|t|^3 - \frac{1}{8}t^4 + O\left(|t|^5\right)$ and $K_1(t) = \frac{1}{8} - \frac{1}{4}t^2 + \frac{1}{3}|t|^3 - \frac{1}{4}t^4 + O\left(|t|^5\right)$. For both autocovariance functions, $\frac{1}{3}|t|^3$ is a principal irregular term. Comparing the models in the spectral domain, the spectral densities both are of the form $2/(\pi\omega^4) + O(\omega^{-6})$ as $\omega \to \infty$, so that the similar high frequency behavior for the two models is readily apparent.

An example with a triangular autocovariance function

Before we conclude that it is always sufficient to find a principal irregular term of an autocovariance function to determine the approximate properties of linear predictors as observations get dense, consider the pair of autocovariance functions: $K_0(t) = e^{-|t|}$ and the triangular autocovariance function $K_1(t) = (1 - |t|)^+$. Both are of the form $1 - |t| + O(t^2)$ as $t \to 0$, so they correspond to processes with similar local behavior. Suppose we observe $Z(j/n)$ for $j = \pm 1, \ldots, \pm n$ and wish to predict $Z(0)$. The BLP (assuming $m_0 = m_1 = 0$) under K_0 is given in Exercise 3 of 3.3 and depends only on $Z(1/n)$ and $Z(-1/n)$. The BLP under K_1 is

$$\frac{2n}{4n+1}\left\{Z\left(\frac{1}{n}\right) + Z\left(-\frac{1}{n}\right)\right\} + \frac{n}{4n+1}\left\{Z\left(\frac{n-1}{n}\right) + Z\left(-\frac{n-1}{n}\right)\right\} - \frac{n+1}{4n+1}\{Z(1) + Z(-1)\} \tag{10}$$

(Exercise 7). The fact that for all n, $Z(\pm 1)$ and $Z(\pm(n-1)/n)$ appear in the BLP, whereas $Z(\pm 2/n), \ldots, Z(\pm(n-2)/n)$ do not, should seem strange, but it is a consequence of the lack of differentiability of K_1 at 1. Furthermore, the role of these relatively distant observations is not negligible, since

$$E_1\left[Z(0)-\frac{2n}{4n+1}\left\{Z\left(\frac{1}{n}\right)+Z\left(-\frac{1}{n}\right)\right\}\right]^2\sim n^{-1}$$

as $n\to\infty$, whereas the mse of the BLP under K_1 is asymptotically $\frac{3}{4}n^{-1}$ (Exercise 7). To compare BLPs under the two models, let $e_j(n)$ be the prediction error using K_j and observations $Z(\pm k/n)$ for $k=1,\ldots,n$. Then as $n\to\infty$, $E_0e_0(n)^2\sim n^{-1}$, $E_1e_0(n)^2\sim n^{-1}$, $E_1e_1(n)^2\sim\frac{3}{4}n^{-1}$ and $E_0e_1(n)^2\sim\frac{5}{4}n^{-1}$ (Exercise 7). Thus, even asymptotically, there are nonnegligible differences between both the actual and presumed performances of the linear predictors under the two models. Note though, that if one uses K_0 when K_1 is the truth, the presumed mse is asymptotically correct; that is, $E_0e_0(n)^2/E_1e_0(n)^2\to 1$ as $n\to\infty$.

The corresponding spectral densities for this situation are $f_0(\omega)=\pi^{-1}(1+\omega^2)^{-1}$ and $f_1(\omega)=\pi^{-1}(1-\cos\omega)/\omega^2$, so that $f_1(\omega)/f_0(\omega)=(1-\cos\omega)\{1+O(\omega^{-2})\}$ as $\omega\to\infty$, whereas in the previous two cases, $f_1(\omega)/f_0(\omega)$ converged to 1 as $\omega\to\infty$. It is instructive to look at the functions corresponding to the prediction errors in this last example. Letting $\hat{e}_j$ be the function corresponding to e_j, Figure 1 plots $|\hat{e}_j(\omega)|^2$ for $j=0,1$ and $n=5$. Now $f_1(2\pi k)=0$ for all nonzero integers k, so that e_1 has a smaller mse than e_0 under f_1 because $|\hat{e}_1(\omega)|^2$ partially matches the oscillations in f_1, being small when $f_1(\omega)$ is large and vice versa.

This example shows that under fixed-domain asymptotics, it is not always the case that two autocovariance functions sharing a common principal irregular term yield asymptotically the same BLPs. However, it is still possible that two spectral densities behaving similarly at high frequen-

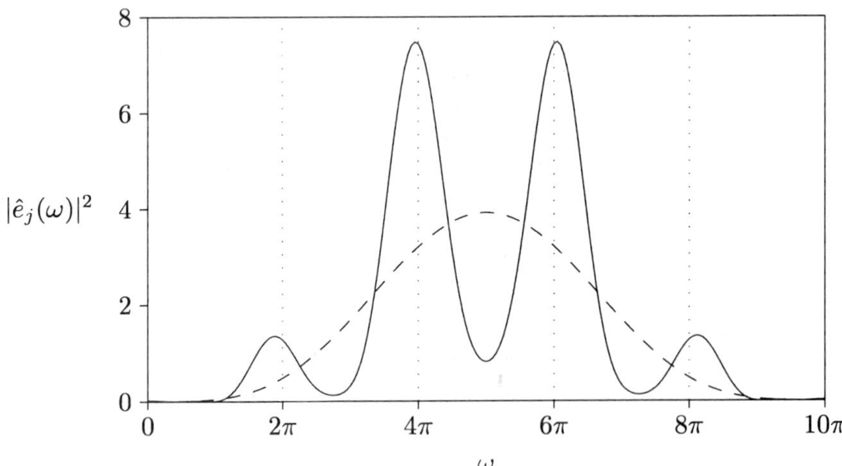

FIGURE 1. For the prediction problem described in this subsection with $n=5$, plots of $|\hat{e}_0(\omega)|^2$ (dashed line) and $|\hat{e}_1(\omega)|^2$ (solid line) as functions of ω. Dotted vertical lines indicate zeroes of f_1.

cies give asymptotically the same BLPs. Indeed, Theorem 10 in Chapter 4 comes very close to proving such a result.

More criticism of Gaussian autocovariance functions

Table 3 gives results for a pair of second-order structures that yield startlingly different linear predictions: $m_0 = m_1 = 0$, $K_0(t) = e^{-|t|}(1 + |t|)$ and $K_1(t) = e^{-t^2/2}$. Under K_0, the process is exactly once mean square differentiable and under K_1 it is infinitely mean square differentiable, so such a result is not surprising. However, as shown in Figure 1 of 2.7, plots of the autocovariance functions give no obvious sign of the radically different natures of the two models. Consider predicting $Z(0)$ based on observing $Z(\pm\delta j)$ for $j = 1, \dots, n$. If K_0 is the truth, then the mse of the BLP goes down with δ but is essentially the same for all $n \geq 5$. If K_1 is the truth, then the mse of the BLP goes down sharply with δ and furthermore, can go down substantially as n increases. In particular, for $\delta = 0.4$, the mse goes from 1.72×10^{-9} to 4.79×10^{-11} when n is increased from 10 to 20. This decrease occurs despite the fact that the added observations at $\pm\delta j$ for $j = 11, \dots, 20$ have correlation of at most 6.25×10^{-5} with the predictand $Z(0)$. Using results in 3.6, it is possible to show via numerical integration that when predicting $Z(0)$ based on observations at $0.4j$ for all nonzero integers j, the mse is 1.94×10^{-11}, which is in turn quite a bit smaller than 4.79×10^{-11}, the mse when $Z(0.4j)$ is observed for $0 < |j| \leq 20$.

Next, consider what happens if K_0 is the truth but K_1 is presumed to be the autocovariance function for Z. The ratio of mses, $E_0e_1^2/E_0e_0^2$, ranges from 1.48 to 3.76 and increases as δ decreases and n increases. The ratio of the presumed and actual mses of the pseudo-BLP, $E_1e_1^2/E_0e_1^2$, ranges from 0.171 down to 7.53×10^{-10}, decreasing sharply with δ but also decreasing robustly as n increases when δ is small. Thus, by acting as if K_1 were the

TABLE 3. Predicting $Z(0)$ based on $Z(\pm\delta j)$ for $j = 1, \dots, n$ with $m_0 = m_1 = 0$, $K_0(t) = e^{-|t|}(1 + |t|)$ and $K_1(t) = e^{-t^2/2}$.

δ	n	$E_0e_0^2$	$E_0e_1^2$	$E_1e_1^2$	$E_1e_0^2$
0.8	5	0.115	0.170	2.90×10^{-2}	7.42×10^{-2}
	10	0.115	0.207	2.51×10^{-2}	7.42×10^{-2}
	20	0.115	0.212	2.49×10^{-2}	7.42×10^{-2}
0.6	5	0.0532	0.0786	6.19×10^{-4}	1.11×10^{-2}
	10	0.0532	0.128	1.94×10^{-4}	1.11×10^{-2}
	20	0.0532	0.172	1.55×10^{-4}	1.11×10^{-2}
0.4	5	0.0169	0.0211	4.81×10^{-7}	5.28×10^{-4}
	10	0.0169	0.0328	1.72×10^{-9}	5.30×10^{-4}
	20	0.0169	0.0636	4.79×10^{-11}	5.30×10^{-4}

truth when K_0 was correct, our predictions will not be all that bad but we may be wildly overoptimistic about their mses. On the other hand, if K_1 is correct but K_0 is presumed to be the autocovariance function of Z, our predictions are severely inefficient for small δ but at least our presumed mses are badly conservative, which is usually a more tolerable mistake than being badly overoptimistic.

I strongly recommend not using autocovariance functions of the form Ce^{-at^2} to model physical processes. If the discussion of the theoretical properties in 2.7 does not convince you of this, the mse of 4.79×10^{-11} in Table 3 for predicting $Z(0)$ when observing $Z(0.4j)$ for $j = \pm 1, \ldots, \pm 20$ should. Considering that $\text{var}\{Z(0)\} = 1$ and that the maximum correlation between $Z(0)$ and the observations is $e^{-0.4^2/2} = 0.923$, this mse is implausibly small for any physical process. Unfortunately, a number of books on geostatistics (Carr 1995; Christakos 1992; Isaaks and Srivastava 1989; Journel and Huijbregts 1978; and Kitanidis 1997) suggest Ce^{-at^2} as a sensible example of an autocovariance function for a mean square differentiable process. Furthermore, the Gaussian model is the only model for differentiable processes available for fitting semivariograms to spatial data in SAS (SAS Institute, Inc. 1997, p. 626), S+SPATIALSTATS (Kaluzny, Vega, Cardoso and Shelly 1998, p. 91) and VARIOWIN (Pannetier 1996, p. 50). Goovaerts (1997) does recognize some of the serious problems with this model but does not give any alternative models for mean square differentiable processes. The Matérn models (Sections 2.7 and 2.10) include a parameter that controls the differentiability of the process and I recommend their adoption as an alternative to Ce^{-at^2} as a model for differentiable processes.

Examples of extrapolation

Let us next reconsider the two pairs of autocovariance functions in Table 2 when there are only observations on one side of 0 so that we are extrapolating rather than interpolating. Table 4 compares these two pairs of second-order structures for predicting $Z(0)$ based on $Z(-\delta j)$, $j = 1, 2, \ldots, 2/\delta$. For the first case $m_0 = m_1 = 0$, $K_0(t) = e^{-|t|}$ and $K_1(t) = \frac{1}{2}e^{-2|t|}$, we see again that $E_0e_1^2/E_0e_0^2$ and $E_1e_1^2/E_0e_1^2$ are both near 1, especially for δ small. However, these ratios are not as close to 1 as in the interpolation case, particularly so for $E_0e_1^2/E_0e_0^2$. Exercise 6 gives analytic results for this problem. Note that the prediction problem itself is not all that much easier in the interpolation case, since, as $\delta \downarrow 0$, $E_0e_0^2 \sim \delta$ when interpolating and $E_0e_0^2 \sim 2\delta$ when extrapolating. What is true is that if we use the model $K(t) = \alpha^{-1}e^{-\alpha|t|}$, or equivalently, $f(\omega) = 1/\{\pi(\alpha^2 + \omega^2)\}$, the value of α is much less critical when interpolating than extrapolating.

The difference between the extrapolation and interpolation problems is more dramatic for the second pair of second-order structures with $m_0 = m_1 = 0$, $K_0(t) = e^{-|t|}(1 + |t|)$ and $K_1(t) = \frac{1}{8}e^{-2|t|}(1 + 2|t|)$. Table 4 shows

TABLE 4. Extrapolating with an incorrect model. Observations at $-\delta j$ for $j = 1, \ldots, 2/\delta$ and predict at 0.

Case 1: $m_0 = m_1 = 0, \quad K_0(t) = e^{-|t|}, \quad K_1(t) = \frac{1}{2}e^{-2|t|}$

δ	$E_0 e_0^2$	$E_0 e_1^2 / E_0 e_0^2 - 1$	$E_1 e_1^2 / E_0 e_1^2 - 1$
0.2	0.3297	0.067	−0.217
0.1	0.1813	0.041	−0.126
0.05	0.0952	0.023	−0.069

Case 2: $m_0 = m_1 = 0, \quad K_0(t) = e^{-|t|}(1 + |t|), \quad K_1(t) = \frac{1}{8}e^{-2|t|}(1 + 2|t|)$

δ	$E_0 e_0^2$	$E_0 e_1^2 / E_0 e_0^2 - 1$	$E_1 e_1^2 / E_0 e_1^2 - 1$
0.2	1.35×10^{-2}	0.392	−0.506
0.1	2.04×10^{-3}	0.251	−0.341
0.05	2.82×10^{-4}	0.143	−0.207

that the values of $E_0 e_1^2 / E_0 e_0^2$ and $E_1 e_1^2 / E_0 e_1^2$ are not nearly as close to 1 when extrapolating as when interpolating.

Let us look more carefully at the transition between interpolation and extrapolation. More specifically, consider what happens when there are observations at $-j/20$ for $j = 1, \ldots, 20$ and at $j/20$ for $j = 1, \ldots, p$ for various values of p. For the exponential autocovariance functions $\alpha^{-1}e^{-\alpha|t|}$, the transition is immediate: for all $p \geq 1$, we get the same results as in Table 2. Table 5 shows that the transition is more gradual for the second case in Tables 2 and 4. We see that once $p = 2$, the mse of the optimal predictor does not change much by adding further observations. The effect on the misspecification of the mse of the pseudo-BLP as measured by $E_1 e_1^2 / E_0 e_1^2 - 1$ settles down at around $p = 3$. However, the loss of efficiency in using the pseudo-BLP as measured by $E_0 e_1^2 / E_0 e_0^2 - 1$ drops dramatically with every increase in p up through $p = 5$. Thus, we see that a prediction location may need to be quite far from the boundary of the observation region to be fully in the "interpolation" setting.

Pseudo-BLPs with spectral densities misspecified at high frequencies

Let us next consider some pairs of spectral densities that both decay algebraically at high frequencies but at different rates. Tables 6 and 7 show numerical results comparing extrapolation and interpolation. Specifically, consider predicting $Z(0)$ based on observations at δj for $j = -2/\delta, \ldots, -1$ for the extrapolation case and $j = \pm 1, \ldots, \pm 2/\delta$ for the interpolation case.

TABLE 5. The transition from extrapolation to interpolation. Observe $Z(0.05j)$ for $j = -20, \ldots, -1$ and $j = 1, \ldots, p$ and predict $Z(0)$ for $m_0 = m_1 = 0, K_0(t) = e^{-|t|}(1 + |t|)$ and $K_1(t) = \frac{1}{8}e^{-2|t|}(1 + 2|t|)$.

p	$E_0 e_0^2$	$E_0 e_1^2 / E_0 e_0^2 - 1$	$E_1 e_1^2 / E_0 e_1^2 - 1$
0	2.815599×10^{-4}	1.427×10^{-1}	-2.069×10^{-1}
1	4.930707×10^{-5}	3.398×10^{-2}	-5.778×10^{-2}
2	3.555431×10^{-5}	1.891×10^{-3}	-5.786×10^{-3}
3	3.485779×10^{-5}	1.219×10^{-4}	-2.798×10^{-3}
4	3.480768×10^{-5}	2.183×10^{-5}	-2.611×10^{-3}
5	3.480417×10^{-5}	7.901×10^{-6}	-2.591×10^{-3}
6	3.480392×10^{-5}	8.644×10^{-6}	-2.591×10^{-3}
20	3.480390×10^{-5}	8.317×10^{-6}	-2.591×10^{-3}

The spectral densities used are $c(\kappa)(\kappa + \omega^2)^{-\kappa}$ for $\kappa = 1$, 2 and 3, where $c(\kappa)$ is chosen to make $\text{var}\{Z(0)\} = 1$. Values for δ of 0.2, 0.1 and 0.05 are considered. For all of these examples, no matter what the value of δ and what two values of ν correspond to the true and presumed spectral densities, $E_0 e_1^2 / E_0 e_0^2$ is larger when extrapolating than interpolating, particularly so for smaller δ. (It is not true that $E_0 e_1^2 / E_0 e_0^2$ is always larger when extrapolating; see Exercise 11 in 3.6 for an example.) Another apparent pattern is that the penalty due to using the wrong spectral density is smaller if the presumed spectral density decays more quickly at high frequencies than the actual spectral density rather than the other way around.

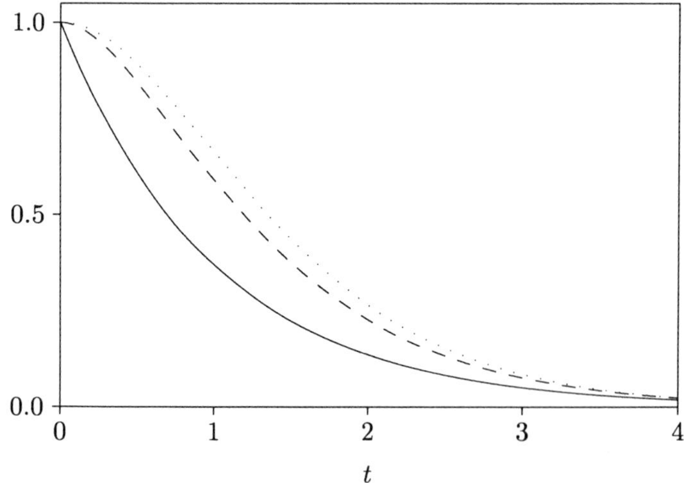

FIGURE 2. Plots of autocovariance functions used in Tables 6 and 7. Solid line corresponds to $\kappa = 1$, dashed line to $\kappa = 2$ and dotted line to $\kappa = 3$.

The reader is strongly encouraged to try other examples of pairs of second-order structures and other locations of observations and predictands, including two-dimensional examples, to gain a better understanding of how changing second-order structures affects linear predictions. When using the autocovariance function $e^{-t^2/2}$, numerical instabilities may limit the examples one can consider. The problem is that for even a moderate number of observations close together, this autocovariance function implies that certain linear combinations of the observations have extremely small variance, so that the covariance matrix of the observations is nearly singular. That is why the smallest value for δ in Table 3 is 0.4.

TABLE 6. Mean squared errors for predicting $Z(0)$ based on $Z(\delta j)$ for $j = 1, \ldots, 2/\delta$ (extrapolation) and based on $Z(\pm\delta j)$ for $j = 1, \ldots, 2/\delta$ (interpolation) under three spectral densities and for $\delta = 0.2$, 0.1 and 0.05. The three spectral densities are $f(\omega) = c(\kappa)(\kappa + \omega^2)^{-\kappa}$ for $\kappa = 1, 2, 3$, where $c(\kappa)$ is chosen to make $\mathrm{var}\{Z(0)\} = 1$. See Figure 2 for plots of the autocovariance functions.

	Presumed value of κ			
True κ	1	2	3	δ
		Extrapolation		
1	0.3295	0.653	1.999	0.2
	0.1813	0.417	1.572	0.1
	0.0952	0.238	1.002	0.05
2	8.72×10^{-2}	3.25×10^{-2}	5.59×10^{-2}	0.2
	2.55×10^{-2}	5.33×10^{-3}	1.09×10^{-2}	0.1
	6.92×10^{-3}	7.65×10^{-4}	1.74×10^{-3}	0.05
3	6.48×10^{-2}	8.75×10^{-3}	4.39×10^{-3}	0.2
	1.80×10^{-2}	7.58×10^{-4}	2.24×10^{-4}	0.1
	4.75×10^{-3}	5.59×10^{-5}	9.02×10^{-6}	0.05
		Interpolation		
1	0.1974	0.2483	0.3195	0.2
	0.0997	0.1256	0.1620	0.1
	0.0500	0.0630	0.0813	0.05
2	1.23×10^{-2}	6.13×10^{-3}	6.79×10^{-3}	0.2
	1.71×10^{-3}	7.83×10^{-4}	8.68×10^{-4}	0.1
	2.25×10^{-4}	9.84×10^{-5}	1.09×10^{-4}	0.05
3	3.40×10^{-3}	2.99×10^{-4}	2.43×10^{-4}	0.2
	2.54×10^{-4}	9.82×10^{-6}	7.90×10^{-6}	0.1
	1.73×10^{-5}	3.11×10^{-7}	2.49×10^{-7}	0.05

Exercises

6 For a process Z on $\mathbb{R}$, consider the two models $(0, K_0)$ and $(0, K_1)$, where $K_i(t) = a_i^{-1} e^{-a_i|t|}$. If Z is observed at $\pm\delta j$ for $j = 1, \ldots, n$, then by a trivial extension of Exercise 3, the BLP of $Z(0)$ under $(0, K_i)$ is

TABLE 7. Some comparisons for the results in Table 6. Unbracketed numbers in off-diagonal entries of the tables are ratios of the mse of the pseudo-BLP to that of the BLP. Numbers in angled brackets are limiting values for these ratios as given by Theorem 3 in 3.6. Numbers in parentheses of diagonal entries of upper table are ratios of mses of the BLPs under extrapolation to the mses of BLPs under interpolation.

True κ	Presumed value of κ: 1	2	3	δ
		Extrapolation		
1	(1.67)	1.98	6.06	0.2
	(1.82)	2.30	8.67	0.1
	(1.90)	2.50	10.53	0.05
		$\langle 2.73 \rangle$	$\langle 12.88 \rangle$	
2	2.68	(5.30)	1.72	0.2
	4.79	(6.80)	2.05	0.1
	9.05	(7.77)	2.28	0.05
	$\langle \infty \rangle$		$\langle 2.55 \rangle$	
3	14.8	2.00	(18.0)	0.2
	80.5	3.38	(28.4)	0.1
	526.9	6.62	(36.2)	0.05
	$\langle \infty \rangle$	$\langle \infty \rangle$		
		Interpolation		
1		1.258	1.619	0.2
		1.260	1.626	0.1
		1.261	1.627	0.05
		$\langle 1.261 \rangle$	$\langle 1.628 \rangle$	
2	2.00		1.107	0.2
	2.18		1.109	0.1
	2.28		1.109	0.05
	$\langle 2.39 \rangle$		$\langle 1.110 \rangle$	
3	14.0	1.229		0.2
	32.1	1.242		0.1
	69.3	1.246		0.05
	$\langle \infty \rangle$	$\langle 1.248 \rangle$		

$\frac{1}{2}\operatorname{sech}(\delta a_i)\{Z(\delta)+Z(-\delta)\}$. Calculate $E_i e_j^2$ and use the result to show that as $\delta \downarrow 0$,

$$E_i e_i^2 = \delta - \frac{a_i^2}{3}\delta^3 + O(\delta^4)$$

and

$$E_0(e_0 - e_1)^2 = \frac{(a_1^2 - a_0^2)^2}{4a_1}\delta^4 + O(\delta^5),$$

so that

$$\frac{E_0 e_1^2}{E_0 e_0^2} = 1 + \frac{(a_1^2 - a_0^2)^2}{4a_0}\delta^3 + O(\delta^4)$$

and

$$\frac{E_1 e_1^2}{E_0 e_1^2} = 1 + \frac{a_0^2 - a_1^2}{3}\delta^2 + O(\delta^3).$$

If, instead, Z is observed at just $-\delta j$ for $j = 1, \ldots, n$, the BLP of $Z(0)$ under $(0, K_j)$ is $e^{-a_j\delta}Z(\delta)$. Show that as $\delta \downarrow 0$,

$$\frac{E_0 e_1^2}{E_0 e_0^2} = 1 + \frac{(a_0 - a_1)^2}{2a_0}\delta + O(\delta^2) \quad \text{and}$$
$$\frac{E_1 e_1^2}{E_0 e_1^2} = 1 + \frac{a_0^2 - a_1^2}{2a_0}\delta + O(\delta^2).$$

7 Suppose Z has known mean 0 and is observed at j/n for $j = \pm 1, \ldots, \pm n$. Verify that (10) gives the BLP for $Z(0)$ under $(0, K_1)$ for $K_1(t) = (1-|t|)^+$. For $K_0(t) = e^{-|t|}$, prove that as $n \to \infty$, $E_0 e_0(n)^2 \sim n^{-1}$, $E_1 e_0(n)^2 \sim n^{-1}$, $E_1 e_1(n)^2 \sim \frac{3}{4}n^{-1}$ and $E_0 e_1(n)^2 \sim \frac{5}{4}n^{-1}$.

8 When comparing linear predictions under two second-order structures with the same covariance functions, show that $E_0 e_0^2 = E_1 e_1^2$ and $E_0 e_1^2 = E_1 e_0^2$, so that $E_0 e_1^2 / E_0 e_0^2 = (E_1 e_1^2 / E_0 e_1^2)^{-1}$. Thus, when only the mean function is misspecified, the effect on the mse due to using the wrong mean is the same as the effect on the evaluation of the mse.

9 For a process Z on $\mathbb{R}$, consider the two models (m_0, K) and (m_1, K), where $K(t) = a^{-1}e^{-a|t|}$ and $m_i(t) \equiv \mu_i$. If Z is observed at $\pm\delta j$ for $j = 1, \ldots, n$, show that the BLP of $Z(0)$ under (m_i, K) is

$$\tfrac{1}{2}\operatorname{sech}(a\delta)\{Z(\delta) + Z(-\delta)\} + \mu_i\{1 - \operatorname{sech}(a\delta)\}.$$

Furthermore, show that as $\delta \downarrow 0$,

$$E_i e_i^2 = \delta - \frac{a^2}{3}\delta^3 + O(\delta^4)$$

and

$$E_0(e_0 - e_1)^2 = (\mu_1 - \mu_0)^2 \frac{a^4}{4}\delta^4 + O(\delta^5).$$

Conclude that

$$\frac{E_0 e_1^2}{E_0 e_0^2} = \left(\frac{E_1 e_1^2}{E_0 e_1^2}\right)^{-1} = 1 + \frac{1}{4}(\mu_1 - \mu_0)^2 a^4 \delta^3 + O(\delta^4).$$

Do similar asymptotic calculations when Z is observed at $-\delta j$ for $j = 1, \ldots, n$. Compare your results for the two cases.

10 Suppose a weakly stationary process Z on $\mathbb{R}$ has known autocovariance function $K(t) = a^{-1}e^{-a|t|}$, unknown constant mean μ and that Z is observed at $-\delta j$ for $j = 1, \ldots, n$.

(i) Find the BLUE of μ and the mse of this estimator. Show that if δn tends to a positive constant as $\delta \downarrow 0$, then the mse of the BLUE tends to a positive constant.

(ii) Find the BLUP or ordinary kriging predictor of $Z(0)$. Find the mse of this predictor and examine its dependence on δ and n, paying particular attention to small values of δ. Show that the BLUP is asymptotically optimal relative to the BLP (which assumes μ is known) as $\delta \downarrow 0$ no matter how n changes with δ (as long as $n > 0$), despite the fact that the BLUE for μ may not be consistent.

(iii) Repeat parts (i) and (ii) when Z is observed at $\pm\delta j$ for $j = 1, \ldots, n$. Compare your results for the two cases.

3.6 Theoretical comparison of extrapolation and interpolation

The previous two sections examined the relationship between the second-order structure and interpolating or extrapolating in some specific instances. This section provides a theoretical basis for supporting the conclusions drawn based on these examples. The approach I use is to consider extrapolation and interpolation based on infinite sequences of observations, for which there are well-known exact results on the properties of BLPs.

More specifically, for a mean 0 weakly stationary process Z on $\mathbb{R}$ and some $\delta > 0$, this section compares predicting $Z(0)$ based on observing $Z(\delta j)$ for all negative integers j with predicting $Z(0)$ based on observing $Z(\delta j)$ for all integers $j \neq 0$. The first setting corresponds to the classical extrapolation problem addressed by Kolmogorov (1941) and Wiener (1949) and the second is an interpolation problem whose general solution is given, for example, in Hannan (1970). Of course, for any given spectrum, the mse when interpolating must be no bigger than when extrapolating, so that interpolating is easier than extrapolating in this sense. However, as I indicated in the previous section, there is a second sense in which interpolation is easier than extrapolation: the actual mse of pseudo-BLPs is generally less sensitive to misspecification of the spectrum at high (see Table 7) or low (compare Tables 2 and 4) frequencies when interpolating than when extrapolating. Considering what happens as δ, the spacing between observations, tends to 0 provides some insight into this phenomenon.

For infinite sequences of observations, it is not appropriate to call what happens as $\delta \downarrow 0$ fixed-domain asymptotics unless we want to allow our fixed domain to be unbounded. However, for many prediction problems, only those observations relatively near the location at which we are predicting the process will have a nonnegligible impact on the prediction. Thus, it will often be the case that results for predicting $Z(0)$ based on observing $Z(\delta j)$ for all integers $j \neq 0$ with δ small will be very similar to those based on observing $Z(\delta j)$ for integers $0 < |j| \leq c/\delta$ for a positive constant c. Exercise 12 can be used to show this is not always the case. Whether or not the observation sequence is infinite, $\delta \downarrow 0$ does correspond to shrinking the distance between neighboring observations, so we might call both of these settings "shrinking interval" asymptotics.

Results on prediction problems when both the observations and the predictands fall on a regular lattice are usually stated in terms of discrete time series observed on the integers. To see the connection between the spectral density of a process Z on $\mathbb{R}$ and the process obtained by observing Z every δ time units, suppose that Z has spectral density f and define $Z_\delta(j) = Z(\delta j)$, so that Z_δ is a process on $\mathbb{Z}$. Then for all integers j and k,

$$\begin{aligned}\operatorname{cov}\{Z_\delta(j), Z_\delta(k)\} &= \int_{-\infty}^{\infty} \exp\{i\omega\delta(j-k)\} f(\omega)\, d\omega \\ &= \int_{-\pi/\delta}^{\pi/\delta} \exp\{i\omega\delta(j-k)\} \sum_{\ell=-\infty}^{\infty} f(\omega + 2\pi\ell\delta^{-1})\, d\omega \\ &= \int_{-\pi}^{\pi} \exp\{i\omega(j-k)\} \tilde{f}^\delta(\omega)\, d\omega,\end{aligned}$$

where

$$\tilde{f}^\delta(\omega) = \delta^{-1} \sum_{\ell=-\infty}^{\infty} f\left(\frac{\omega + 2\pi\ell}{\delta}\right). \tag{11}$$

Thus, we can view $\tilde{f}^\delta$ as the spectral density on $(-\pi, \pi]$ of the process Z_δ on $\mathbb{Z}$ and, indeed, weakly stationary stochastic processes on the integers are generally taken to have spectral distributions on $(-\pi, \pi]$. To avoid any possible ambiguities, I always use a $\tilde{}$ to indicate a spectral density on $(-\pi, \pi]$ for a process on $\mathbb{Z}$.

An interpolation problem

Dropping the δs for now, let Z be a mean 0 weakly stationary process on the integers with spectral density $\tilde{f}$ on $(-\pi, \pi]$ and consider the interpolation problem of finding the BLP of $Z(0)$ based on observing $Z(j)$ for all integers $j \neq 0$. Since the BLP will obviously have mean 0 ($\lambda_0 = 0$ in the notation of 1.2), we can restrict attention to predictors with mean 0 in order to find the BLP. The space of linear predictors with mean 0 is equivalent to the closed real linear manifold $\mathcal{L}_{-0}(\tilde{f})$ of functions $\exp(i\omega j)$ on $(-\pi, \pi]$ for integers $j \neq 0$ under the inner product defined by the spectral density $\tilde{f}$. The BLP of $Z(0)$ corresponds to the function $\hat{H} \in \mathcal{L}_{-0}(\tilde{f})$ satisfying $\int_{-\pi}^{\pi} \{1 - \hat{H}(\omega)\} \exp(-i\omega j)\tilde{f}(\omega)d\omega = 0$ for all integers $j \neq 0$. By the completeness and orthogonality of the functions $\{\exp(i\omega j)\}_{j\in\mathbb{Z}}$ in the space of square integrable functions on $(-\pi, \pi]$ (see, for example, Akhiezer and Glazman 1981), we must have $\{1 - \hat{H}(\omega)\}\tilde{f}(\omega)$ constant almost everywhere on $(-\pi, \pi]$. Suppose $1/\tilde{f}$ is integrable on $(-\pi, \pi]$, so that, in particular $\tilde{f}$ is positive almost everywhere on $(-\pi, \pi]$. It follows that $\hat{H}$ must be of the form $\hat{H}(\omega) = 1 + b/\tilde{f}(\omega)$ for almost every $\omega \in (-\pi, \pi]$ for some constant b. But $\hat{H} \in \mathcal{L}_{-0}(\tilde{f})$ implies $\int_{-\pi}^{\pi} \hat{H}(\omega)d\omega = 0$, so

$$\hat{H}(\omega) = 1 - \frac{2\pi}{\tilde{f}(\omega) \int_{-\pi}^{\pi} \tilde{f}(\nu)^{-1} d\nu}. \tag{12}$$

The mean square prediction error is then $4\pi^2 / \int_{-\pi}^{\pi} \tilde{f}(\omega)^{-1} d\omega$. Furthermore, when $\tilde{f}(\omega)^{-1}$ is not integrable, the mean square prediction error is 0 (Exercise 13). Hannan (1970, page 164) gives a generalization of these results to vector-valued time series. Defining e_i to be the prediction error under $\tilde{f}_i$,

$$E_0 e_1^2 = \frac{4\pi^2 \int_{-\pi}^{\pi} \{\tilde{f}_0(\omega)/\tilde{f}_1(\omega)^2\}\, d\omega}{\left\{\int_{-\pi}^{\pi} \tilde{f}_1(\omega)^{-1} d\omega\right\}^2}, \tag{13}$$

which we need to assess the effect of using the wrong spectral density for prediction.

An extrapolation problem

Next consider the classical extrapolation problem of finding the BLP of $Z(0)$ when observing $Z(j)$ for all negative integers j. This problem is mathematically more difficult than the interpolation case and I only present its solution; details are provided in many sources including Hannan (1970), Priestley (1981) and Yaglom (1962). Let $\mathcal{L}_{-}(\tilde{f})$ be the closed real linear manifold of $\exp(i\omega j)$ for integers $j < 0$ with respect to the inner product defined by $\tilde{f}$ and define

$$\sigma^2 = 2\pi \exp\left[\frac{1}{2\pi} \int_{-\pi}^{\pi} \log\{\tilde{f}(\omega)\}\, d\omega\right].$$

Theorem 2 (Hannan 1970)). *Suppose Z is a mean 0 process on the integers with spectral density $\tilde{f}$ satisfying $\sigma^2 > 0$. The function in $\mathcal{L}_{-}(\tilde{f})$*

corresponding to the BLP of $Z(0)$ based on $Z(j)$ for $j < 0$ is

$$\hat{H}(\omega) = \frac{\sum_{j=1}^{\infty} a(j)\exp(-i\omega j)}{\sum_{j=0}^{\infty} a(j)\exp(-i\omega j)},$$

where the $a(j)$s satisfy

$$\frac{\sigma}{\sqrt{2\pi}}\sum_{j=0}^{\infty} a(j)z^j = \exp\left\{c(0) + 2\sum_{j=1}^{\infty} c(j)z^j\right\},$$

$$c(j) = \frac{1}{4\pi}\int_{-\pi}^{\pi} \exp(-i\omega j)\log \tilde{f}(\omega)\,d\omega$$

for $j \neq 0$ and

$$\frac{\sigma}{\sqrt{2\pi}} = \exp\{c(0)\}.$$

In addition, if $\sigma^2 = 0$, then perfect prediction is possible.

For our purposes, we only need this result to obtain $E_0 e_1^2$. Using the subscript k on σ_k, $a_k(j)$ and $c_k(j)$ to indicate that these quantities are defined in terms of $\tilde{f}_k$, we have

$$\begin{aligned} E_0 e_1^2 &= \int_{-\pi}^{\pi} \tilde{f}_0(\omega)\left|\sum_{j=0}^{\infty} a_1(j)\exp(-i\omega j)\right|^{-2} d\omega \\ &= \frac{\sigma_1^2}{2\pi}\int_{-\pi}^{\pi} \tilde{f}_0(\omega)\exp\left\{-2\sum_{j=-\infty}^{\infty} c_1(j)\exp(-i\omega j)\right\}d\omega \\ &= \exp\left\{\frac{1}{2\pi}\int_{-\pi}^{\pi} \log \tilde{f}_1(\omega)\,d\omega\right\}\int_{-\pi}^{\pi}\frac{\tilde{f}_0(\omega)}{\tilde{f}_1(\omega)}\,d\omega, \end{aligned}$$

where the last step uses the fact that the $c_1(j)$s are the coefficients in the Fourier series for $\frac{1}{2}\log \tilde{f}_1$. Taking $\tilde{f}_1 = \tilde{f}_0$ recovers the well-known Kolmogorov formula of $E_0 e_0^2 = \sigma_0^2$.

Asymptotics for BLPs

Suppose Z is a mean 0 weakly stationary process on $\mathbb{R}$ with spectral density f_0 (on $\mathbb{R}$) but we instead presume that f_1 is the spectral density and we wish to predict $Z(0)$. Define $e_i(\text{in}, \delta)$ to be the prediction error under f_i with observations at δj for integers $j \neq 0$ and $e_i(\text{ex}, \delta)$ the prediction error under f_i with observations at δj for integers $j < 0$. Suppose that for some $\alpha > 1$, $f_0(\omega) \sim c|\omega|^{-\alpha}$ as $|\omega| \to \infty$ so that there exists T such that

$\frac{1}{2}c \le f_0(\omega)|\omega|^\alpha \le 2c$ for all $|\omega| \ge T$. Consider approximating

$$E_0 e_0(\mathrm{ex}, \delta)^2 = 2\pi c\delta^{\alpha-1} \exp\left\{\frac{1}{2\pi}\int_{-\pi}^{\pi} \log \frac{\tilde{f}_0^\delta(\omega)}{c\delta^{\alpha-1}}\, d\omega\right\} \tag{14}$$

as $\delta \downarrow 0$, where $\tilde{f}_i^\delta$ is defined as in (11) with f_i in place of f. For $\delta \le \pi/T$, we have

$$\int_{-\pi}^{\pi} \log \frac{\tilde{f}_0^\delta(\omega)}{c\delta^{\alpha-1}}\, d\omega = \int_{-\delta T}^{\delta T} \log \frac{\tilde{f}_0^\delta(\omega)}{c\delta^{\alpha-1}}\, d\omega + \int_{-\pi}^{\pi} \log \frac{\tilde{f}_0^\delta(\omega)}{c\delta^{\alpha-1}} 1\{|\omega| > \delta T\}\, d\omega. \tag{15}$$

For two functions a and b on a set T, I write $a(t) \ll b(t)$ if there exists C finite such that $|a(t)| \le Cb(t)$ for all $t \in T$. The first integral on the right side of (15) tends to 0 as $\delta \downarrow 0$, which follows from

$$\log \frac{\tilde{f}_0^\delta(\omega)}{c\delta^{\alpha-1}} \ll \log\left\{1 + \delta^{-\alpha} f_0(\omega/\delta)\right\}$$

for all δ sufficiently small. For $\alpha > 1$, define

$$\eta_\alpha(\omega) = \sum_{j=-\infty}^{\infty} |\omega + 2\pi j|^{-\alpha}. \tag{16}$$

The second integral on the right side of (15) converges to $\int_{-\pi}^{\pi} \log \eta_\alpha(\omega)\, d\omega$ since the integrand converges to $\log \eta_\alpha(\omega)$ for all $\omega \in (-\pi, \pi]$ other than $\omega = 0$ and is dominated by the integrable function $|\log \eta_\alpha(\omega)|$. Applying these results to (14) gives

$$E_0 e_0(\mathrm{ex}, \delta)^2 \sim 2\pi c\delta^{\alpha-1} \exp\left\{\frac{1}{2\pi}\int_{-\pi}^{\pi} \log \eta_\alpha(\omega)\, d\omega\right\}. \tag{17}$$

Similarly (Exercise 14),

$$E_0 e_0(\mathrm{in}, \delta)^2 \sim \frac{4\pi^2 c\delta^{\alpha-1}}{\int_{-\pi}^{\pi} \eta_\alpha(\omega)^{-1} d\omega}. \tag{18}$$

Of course, the right side of (18) must be no greater than the right side of (17), which can be directly verified by Jensen's inequality. A less trivial consequence of (17) and (18) is that the mse is of order $\delta^{\alpha-1}$ whether extrapolating or interpolating.

We can now look at the behavior of the prediction errors in the spectral domain. Let F be a positive finite measure on $\mathbb{R}$ with density f and let $\mathcal{L}(F)$ be the class of functions square integrable with respect to F. Suppose $V_\delta(\omega; \mathrm{ex})$ is the function in $\mathcal{L}(F)$ corresponding to the prediction error for the extrapolation problem with spacing δ and spectral density f, so that $Ee(\mathrm{ex}, \delta)^2 = \int_{-\infty}^{\infty} |V_\delta(\omega; \mathrm{ex})|^2 f(\omega) d\omega$. It follows from Theorem 2 that

$$|V_\delta(\omega; \mathrm{ex})|^2 = \exp\left\{\frac{1}{2\pi}\int_{-\pi}^{\pi} \log \tilde{f}^\delta(\nu)\, d\nu\right\} \frac{1}{\tilde{f}^\delta(\delta\omega)} \tag{19}$$

(Exercise 15). For two positive functions a and b on some set D, write $a(x) \asymp b(x)$ if there exist positive finite constants c_0 and c_1 such that $c_0 \leq a(x)/b(x) \leq c_1$ for all $x \in D$. If $f(\omega) \asymp (1+|\omega|)^{-\alpha}$, then for $|\omega| < \pi\delta^{-1}$,

$$|V_\delta(\omega; \text{ex})|^2 \asymp \delta^\alpha (1+|\omega|)^\alpha$$

and for $0 < T < \pi\delta^{-1}$,

$$\frac{\int_{-T}^{T} |V_\delta(\omega; \text{ex})|^2 f(\omega)\, d\omega}{Ee(\text{ex}, \delta)^2} \asymp \delta T \tag{20}$$

(Exercise 15). For the interpolation problem, let $V_\delta(\omega; \text{in})$ be the function in $\mathcal{L}(F)$ corresponding to the prediction error. From (12),

$$|V_\delta(\omega; \text{in})|^2 = \frac{4\pi^2}{\left\{\tilde{f}^\delta(\delta\omega) \int_{-\pi}^{\pi} \tilde{f}^\delta(\omega)^{-1} d\omega\right\}^2},$$

so that if $f(\omega) \asymp (1+|\omega|)^{-\alpha}$, then for $|\omega| < \pi\delta^{-1}$,

$$|V_\delta(\omega; \text{in})|^2 \asymp \delta^{2\alpha}(1+|\omega|)^{2\alpha} \tag{21}$$

and for $0 < T < \pi\delta^{-1}$,

$$\frac{\int_{-T}^{T} |V_\delta(\omega; \text{in})|^2 f(\omega)\, d\omega}{Ee(\text{in}, \delta)^2} \asymp \delta^{\alpha+1}(T + T^{\alpha+1}) \tag{22}$$

(Exercise 16). Equations (20) and (22) agree with (8) and (9) in 3.4 for $f(\omega) = (1+\omega^2)^{-1}$ and support the numerical results in Table 1 of 3.4 for $f(\omega) = (2+\omega^2)^{-2}$. For fixed T, when extrapolating, the rate of convergence of the fraction of the variance of the prediction error attributable to the frequencies $[-T, T]$ is linear in δ as $\delta \downarrow 0$ irrespective of α. In contrast, when interpolating, this rate of convergence is of order $\delta^{\alpha+1}$ as $\delta \downarrow 0$. Thus, the low frequencies make much less of a contribution when interpolating, especially if the process is smooth.

Inefficiency of pseudo-BLPs with misspecified high frequency behavior

Let us next look at the behavior of interpolations and extrapolations when the spectral density is misspecified. Define

$$\begin{aligned} r_{01}(\text{ex}, \delta) &= \frac{E_0 e_1(\text{ex}, \delta)^2}{E_0 e_0(\text{ex}, \delta)^2} \\ &= \frac{1}{2\pi} \exp\left\{\frac{1}{2\pi}\int_{-\pi}^{\pi} \log \frac{\tilde{f}_1^\delta(\omega)}{\tilde{f}_0^\delta(\omega)}\, d\omega\right\} \int_{-\pi}^{\pi} \frac{\tilde{f}_0^\delta(\omega)}{\tilde{f}_1^\delta(\omega)}\, d\omega \end{aligned}$$

and

$$r_{01}(\text{in},\delta) = \frac{E_0 e_1(\text{in},\delta)^2}{E_0 e_0(\text{in},\delta)^2} = \frac{\int_{-\pi}^{\pi} \tilde{f}_0^\delta(\omega)\tilde{f}_1^\delta(\omega)^{-2}d\omega \int_{-\pi}^{\pi} \tilde{f}_0^\delta(\omega)^{-1}d\omega}{\left\{\int_{-\pi}^{\pi} \tilde{f}_1^\delta(\omega)^{-1}d\omega\right\}^2}.$$

The following theorem describes the asymptotic behavior of $r_{01}(\text{ex},\delta)$ and $r_{01}(\text{in},\delta)$ when f_0 and f_1 both decay algebraically at high frequencies but at different rates. In particular, this theorem allows us to make sense of the numerical results in Tables 6 and 7. Define L^1_{loc} to be the class of real-valued functions on $\mathbb{R}$ that are integrable on all bounded intervals. All results in Theorem 3 are limits as $\delta \downarrow 0$.

Theorem 3. *Suppose $f_i(\omega) \sim c_i|\omega|^{-\alpha_i}$ as $|\omega| \to \infty$. If $f_0/f_1 \in L^1_{\text{loc}}$, then for $\alpha_1 > \alpha_0 - 1$,*

$$r_{01}(\text{ex},\delta) \to \frac{1}{2\pi}\exp\left\{\frac{1}{2\pi}\int_{-\pi}^{\pi}\log\frac{\eta_{\alpha_1}(\omega)}{\eta_{\alpha_0}(\omega)}d\omega\right\}\int_{-\pi}^{\pi}\frac{\eta_{\alpha_0}(\omega)}{\eta_{\alpha_1}(\omega)}d\omega \tag{23}$$

for $\alpha_1 = \alpha_0 - 1$,

$$r_{01}(\text{ex},\delta) \sim \frac{1}{\pi}\exp\left\{\frac{1}{2\pi}\int_{-\pi}^{\pi}\log\frac{\eta_{\alpha_1}(\omega)}{\eta_{\alpha_0}(\omega)}d\omega\right\}\log\delta^{-1} \tag{24}$$

and for $\alpha_1 < \alpha_0 - 1$,

$$r_{01}(\text{ex},\delta) \sim \left[\frac{c_1}{2\pi c_0}\exp\left\{\frac{1}{2\pi}\int_{-\pi}^{\pi}\log\frac{\eta_{\alpha_1}(\omega)}{\eta_{\alpha_0}(\omega)}d\omega\right\}\int_{-\infty}^{\infty}\frac{f_0(\omega)}{f_1(\omega)}d\omega\right]\delta^{1+\alpha_1-\alpha_0}. \tag{25}$$

Furthermore, if f_0/f_1^2 and $1/f_0$ are in L^1_{loc}, then for $\alpha_1 > (\alpha_0 - 1)/2$,

$$r_{01}(\text{in},\delta) \to \frac{\int_{-\pi}^{\pi}\eta_{\alpha_0}(\omega)\eta_{\alpha_1}(\omega)^{-2}d\omega\int_{-\pi}^{\pi}\eta_{\alpha_0}(\omega)^{-1}d\omega}{\left\{\int_{-\pi}^{\pi}\eta_{\alpha_1}(\omega)^{-1}d\omega\right\}^2}, \tag{26}$$

for $\alpha_1 = (\alpha_0 - 1)/2$,

$$r_{01}(\text{in},\delta) \sim \frac{2\int_{-\pi}^{\pi}\eta_{\alpha_0}(\omega)^{-1}d\omega}{\left\{\int_{-\pi}^{\pi}\eta_{\alpha_1}(\omega)^{-1}d\omega\right\}^2}\log\delta^{-1} \tag{27}$$

and for $\alpha_1 < (\alpha_0 - 1)/2$,

$$r_{01}(\text{in},\delta) \sim \frac{c_1^2\int_{-\infty}^{\infty}f_0(\omega)f_1(\omega)^{-2}d\omega\int_{-\pi}^{\pi}\eta_{\alpha_0}(\omega)^{-1}d\omega}{c_0\left\{\int_{-\pi}^{\pi}\eta_{\alpha_1}(\omega)^{-1}d\omega\right\}^2}\delta^{1+2\alpha_1-\alpha_0}. \tag{28}$$

Before proving these results, some comments are in order. For both interpolation and extrapolation, we can use f_1 rather than the correct f_0 and still get the optimal rate of convergence for the mse as long as α_1 is not too much smaller than α_0; that is, the process is not too much rougher under f_1 than under f_0. However, for extrapolation, we need $\alpha_1 > \alpha_0 - 1$ for this to hold, whereas for interpolation, we only need $\alpha_1 > (\alpha_0 - 1)/2$. A second point is that whenever the optimal rate is obtained, the limit for either $r_{01}(\text{ex}, \delta)$ or $r_{01}(\text{in}, \delta)$ depends on f_0 and f_1 only through α_0 and α_1 and is independent of the behavior of f_0 and f_1 on any bounded set. Thus, the range of problems for which the low frequency behavior has asymptotically negligible effect as $\delta \downarrow 0$ is larger when interpolating. When $\alpha_1 > \alpha_0 - 1$, so that both interpolation and extrapolation give the asymptotically best rate, Table 8 shows that the limit of $r_{01}(\text{in}, \delta)$ tends to be much smaller than $r_{01}(\text{ex}, \delta)$. A reasonable conjecture is that for all α_0 and α_1, $\lim_{\delta \downarrow 0} r_{01}(\text{ex}, \delta)/r_{01}(\text{in}, \delta) \geq 1$. We do know that this limit is $+\infty$ whenever $\alpha_1 \leq \alpha_0 - 1$.

Let us now return to Tables 6 and 7 and see how these results relate to Theorem 3. First, as I have already noted, Theorem 3 does not directly apply to the setting in Tables 6 and 7 in which there are only a finite number of observations. However, numerical calculations show that the results in these tables do not noticeably change by increasing n. For interpolation, Theorem 3 suggests that the ratios $E_0 e_1^2 / E_0 e_0^2$ should tend to a finite constant except when $f_0(\omega) = (3 + \omega^2)^{-3}$ and $f_1(\omega) = (1 + \omega^2)^{-1}$, in which case, it should be proportional to δ^{-1}. The numerical results fit these patterns well, particularly for $\alpha_1 > \alpha_0$, when the dependence of $E_0 e_1^2 / E_0 e_0^2$ on δ is extremely weak. For extrapolating, Theorem 3 suggests the ratios tend to a finite constant in those entries above the main diagonal of the table. When $f_0(\omega) = (3 + \omega^2)^{-3}$ and $f_1(\omega) = (2 + \omega^2)^{-2}$, or $f_0(\omega) = (2+\omega^2)^{-2}$ and $f_1(\omega) = (1+\omega^2)^{-1}$, then the ratio should grow like δ^{-1}. When $f_0(\omega) = (3 + \omega^2)^{-3}$ and $f_1(\omega) = (1 + \omega^2)^{-1}$, the ratio should grow like δ^{-3}. Although the numerical results roughly correspond to these patterns, the agreement is not nearly as good as when interpolating. It appears that the asymptotics "kick in" for larger values of δ when interpolating than when extrapolating, providing another argument for the greater relevance of shrinking interval asymptotics for interpolation problems.

PROOF OF THEOREM 3. The proofs of (23) and (26) are similar to that of (17) (Exercise 17). When $\alpha_1 < \alpha_0 - 1$, (25) follows from

$$\delta^{\alpha_0 - \alpha_1} \exp\left\{ \frac{1}{2\pi} \int_{-\pi}^{\pi} \log \frac{\tilde{f}_1^\delta(\omega)}{\tilde{f}_0^\delta(\omega)}\, d\omega \right\} \to \frac{c_1}{c_0} \exp\left\{ \frac{1}{2\pi} \int_{-\pi}^{\pi} \log \frac{\eta_{\alpha_1}(\omega)}{\eta_{\alpha_0}(\omega)}\, d\omega \right\}$$

and

$$\begin{aligned} \delta^{-1} \int_{-\pi}^{\pi} \frac{\tilde{f}_0^\delta(\omega)}{\tilde{f}_1^\delta(\omega)}\, d\omega &= \int_{-\pi/\delta}^{\pi/\delta} \frac{\sum_{j=-\infty}^{\infty} f_0(\omega + 2\pi\delta^{-1} j)}{\sum_{j=-\infty}^{\infty} f_1(\omega + 2\pi\delta^{-1} j)}\, d\omega \\ &\to \int_{-\infty}^{\infty} \frac{f_0(\omega)}{f_1(\omega)}\, d\omega \end{aligned}$$

TABLE 8. Limiting values of $r_{01}(\text{in}, \delta)$ and $r_{01}(\text{ex}, \delta)$ as given by Theorem 3 for various values of α_0 and α_1. For each α_0, the largest value for α_1 is $\alpha_0 + 4.8$, which facilitates comparisons of how these limits depend on $\alpha_1 - \alpha_0$ as α_0 varies.

	α_0					
	2		4		6	
α_1	in	ex	in	ex	in	ex
1.2	1.655	3.878	$+\infty$	$+\infty$	$+\infty$	$+\infty$
1.6	1.053	1.171	16.55	·	·	·
2.0	1	1	2.392	·	·	·
2.4	1.022	1.081	1.426	·	$+\infty$	·
2.8	1.070	1.285	1.154	$+\infty$	4.645	·
3.2	1.128	1.606	1.050	2.302	2.110	·
3.6	1.193	2.071	1.010	1.122	1.498	·
4.0	1.261	2.728	1	1	1.248	·
4.4	1.331	3.649	1.007	1.068	1.123	·
4.8	1.404	4.939	1.024	1.247	1.057	$+\infty$
5.2	1.478	6.749	1.048	1.534	1.021	2.265
5.6	1.552	9.294	1.077	1.952	1.005	1.116
6.0	1.628	12.88	1.110	2.546	1	1
6.4	1.704	17.95	1.144	3.381	1.004	1.067
6.8	1.781	25.13	1.181	4.552	1.013	1.241
7.2			1.220	6.196	1.027	1.522
7.6			1.260	8.507	1.044	1.931
8.0			1.300	11.77	1.064	2.510
8.4			1.342	16.31	1.085	3.324
8.8			1.384	22.90	1.109	4.464
9.2					1.134	6.062
9.6					1.159	8.309
10.0					1.186	11.47
10.4					1.214	15.94
10.8					1.242	22.28

as $\delta \downarrow 0$. Similarly, (28) follows for $\alpha_1 < (\alpha_0 - 1)/2$ by showing

$$\delta^{\alpha_i - 1} \int_{-\pi}^{\pi} \tilde{f}_i^\delta(\omega)^{-1} d\omega \to \frac{1}{c_i} \int_{-\pi}^{\pi} \eta_{\alpha_i}(\omega)^{-1} d\omega$$

and

$$\delta^{-2} \int_{-\pi}^{\pi} \frac{\tilde{f}_0^\delta(\omega)}{\tilde{f}_1^\delta(\omega)^2} d\omega \to \int_{-\infty}^{\infty} \frac{f_0(\omega)}{f_1(\omega)^2} d\omega$$

as $\delta \downarrow 0$. Equation (24) follows by showing that when $\alpha_1 = \alpha_0 - 1$,

$$\int_{-\pi}^{\pi} \frac{\tilde{f}_0^\delta(\omega)}{\tilde{f}_1^\delta(\omega)}\, d\omega \sim 2\frac{c_0}{c_1}\delta \log \delta^{-1},$$

which is a consequence of

$$\int_{-\pi}^{\pi} \left| \frac{\tilde{f}_0^\delta(\omega)}{\tilde{f}_1^\delta(\omega)} - \frac{c_0/c_1}{1+|\omega|/\delta} \right| d\omega = o(\delta \log \delta^{-1}) \tag{29}$$

as $\delta \downarrow 0$ (Exercise 18). The proof of (27) is similar (Exercise 18). □

Presumed mses for pseudo-BLPs with misspecified high frequency behavior

As described in 3.1, another measure of the effect of using f_1 rather than the correct f_0 is the ratio of the mse of the pseudo-BLP evaluated under f_1 to the mse of the pseudo-BLP under the true spectral density f_0. Define

$$s_{01}(\text{ex}, \delta) = \frac{E_1 e_1(\text{ex}, \delta)^2}{E_0 e_1(\text{ex}, \delta)^2} = \frac{2\pi}{\int_{-\pi}^{\pi} \tilde{f}_0^\delta(\omega) \tilde{f}_1^\delta(\omega)^{-1} d\omega} \tag{30}$$

and

$$s_{01}(\text{in}, \delta) = \frac{E_1 e_1(\text{in}, \delta)^2}{E_0 e_1(\text{in}, \delta)^2} = \frac{\int_{-\pi}^{\pi} \tilde{f}_1^\delta(\omega)^{-1} d\omega}{\int_{-\pi}^{\pi} \tilde{f}_0^\delta(\omega) \tilde{f}_1^\delta(\omega)^{-2} d\omega}. \tag{31}$$

How to do the comparisons is now not so simple since, as opposed to the results in (23) and (26) for $r_{01}(\text{ex}, \delta)$ and $r_{01}(\text{in}, \delta)$, the values of c_0 and c_1 matter asymptotically for all α_0 and α_1. One informative choice is to allow c_1 to depend on δ and take $c_1(\delta)$ to satisfy $c_1(\delta)(2\pi/\delta)^{-\alpha_1} = c_0(2\pi/\delta)^{-\alpha_0}$; that is, for δ small, make the two spectral densities nearly the same at frequency $2\pi/\delta$, which is arguably the highest frequency about which we get information for observations spaced δ apart since $\exp(i2\pi t/\delta)$ makes one complete cycle in δ units of t. For observations spaced δ apart, $2\pi/\delta$ is known as the Nyquist frequency (Priestley 1981, Yaglom 1987a). All results in Theorem 4 are limits as $\delta \downarrow 0$.

Theorem 4. *Suppose $f_0 \sim c_0|\omega|^{-\alpha_0}$ and $f_1 \sim |\omega|^{-\alpha_1}$ as $|\omega| \to \infty$. For the purposes of this theorem only, define*

$$\tilde{f}_1^\delta(\omega) = c_1(\delta)\delta^{-1} \sum_{j=-\infty}^{\infty} f_1\left(\frac{\omega + 2\pi j}{\delta}\right)$$

in (30) and (31), where $c_1(\delta) = c_0(2\pi/\delta)^{\alpha_1 - \alpha_0}$. If $f_0/f_1 \in L^1_{\text{loc}}$, then for $\alpha_1 > \alpha_0 - 1$,

$$s_{01}(\text{ex}, \delta) \to \frac{(2\pi)^{1+\alpha_1-\alpha_0}}{\int_{-\pi}^{\pi} \eta_{\alpha_0}(\omega)\eta_{\alpha_1}(\omega)^{-1} d\omega},$$

for $\alpha_1 < \alpha_0 - 1$,

$$s_{01}(\mathrm{ex}, \delta) \sim \frac{(2\pi)^{1+\alpha_1-\alpha_0} c_0}{\int_{-\infty}^{\infty} f_0(\omega) f_1(\omega)^{-1} d\omega} \delta^{\alpha_0-\alpha_1-1}$$

and for $\alpha_1 = \alpha_0 - 1$, $s_{01}(\mathrm{ex}, \delta) \sim \frac{1}{2}/\log \delta^{-1}$. *If* f_0/f_1^2 *and* $1/f_1$ *are in* L^1_{loc}, *then for* $\alpha_1 > (\alpha_0 - 1)/2$,

$$s_{01}(\mathrm{in}, \delta) \to \frac{(2\pi)^{\alpha_1-\alpha_0} \int_{-\pi}^{\pi} \eta_{\alpha_1}(\omega)^{-1} d\omega}{\int_{-\pi}^{\pi} \eta_{\alpha_0}(\omega) \eta_{\alpha_1}(\omega)^{-2} d\omega},$$

for $\alpha_1 < (\alpha_0 - 1)/2$,

$$s_{01}(\mathrm{in}, \delta) \sim \frac{(2\pi)^{\alpha_1-\alpha_0} c_0 \int_{-\pi}^{\pi} \eta_{\alpha_1}(\omega)^{-1} d\omega}{\int_{-\infty}^{\infty} f_0(\omega) f_1(\omega)^{-2} d\omega} \delta^{\alpha_0-2\alpha_1-1}$$

and for $\alpha_1 = (\alpha_0 - 1)/2$, $s_{01}(\mathrm{in}, \delta) \sim \frac{1}{2}(2\pi)^{\alpha_1-\alpha_0}/\log \delta^{-1}$.

The proof of Theorem 4 is left as Exercise 19.

We see that in all cases, $E_0 e_0(\mathrm{ex}, \delta)^2 \asymp E_1 e_1(\mathrm{ex}, \delta)^2$ and $E_0 e_0(\mathrm{in}, \delta)^2 \asymp E_1 e_1(\mathrm{in}, \delta)^2$, which provides some support for choosing $c_1(\delta) \asymp \delta^{\alpha_0-\alpha_1}$. Theorems 3 and 4 together imply that when interpolating, for example, the pseudo-BLPs have the optimal rate of convergence and the presumed mses are of the same order of magnitude as the actual mses when $\alpha_1 > (\alpha_0 - 1)/2$. However, when $\alpha_1 \le (\alpha_0 - 1)/2$, pseudo-BLPs have suboptimal rates of convergence but the presumed mses still converge to 0 at the faster rate obtained by the mse for the BLP.

Pseudo-BLPs with correctly specified high frequency behavior

We next develop some asymptotic results that elucidate the numerical results given in Tables 2 and 4 of the previous section. Suppose f_0 and f_1 are of the form $f_i(\omega) = c|\omega|^{-\alpha} + d_i|\omega|^{-\beta} + o(|\omega|^{-\beta})$ as $|\omega| \to \infty$, where $\beta > \alpha > 1$, so that $f_0(\omega)/f_1(\omega) \to 1$ as $|\omega| \to \infty$. Furthermore, the larger the value of $\beta - \alpha$, the more similar the two spectral densities are at high frequencies. Thus, to the extent that the behavior of the BLPs is dominated by the high frequency behavior of the spectrum, larger values of $\beta - \alpha$ should correspond to smaller effects from using f_1 rather than f_0. Again, all results in Theorems 5 and 6 are limits as $\delta \downarrow 0$.

Theorem 5. *Suppose that for* $i = 0, 1$, $f_i(\omega) \asymp (1 + |\omega|)^{-\alpha}$ *and* $f_i(\omega) = c|\omega|^{-\alpha} + d_i|\omega|^{-\beta} + o(|\omega|^{-\beta})$ *as* $|\omega| \to \infty$, *where* $\beta > \alpha > 1$. *For* $\beta < \alpha + \frac{1}{2}$,

$$r_{01}(\mathrm{ex}, \delta) - 1 \sim \frac{(d_1 - d_0)^2}{4\pi c^2} \left[\int_{-\pi}^{\pi} \frac{\eta_\beta(\omega)^2}{\eta_\alpha(\omega)^2} d\omega - \frac{1}{2\pi} \left\{ \int_{-\pi}^{\pi} \frac{\eta_\beta(\omega)}{\eta_\alpha(\omega)} d\omega \right\}^2 \right] \delta^{2(\beta-\alpha)} \quad (32)$$

and for $\beta > \alpha + \frac{1}{2}$,

$$r_{01}(\mathrm{ex}, \delta) - 1 \sim \frac{\delta}{2\pi} \int_{-\infty}^{\infty} \left\{ \frac{f_0(\omega)}{f_1(\omega)} - 1 - \log \frac{f_0(\omega)}{f_1(\omega)} \right\} d\omega. \tag{33}$$

Furthermore, for $\beta < \frac{3}{2}\alpha + \frac{1}{2}$,

$$\begin{aligned} r_{01}(\mathrm{in}, \delta) - 1 \sim \frac{(d_1 - d_0)^2}{c^2} & \left[\int_{-\pi}^{\pi} \eta_\alpha(\omega)^{-1} d\omega \int_{-\pi}^{\pi} \eta_\beta(\omega)^2 \eta_\alpha(\omega)^{-3} d\omega \right. \\ & \left. - \left\{ \int_{-\pi}^{\pi} \eta_\beta(\omega) \eta_\alpha(\omega)^{-2} d\omega \right\}^2 \right] \\ & \times \left\{ \int_{-\pi}^{\pi} \eta_\alpha(\omega)^{-1} d\omega \right\}^{-2} \delta^{2(\beta - \alpha)} \end{aligned} \tag{34}$$

and for $\beta > \frac{3}{2}\alpha + \frac{1}{2}$,

$$r_{01}(\mathrm{in}, \delta) - 1 \sim \left\{ c \int_{-\infty}^{\infty} \frac{\{f_0(\omega) - f_1(\omega)\}^2}{f_0(\omega) f_1(\omega)^2} d\omega \Big/ \int_{-\pi}^{\pi} \eta_\alpha(\omega)^{-1} d\omega \right\} \delta^{\alpha+1}. \tag{35}$$

Theorem 6. *Under the same conditions on* f_0 *and* f_1 *as in Theorem 5, for* $\beta < \alpha + 1$,

$$s_{01}(\mathrm{ex}, \delta) - 1 \sim \left\{ \frac{d_1 - d_0}{2\pi c} \int_{-\pi}^{\pi} \frac{\eta_\beta(\omega)}{\eta_\alpha(\omega)} d\omega \right\} \delta^{\beta - \alpha}, \tag{36}$$

and for $\beta > \alpha + 1$,

$$s_{01}(\mathrm{ex}, \delta) - 1 \sim \frac{\delta}{2\pi} \int_{-\infty}^{\infty} \frac{f_1(\omega) - f_0(\omega)}{f_0(\omega)} d\omega. \tag{37}$$

Furthermore, for $\beta < 2\alpha + 1$,

$$s_{01}(\mathrm{in}, \delta) - 1 \sim \frac{(d_1 - d_0) \int_{-\pi}^{\pi} \eta_\beta(\omega) \eta_\alpha(\omega)^{-2} d\omega}{c \int_{-\pi}^{\pi} \eta_\alpha(\omega)^{-1} d\omega} \delta^{\beta - \alpha}, \tag{38}$$

and for $\beta > 2\alpha + 1$,

$$s_{01}(\mathrm{in}, \delta) - 1 \sim \frac{\int_{-\infty}^{\infty} \{|\omega|^\alpha - c f_0(\omega) f_1(\omega)^{-2}\} d\omega}{\int_{-\infty}^{\infty} \eta_\alpha(\omega)^{-1} d\omega} \delta^{\alpha+1}. \tag{39}$$

The proofs of (32) and (33) are given at the end of this section; (34)–(39) are left as exercises. Note that the conditions given imply that all of the integrals in (32)–(39) are well defined and finite. The condition $f_i(\omega) \asymp (1 + |\omega|)^{-\alpha}$ for all ω and $i = 0, 1$ is stronger than necessary but simplifies the proofs considerably.

The results for interpolation and extrapolation have a number of features in common and one major difference. In both cases, the relative increase in mse due to using f_1 rather than f_0, given by $r_{01}(\cdot, \delta) - 1$, is of order $\delta^{2(\beta-\alpha)}$ when β is not too large. Furthermore, again when β is not too large, for the predictor obtained under f_1, the relative misspecification of its mse obtained by using f_1 rather than f_0 to evaluate its mse, given by $s_{01}(\cdot, \delta)-1$, is of order $\delta^{\beta-\alpha}$. Thus, for β not too large, using f_1 rather than f_0 has a much smaller effect on the actual mse of the predictor than on its presumed mse. The major difference between interpolation and prediction is in what it means for β to be too large. For example, $r_{01}(\text{ex}, \delta) - 1$ is only of order δ and not $\delta^{2(\beta-\alpha)}$ for $\beta > \alpha + \frac{1}{2}$, whereas $r_{01}(\text{in}, \delta) - 1$ is of order $\delta^{2(\beta-\alpha)}$ for $\beta < \frac{3}{2}\alpha + \frac{1}{2}$. Similarly, assuming $\int_{-\infty}^{\infty} \{f_1(\omega) - f_0(\omega)\} f_0(\omega)^{-1} d\omega \neq 0$, $s_{01}(\text{ex}, \delta) - 1$ is only of order δ and not $\delta^{\beta-\alpha}$ for $\beta > \alpha + 1$, whereas $s_{01}(\text{in}, \delta) - 1$ is of order $\delta^{\beta-\alpha}$ for $\beta < 2\alpha + 1$.

Comparing the asymptotic results in Theorems 3 and 4 to the numerical ones in Table 7 shows that the asymptotics tend to give much more accurate approximations for finite δ when interpolating than when extrapolating. A similar result holds when comparing Theorems 5 and 6 to the numerical outcomes in Tables 2 and 4. For the first pair of spectral densities in Tables 2 and 4, $\alpha = 2$ and $\beta = 4$; for the second pair, $\alpha = 4$ and $\beta = 6$. Theorems 5 and 6 thus suggest that when extrapolating, both $E_0 e_1^2 / E_0 e_0^2 - 1$ and $E_1 e_1^2 / E_0 e_1^2 - 1$ should be of order δ for either pair of spectral densities. When interpolating, $E_0 e_1^2 / E_0 e_0^2 - 1$ should be of order δ^3 for the first pair and order δ^4 for the second, and $E_1 e_1^2 / E_0 e_1^2 - 1$ should be of order δ^2 for both pairs. Although these rates qualitatively agree with the numerical outcomes, there is again evidence that for the values of δ considered in Tables 2 and 4, the asymptotic formulae give considerably more accurate results for interpolation than extrapolation.

One general conclusion from Theorems 5 and 6 is that as $\delta \downarrow 0$, the low frequency behavior of the spectral density has asymptotically negligible impact on both interpolations and extrapolations. This result has important implications for the modeling of frequently observed processes when the goal is prediction. In particular, it implies that when neighboring observations are strongly correlated, one's focus in choosing models and methods of estimation should be to get the high frequency behavior as accurately as possible and not worry so much about the low frequency behavior. The theoretical and numerical results also suggest that this strategy of focusing on the high frequencies is likely to work better when interpolating than extrapolating. Indeed, when extrapolating more than a small distance from the last observation, getting the high frequency behavior of the spectral density correct does not guarantee asymptotically optimal predictions. For example, suppose $f_0(\omega) = 1/\{\pi(1+\omega^2)\}$ and $f_1(\omega) = 1/\{\pi(4+\omega^2)\}$ as in the first example in Tables 2 and 4 and that we wish to predict $Z(t)$ for some $t > 0$ based on observing $Z(\delta j)$ for integers $j \leq 0$. Then independent of δ,

$$\begin{aligned} \frac{E_0 e_1^2}{E_0 e_0^2} &= \frac{1 - 2e^{-3t} + e^{-4t}}{1 - e^{-2t}} \quad \text{and} \\ \frac{E_1 e_1^2}{E_0 e_1^2} &= \frac{1 - e^{-4t}}{2(1 - 2e^{-3t} + e^{-4t})}. \end{aligned} \tag{40}$$

For $t = 0.5$, this gives $E_0 e_1^2 / E_0 e_0^2 = 1.0901$ and $E_1 e_1^2 / E_0 e_1^2 = 0.6274$. This example is reconsidered in 4.4.

PROOF OF (32) AND (33). Let us first consider why these results are plausible. Write

$$\frac{\tilde{f}_0^\delta(\omega)}{\tilde{f}_1^\delta(\omega)} = 1 + \delta^{\beta-\alpha} R(\omega) + S_\delta(\omega),$$

where

$$R(\omega) = \frac{d_0 - d_1}{c} \cdot \frac{\eta_\beta(\omega)}{\eta_\alpha(\omega)},$$

so that for any fixed $\omega \neq 0$ in $(-\pi, \pi]$, $S_\delta(\omega) = o(\delta^{\beta-\alpha})$. If $S_\delta(\omega)$ were $o(\delta^{\beta-\alpha})$ uniformly in $\omega \in (-\pi, \pi]$, then we could say

$$\begin{aligned} &\exp\left\{-\frac{1}{2\pi}\int_{-\pi}^{\pi} \log \frac{\tilde{f}_0^\delta(\omega)}{\tilde{f}_1^\delta(\omega)}\, d\omega\right\} \\ &= \exp\left[-\frac{1}{2\pi}\int_{-\pi}^{\pi} \left\{\delta^{\beta-\alpha} R(\omega) + S_\delta(\omega) - \tfrac{1}{2}\delta^{2(\beta-\alpha)} R(\omega)^2\right\} d\omega \right. \\ &\qquad \left. + o\big(\delta^{2(\beta-\alpha)}\big)\right] \\ &= 1 - \frac{\delta^{\beta-\alpha}}{2\pi}\int_{-\pi}^{\pi} R(\omega)\, d\omega - \frac{1}{2\pi}\int_{-\pi}^{\pi} S_\delta(\omega)\, d\omega + \frac{\delta^{2(\beta-\alpha)}}{4\pi}\int_{-\pi}^{\pi} R(\omega)^2 d\omega \\ &\quad + \frac{\delta^{2(\beta-\alpha)}}{8\pi^2}\left\{\int_{-\pi}^{\pi} R(\omega)\, d\omega\right\}^2 + o\big(\delta^{2(\beta-\alpha)}\big), \end{aligned} \tag{41}$$

suggesting

$$r_{01}(\text{ex}, \delta) - 1 \sim \frac{\delta^{2(\beta-\alpha)}}{4\pi}\left[\int_{-\pi}^{\pi} R(\omega)^2 d\omega - \frac{1}{2\pi}\left\{\int_{-\pi}^{\pi} R(\omega) d\omega\right\}^2\right],$$

which is the same as (32). This argument overlooks what happens for ω near 0 and gives a wrong result for $\beta > \alpha + \frac{1}{2}$. However, this calculation does correctly suggest that frequencies not too near the origin contribute a term of order $\delta^{2(\beta-\alpha)}$ to $r_{01}(\text{ex}, \delta) - 1$. For $\beta > \alpha + \frac{1}{2}$, $\delta^{2(\beta-\alpha)} = o(\delta)$, which suggests the following heuristic justification for (33).

$r_{01}(\text{ex}, \delta)$

$$
\begin{aligned}
&= \exp\left\{-\frac{1}{2\pi}\int_{-\pi}^{\pi}\log\frac{\tilde{f}_0^\delta(\omega)}{\tilde{f}_1^\delta(\omega)}\,d\omega\right\}\left[1+\frac{1}{2\pi}\int_{-\pi}^{\pi}\left\{\frac{\tilde{f}_0^\delta(\omega)}{\tilde{f}_1^\delta(\omega)}-1\right\}d\omega\right]\\
&= \exp\left\{-\frac{1}{2\pi}\int_{-\delta^{1/2}}^{\delta^{1/2}}\log\frac{\tilde{f}_0^\delta(\omega)}{\tilde{f}_1^\delta(\omega)}\,d\omega\right\}\\
&\quad\times\left[1+\frac{1}{2\pi}\int_{-\delta^{1/2}}^{\delta^{1/2}}\left\{\frac{\tilde{f}_0^\delta(\omega)}{\tilde{f}_1^\delta(\omega)}-1\right\}d\omega\right]+o(\delta)\\
&= \exp\left\{-\frac{\delta}{2\pi}\int_{-\delta^{-1/2}}^{\delta^{-1/2}}\log\frac{f_0(\omega)}{f_1(\omega)}\,d\omega\right\}\\
&\quad\times\left[1+\frac{\delta}{2\pi}\int_{-\delta^{-1/2}}^{\delta^{-1/2}}\left\{\frac{f_0(\omega)}{f_1(\omega)}-1\right\}d\omega\right]+o(\delta)\\
&= 1+\frac{\delta}{2\pi}\int_{-\delta^{-1/2}}^{\delta^{-1/2}}\left\{\frac{f_0(\omega)}{f_1(\omega)}-1-\log\frac{f_0(\omega)}{f_1(\omega)}\right\}d\omega+o(\delta)\\
&= 1+\frac{\delta}{2\pi}\int_{-\infty}^{\infty}\left\{\frac{f_0(\omega)}{f_1(\omega)}-1-\log\frac{f_0(\omega)}{f_1(\omega)}\right\}d\omega+o(\delta),
\end{aligned}
\tag{42}
$$

where I have used $\tilde{f}_i^\delta(\omega)\sim\delta^{-1}f_i(\omega/\delta)$ uniformly for $|\omega|<\delta^{1/2}$ as $\delta\downarrow 0$. Note that it is important to combine the two integrals before letting the limits of integration go to $\pm\infty$ since $\left|\int_{-\infty}^{\infty}[\{f_0(\omega)/f_1(\omega)\}-1]\,d\omega\right|$ and $\left|\int_{-\infty}^{\infty}\log\{f_0(\omega)/f_1(\omega)\}\,d\omega\right|$ are $+\infty$ for $\beta\le\alpha+1$ but $\int_{-\infty}^{\infty}[\{f_0(\omega)/f_1(\omega)\}-1-\log\{f_0(\omega)/f_1(\omega)\}]d\omega$ is finite (and nonnegative) for all $\beta>\alpha+\frac{1}{2}$.

To provide a rigorous proof of (32), we need to consider the behavior of $S_\delta(\omega)$ more carefully for $\beta<\alpha+\frac{1}{2}$. First, for $|\omega|<\delta$, $S_\delta(\omega)\ll 1+|\delta/\omega|^{\beta-\alpha}$ so that $\int_{-\delta}^{\delta}|S_\delta(\omega)|d\omega\ll\delta$. Next, define

$$
\rho_i^\delta(\omega)=\delta\tilde{f}_i^\delta(\omega)-c\delta^\alpha\eta_\alpha(\omega)-d_i\delta^\beta\eta_\beta(\omega),
$$

so that $|\omega/\delta|^\beta\rho_i^\delta(\omega)\ll 1$ and $\lim_{\delta\downarrow 0}|\omega/\delta|^\beta\rho_i^\delta(\omega)=0$ for all $\omega\ne 0$ in $(-\pi,\pi]$. Then for $\delta\le|\omega|\le\pi$,

$$
\begin{aligned}
S_\delta(\omega) &= \frac{\rho_0^\delta(\omega)-\rho_1^\delta(\omega)}{\delta\tilde{f}_1^\delta(\omega)}\\
&\quad+\frac{d_1-d_0}{c\delta\eta_\alpha(\omega)\tilde{f}_1^\delta(\omega)}\left\{\delta^{2\beta-\alpha}\eta_\beta(\omega)^2+\delta^{\beta-\alpha}\eta_\beta(\omega)\rho_1^\delta(\omega)\right\}\\
&\ll\left|\frac{\omega}{\delta}\right|^\alpha|\rho_0^\delta(\omega)-\rho_1^\delta(\omega)|+\left|\frac{\omega}{\delta}\right|^{2(\alpha-\beta)}+\left|\frac{\omega}{\delta}\right|^{2(\alpha-\beta)}\left|\frac{\omega}{\delta}\right|^\beta|\rho_1^\delta(\omega)|\\
&\ll\left|\frac{\omega}{\delta}\right|^{\alpha-\beta}\left|\frac{\omega}{\delta}\right|^\beta|\rho_0^\delta(\omega)-\rho_1^\delta(\omega)|+\left|\frac{\omega}{\delta}\right|^{2(\alpha-\beta)}
\end{aligned}
$$

by using $|\omega|^\gamma \eta_\gamma(\omega) \asymp 1$ on $|\omega| \leq \pi$ for all $\gamma > 1$. Thus,

$$\begin{aligned}
&\int_{\delta\leq|\omega|\leq\pi} |S_\delta(\omega)|\, d\omega \\
&\quad \ll \delta^{\beta-\alpha} \int_\delta^\pi |\omega|^{\alpha-\beta} \left|\frac{\omega}{\delta}\right|^\beta |\rho_0^\delta(\omega) - \rho_1^\delta(\omega)|\, d\omega + \delta^{2(\beta-\alpha)} \\
&\quad = o(\delta^{\beta-\alpha}),
\end{aligned}$$

so that $\int_{-\pi}^\pi |S_\delta(\omega)| d\omega = o(\delta^{\beta-\alpha})$. Now

$$\int_{-\pi}^\pi \log \frac{\tilde f_0^\delta(\omega)}{\tilde f_1^\delta(\omega)}\, d\omega = 2\int_0^\delta \log \frac{\tilde f_0^\delta(\omega)}{\tilde f_1^\delta(\omega)}\, d\omega + 2\int_\delta^\pi \log \frac{\tilde f_0^\delta(\omega)}{\tilde f_1^\delta(\omega)}\, d\omega.$$

Since $\tilde f_0^\delta(\omega) \asymp \tilde f_1^\delta(\omega)$, the first term on the right side is $O(\delta)$ and

$$\begin{aligned}
&\int_\delta^\pi \log \frac{\tilde f_0^\delta(\omega)}{\tilde f_1^\delta(\omega)}\, d\omega - \int_0^\pi \left\{\delta^{\beta-\alpha} R(\omega) + S_\delta(\omega) - \tfrac{1}{2}\delta^{2(\beta-\alpha)} R(\omega)^2\right\} d\omega \\
&\quad \ll \int_0^\delta \left\{\delta^{\beta-\alpha}|R(\omega)| + |S_\delta(\omega)| + \delta^{2(\beta-\alpha)} R(\omega)^2\right\} d\omega \\
&\qquad + \int_\delta^\pi \left\{\delta^{3(\beta-\alpha)}|R(\omega)|^3 + \delta^{\beta-\alpha}|R(\omega)S_\delta(\omega)| + S_\delta(\omega)^2 + |S_\delta(\omega)|^3\right\} d\omega \\
&\quad = o\left(\delta^{2(\beta-\alpha)}\right)
\end{aligned}$$

for $\beta < \alpha + \frac{1}{2}$ by straightforward calculations using the properties of $R(\omega)$ and $S_\delta(\omega)$. Thus, the heuristic calculation in (41) is correct for $\beta < \alpha + \frac{1}{2}$ and (32) follows.

To prove (42) for $\beta > \alpha + \frac{1}{2}$, first note that as $\delta \downarrow 0$,

$$\int_{-\delta^{1/2}}^{\delta^{1/2}} \log\left\{\frac{\tilde f_0^\delta(\omega)}{\tilde f_1^\delta(\omega)} \frac{f_1(\omega/\delta)}{f_0(\omega/\delta)}\right\} d\omega \ll \int_0^\delta \delta^\alpha\, d\omega + \int_\delta^{\delta^{1/2}} \omega^\alpha\, d\omega = o(\delta)$$

and

$$\begin{aligned}
&\int_{\delta^{1/2}}^\pi \left|\log \frac{\tilde f_0^\delta(\omega)}{\tilde f_1^\delta(\omega)} - \delta^{\beta-\alpha}R(\omega) - S_\delta(\omega)\right| d\omega \\
&\quad \ll \int_{\delta^{1/2}}^\pi \left\{\delta^{2(\beta-\alpha)} R(\omega)^2 + S_\delta(\omega)^2\right\} d\omega = o(\delta).
\end{aligned}$$

Moreover,

$$\int_{-\delta^{1/2}}^{\delta^{1/2}} \log \frac{f_0(\omega/\delta)}{f_1(\omega/\delta)}\, d\omega = \delta \int_{-\delta^{-1/2}}^{\delta^{-1/2}} \log \frac{f_0(\omega)}{f_1(\omega)}\, d\omega$$

so that

$$\int_{-\pi}^{\pi} \log \frac{\tilde{f}_0^\delta(\omega)}{\tilde{f}_1^\delta(\omega)}\, d\omega - \delta \int_{-\delta^{-1/2}}^{\delta^{-1/2}} \log \frac{f_0(\omega)}{f_1(\omega)}\, d\omega$$
$$-2\int_{\delta^{1/2}}^{\pi} \{\delta^{\beta-\alpha} R(\omega) + S_\delta(\omega)\} d\omega = o(\delta).$$

Similarly,

$$\int_{-\pi}^{\pi} \left\{ \frac{f_0^\delta(\omega)}{f_1^\delta(\omega)} - 1 \right\} d\omega - \delta \int_{-\delta^{-1/2}}^{\delta^{-1/2}} \left\{ \frac{f_0(\omega)}{f_1(\omega)} - 1 \right\} d\omega$$
$$-2\int_{\delta^{1/2}}^{\pi} \{\delta^{\beta-\alpha} R(\omega) + S_\delta(\omega)\} d\omega = o(\delta).$$

Now, $\int_{-\delta^{-1/2}}^{\delta^{-1/2}} \log\{f_0(\omega)/f_1(\omega)\}\, d\omega = o(\delta^{-1/2})$, so

$$\begin{aligned}
r_{01}(\text{ex}, \delta) = & \left[1 - \frac{\delta}{2\pi} \int_{-\delta^{-1/2}}^{\delta^{-1/2}} \log \frac{f_0(\omega)}{f_1(\omega)}\, d\omega \right. \\
& \left. - \frac{1}{\pi} \int_{\delta^{1/2}}^{\pi} \{\delta^{\beta-\alpha} R(\omega) + S_\delta(\omega)\}\, d\omega + o(\delta) \right] \\
& \times \left[1 + \frac{\delta}{2\pi} \int_{-\delta^{-1/2}}^{\delta^{-1/2}} \left\{ \frac{f_0(\omega)}{f_1(\omega)} - 1 \right\} d\omega \right. \\
& \left. + \frac{1}{\pi} \int_{\delta^{1/2}}^{\pi} \{\delta^{\beta-\alpha} R(\omega) + S_\delta(\omega)\}\, d\omega + o(\delta) \right] \\
= & 1 + \frac{\delta}{2\pi} \int_{-\delta^{-1/2}}^{\delta^{-1/2}} \left\{ \frac{f_0(\omega)}{f_1(\omega)} - 1 - \log \frac{f_0(\omega)}{f_1(\omega)} \right\} d\omega \\
& - \left[\frac{1}{\pi} \int_{\delta^{1/2}}^{\pi} \{\delta^{\beta-\alpha} R(\omega) + S_\delta(\omega)\}\, d\omega \right]^2 + o(\delta) \\
= & 1 + \frac{\delta}{2\pi} \int_{-\infty}^{\infty} \left\{ \frac{f_0(\omega)}{f_1(\omega)} - 1 - \log \frac{f_0(\omega)}{f_1(\omega)} \right\} d\omega + o(\delta)
\end{aligned}$$

proving (33). □

Exercises

11 Give an example of a pair of autocovariance functions for a mean 0 weakly stationary process for which $r_{01}(\text{in}, 1) = \infty$ but $r_{01}(\text{ex}, 1) < \infty$.

12 Suppose Z is a mean 0 weakly stationary process with triangular autocovariance function $K(t) = (1 - |t|)^+$. For n a positive integer, find the BLP for $Z(0)$ when Z is observed at j/n for all integers $j \neq 0$. Do the same when Z is observed at j/n for all integers $j < 0$.

13 For a mean 0 weakly stationary time series on $\mathbb{Z}$ with spectral density $\tilde{f}$ on $(-\pi, \pi]$, show that perfect prediction of $Z(0)$ is possible if $Z(j)$ is observed for all $j \neq 0$ when $1/\tilde{f}$ is not integrable.

14 Prove (18).

15 Verify (19) and (20).

16 Verify (22).

17 Prove (23) and (26).

18 Prove (29). Show that (24) follows. Prove (27).

19 Prove Theorem 4.

20 Develop an asymptotic theory for interpolation and extrapolation similar to that in Theorems 5 and 6 when $f_0 = f_1$, f_0 satisfies the conditions in Theorem 5, $m_0(t) = 0$ for all t and $m_1(t)$ is some given function.

21 State and prove asymptotic results for the borderline cases not treated in Theorems 5 and 6. Specifically, consider $r_{01}(\text{ex}, \delta)$ when $\beta = \alpha + \frac{1}{2}$, $r_{01}(\text{in}, \delta)$ when $\beta = \frac{3}{2}\alpha + \frac{1}{2}$, $s_{01}(\text{ex}, \delta)$ when $\beta = \alpha + 1$ and $s_{01}(\text{in}, \delta)$ when $\beta = 2\alpha + 1$.

22 Write a program that efficiently calculates $\eta_\alpha(\omega)$ as defined in (16) correctly to six significant digits for all $\alpha > 1$ and $0 < \omega < \pi$. Note that just truncating the infinite sum in the definition of η_α is not an efficient method, particularly for α near 1.

23 It is possible to give a closed form expression for η_α when α is a positive even integer (see, for example, Carrier, Crook and Pearson (1966, p. 97)). Do so for $\alpha = 2$ and $\alpha = 4$.

24 Write a program to calculate via numerical integration the ratios (23) and (26) in Theorem 3. Note that special care is needed in (23) when $\alpha_1 + 1 - \alpha_0$ is near 0 and in (26) when $2\alpha_1 + 1 - \alpha_0$ is near 0.

25 Suppose that $EZ(t) = \boldsymbol{\beta}^T \mathbf{m}(t)$, where $\mathbf{m}$ is a known vector-valued function and $\boldsymbol{\beta}$ is an unknown vector. Assume Z has spectral density f satisfying $f(\omega) \asymp (1 + |\omega|)^{-\alpha}$ for some $\alpha > 1$. For the interpolation problem considered in this section, find conditions on $\mathbf{m}$ under which the BLUP and the BLP are identical. Do the same for the extrapolation problem. For definiteness, assume $\delta = 1$ in each case.

26 Prove (34) and (35). As a way to get started on (34), consider what would happen if $f_i(\omega) = c|\omega|^{-\alpha} + d_i|\omega|^{-\beta}$ for all ω and $i = 0, 1$.

27 Prove Theorem 6. As a way to get started on (36) and (38), consider what would happen if $f_i(\omega) = c|\omega|^{-\alpha} + d_i|\omega|^{-\beta}$ for all ω and $i = 0, 1$. As a way to get started on (35), approximate $\tilde{f}_i^\delta(\omega)$ by $\delta^{-1} f_i(\omega/\delta)$.

3.7 Measurement errors

Until now, we have been assuming that the random field Z is continuous and that it can be observed at specified locations without error. This scenario is, of course, an idealization. For many physical quantities, the whole notion of its value at a single point is ill defined. For example, a concentration of a substance is inherently an average over some region of space (see Vanmarcke (1983) and Cressie (1996) for further discussion). So, when we say $Z(\mathbf{x})$, we often are referring to an average over some region containing $\mathbf{x}$ whose dimensions are small compared to the distance between neighboring observations. We may also be implicitly assuming that this average is insensitive to modest changes in the region over which we are taking an average. Even if the quantity of interest can, for all practical purposes, be taken to be well defined pointwise, there are inevitably errors, however tiny, in the measured values of both $\mathbf{x}$ and $Z(\mathbf{x})$.

The usual approach in spatial statistics is to ignore errors in $\mathbf{x}$ and to assume that errors in $Z(\mathbf{x})$ are additive or perhaps multiplicative. Diggle, Tawn and Moyeed (1998) consider more general error structures, but still of the form that the conditional distribution of the observation at $\mathbf{x}$ given the actual value of $Z(\mathbf{x})$ does not depend on Z at any other location. Chan, Hall and Poskitt (1995) note that if the height of a surface is measured by a stylus, then the error at a particular $\mathbf{x}$ can naturally depend on values of Z in a neighborhood of $\mathbf{x}$. In this section, I consider the simplest possible setting for measurement errors: observation locations $\mathbf{x}_1, \ldots, \mathbf{x}_n$ are known without error and the observations are

$$Y_i = Z(\mathbf{x}_i) + U_i \quad \text{for} \quad i = 1, \ldots, n,$$

where the U_is are independent and identically distributed with mean 0 and common variance σ^2 and independent of Z. I further assume the U_is are $N(0, \sigma^2)$ when considering estimation of covariance structures.

In practice, it is commonly found that even the closest observations in space differ by far more than the technical errors in the measurement process; see Laslett, McBratney, Pahl and Hutchinson (1987) for a convincing example of this phenomenon. Such variation is called the nugget effect in the geostatistical literature (Cressie 1993, pp. 59–60). Standard practice is to model both measurement errors and nugget effects by introducing a discontinuity at the origin in the autocovariance function of the observations. As Cressie (1993, pp. 127–130) points out, whether we consider this local variation in the observations due to measurement error or a nugget effect does have an impact on the evaluation of mses when predicting the random field at a point. However, I suspect that when there is a substantial nugget effect, its magnitude must significantly vary with the region over which the observations are averages, in which case, it is not clear to me that prediction at a point is meaningful. Furthermore, when predicting area averages over sufficiently large regions, there will be effectively no difference in either the prediction or the evaluation of its mse depending on whether the spatial discontinuity in the observations is attributed to measurement error, a nugget effect or some combination of the two. For the purposes of this work, I ignore the distinction and presume that any discontinuity at the origin in the autocovariance function of the observations is due to measurement error.

Given that measurement errors are inevitable, an essential first issue is to determine when they matter. In particular, for what sorts of prediction problems can we act as if $\sigma^2 = 0$ and still get nearly optimal predictions and accurate assessments of mse even though σ^2 is in fact positive but small? For any $\sigma^2 > 0$ and any finite set of observations, the answer to this question will depend on the predictand. For example, suppose model 0 is the correct model, which includes a measurement error with positive variance, and model 1 is the same model except that σ^2 is taken to be 0. Let $e_i(\mathbf{x})$ be the error of the BLP of $Z(\mathbf{x})$ under model i. Then if $\mathbf{x}$ is an observation location, clearly $E_1 e_1(\mathbf{x})^2 = 0$ and $E_0 e_1(\mathbf{x})^2 > 0$, so that the presumed mse of the pseudo-BLP under model 1 is infinitely too small. Thus, we cannot have uniformly good assessments of mse when ignoring measurement errors, no matter how small the value of σ^2. However, if we consider predicting Z at locations not too near any of the observations, we may be able to obtain nearly optimal predictions and accurate assessments of mse when ignoring a sufficiently small measurement error.

Some asymptotic theory

One way in which we can investigate this issue is to consider the interpolation problem addressed in 3.6: a mean 0 weakly stationary process Z on $\mathbb{R}$ with spectral density f is observed at δj for $j \neq 0$ and we wish to predict $Z(0)$. Suppose the observations are $Y_\delta(j) = Z(\delta j) + U_j$ for $j \neq 0$, where the U_js have mean 0, are uncorrelated and have common variance σ_δ^2. Allowing the error variance to depend on δ may appear unconventional to some, but it provides us with a way of assessing how the smoothness of Z relates to the level of measurement errors that can be safely ignored. Define Z_δ and $\tilde{f}^\delta$ as in 3.6. The spectral density of $Y_\delta(j)$ on $(-\pi, \pi]$ is then $\tilde{f}^\delta(\omega) + (2\pi)^{-1}\sigma_\delta^2$. It follows that the mse of the BLP of $Z(0)$ based on $Y_\delta(j)$ for $j \neq 0$ is

$$\frac{4\pi^2}{\displaystyle\int_{-\pi}^{\pi} \left\{ \tilde{f}^\delta(\omega) + \frac{1}{2\pi}\sigma_\delta^2 \right\}^{-1} d\omega} - \sigma_\delta^2.$$

Next, suppose that $f(\omega) \sim c\omega^{-\alpha}$ as $\omega \to \infty$ for some $c > 0$ and some $\alpha > 1$. Then for any fixed $\omega \in (-\pi, \pi]$ other than $\omega = 0$, $\tilde{f}^\delta(\omega) \sim c\delta^{\alpha-1}\eta_\alpha(\omega)$ as $\delta \downarrow 0$, where η_α is defined as in (16). Thus, a plausible conjecture is that if $\sigma_\delta^2 = O(\delta^{\alpha-1})$ as $\delta \downarrow 0$, the measurement error will have asymptotically negligible effect on predicting $Z(0)$ and assessing the mse of prediction. To prove this assertion, define $\tilde{f}_0^\delta(\omega) = \tilde{f}^\delta(\omega) + (2\pi)^{-1}\sigma_\delta^2$ and $\tilde{f}_1^\delta(\omega) = \tilde{f}^\delta(\omega)$ as functions on $(-\pi, \pi]$, so that $\tilde{f}_0^\delta$ is the actual spectral density for the Y_δ process and $\tilde{f}_1^\delta$ is the presumed spectral density obtained by ignoring the measurement errors. Take $e_i(\delta)$ to be the error of the BLP of $Z(0)$ under model i with observations $Y_\delta(j)$ for all $j \neq 0$.

Theorem 7. *Suppose $f(\omega) \sim c\omega^{-\alpha}$ as $\omega \to \infty$ for some $c > 0$ and $\alpha > 1$. If $\sigma_\delta^2 = o(\delta^{\alpha-1})$ as $\delta \downarrow 0$, then*

$$E_0 e_0(\delta)^2 \sim E_0 e_1(\delta)^2 \sim E_1 e_1(\delta)^2 \sim \frac{4\pi^2 c\delta^{\alpha-1}}{\int_{-\pi}^{\pi} \eta_\alpha(\omega)^{-1} d\omega}.$$

The proof of Theorem 7 is similar to the proofs in 3.6 and is left as an exercise (Exercise 28). The next result gives asymptotic approximations to $E_0 e_0(\delta)^2$ for various circumstances for which σ_δ^2 is not $o(\delta^{\alpha-1})$.

Theorem 8. *Suppose f is as in Theorem 7. As $\delta \downarrow 0$, if $\sigma_\delta^2 = b\delta^{\alpha-1}$ for some fixed $b > 0$, then*

$$E_0 e_0(\delta)^2 \sim b\delta^{\alpha-1} \left[\frac{2\pi}{\int_{-\pi}^{\pi} \left\{ 1 + \frac{2\pi c}{b} \eta_\alpha(\omega) \right\}^{-1} d\omega} - 1 \right], \tag{43}$$

if $\sigma_\delta^2/\delta^{\alpha-1} \to \infty$ and $\delta\sigma_\delta^2 \to 0$ as $\delta \downarrow 0$, then

$$E_0 e_0(\delta)^2 \sim \frac{(2\pi c)^{1/\alpha}}{\alpha \sin\left(\frac{\pi}{\alpha}\right)} \left(\frac{\sigma_\delta^2}{\delta} \right)^{1-1/\alpha}, \tag{44}$$

if $\sigma_\delta^2 = r/\delta$ for some fixed $r > 0$, then

$$E_0 e_0(\delta)^2 \to r \int_{-\infty}^{\infty} \frac{f(\omega)}{r + 2\pi f(\omega)} d\omega \tag{45}$$

and if $\delta\sigma_\delta^2 \to \infty$ as $\delta \downarrow 0$, then

$$E_0 e_0(\delta)^2 \to \int_{-\infty}^{\infty} f(\omega)\, d\omega. \tag{46}$$

The proof of Theorem 8 is left as a series of exercises (Exercises 29–31).

Theorems 7 and 8 imply that the measurement error has asymptotically negligible impact on predictions of $Z(0)$ if and only if $\sigma_\delta^2 = o(\delta^{\alpha-1})$. Thus, the smoother Z is, the smaller σ_δ^2 needs to be before it can be ignored. Such a result makes sense, since even small measurement errors can make it quite difficult to extract information about the derivatives of Z and hence seriously degrade predictions that exploit the existence of these derivatives. As a consequence of this result, one should be quite reluctant to leave out a measurement error term from a model for observations from a highly smooth random field, even if the measurement errors are expected to be quite small.

Equation (44) includes the case $\sigma_\delta^2 = \sigma^2 > 0$ independent of δ, which is the most common asymptotic regime to consider (see, for example, Stein (1993b)). However, the more general result for $\sigma_\delta^2/\delta^{\alpha-1} \to \infty$ and $\delta\sigma_\delta^2 \to 0$ is no more difficult to obtain.

As σ_δ^2 increases, we have to average a greater number of observations in order to reduce the contribution of the measurement error to the prediction error. Hence, we should expect that if σ_δ^2 is sufficiently large, even if δ is small, observations outside any fixed bounded interval will have a nontrivial impact on the prediction of $Z(0)$. The case $\delta\sigma_\delta^2 = r > 0$ given by (45) addresses this setting. Here we see that the low frequency behavior of f does affect the asymptotic mse. Furthermore, Exercise 32 asks you to show that the BLP makes nonnegligible use of observations outside any fixed neighborhood of the origin. In contrast, I conjecture that (43) and (44) still hold if $Y_\delta(j)$ is observed only for j satisfying $0 < \delta|j| < a$ for any fixed $a > 0$.

Finally, (46) just says that if the noise is too large, one might as well predict $Z(0)$ by 0, which has mse $\operatorname{var}\{Z(0)\} = \int_{-\infty}^{\infty} f(\omega)\,d\omega$.

Exercises

28 Prove Theorem 7.

29 Prove (43).

30 Prove (44).

31 Prove (45) and (46). Show that (46) still holds if one observes $Y_\delta(j)$ for all integers j.

32 Suppose that f is as in Theorem 7 and that observations are restricted to those $Y_\delta(j)$ for which j satisfies $0 < \delta|j| < a$ for some fixed $a > 0$. Show that if $\sigma_\delta^2 = r/\delta$ for some fixed $r > 0$,

$$\liminf_{\delta \downarrow 0} E_0 e_0(\delta)^2 > r \int_{-\infty}^{\infty} \frac{f(\omega)}{r + 2\pi f(\omega)}\,d\omega.$$

3.8 Observations on an infinite lattice

Section 3.6 gave some theory for predicting a mean 0 weakly stationary process Z at the origin based on observing Z at δj for all integers $j \neq 0$ or for all integers $j < 0$. The arguments were based on having exact results for the spectral representation of the optimal predictions in these two settings. There is no easy way to extend these results if we wanted to predict, for example, $Z(2^{-1/2}\delta)$, or $\int_0^1 Z(t)\,dt$. If, however, we observe Z at δj for all integers j, then all possible linear prediction problems become in a sense trivial. Furthermore, the same approach works just as well for a random field in any number of dimensions observed at $\delta\mathbf{j}$ for all $\mathbf{j} \in \mathbb{Z}^d$, the d-dimensional integer lattice. For this setting, this section provides a simple bound on the fraction of the variance of the prediction error attributable to some set of frequencies. This result is then used to obtain bounds on the effects on prediction of using the wrong second-order structure. The bounds in this section are uniform over all possible linear predictions.

Characterizing the BLP

Suppose Z is a real mean 0 weakly stationary random field on $\mathbb{R}^d$ with spectrum F and we observe Z at $\delta\mathbf{j}$ for all $\mathbf{j} \in \mathbb{Z}^d$. Let $\mathcal{H}(F)$ be the closed real linear manifold of the random variables $Z(\mathbf{x})$, $\mathbf{x} \in \mathbb{R}^d$, with respect to the inner product defined by F and let $\mathcal{H}_\delta(F)$ be the subspace of $\mathcal{H}(F)$ generated by $Z(\delta\mathbf{j})$ for $\mathbf{j} \in \mathbb{Z}^d$. Similarly, let $\mathcal{L}(F)$ be the closed real linear manifold of the functions $\exp(i\boldsymbol{\omega}^T\mathbf{x})$ for $\mathbf{x} \in \mathbb{R}^d$ with respect to the inner product defined by F and $\mathcal{L}_\delta(F)$ the subspace of $\mathcal{L}(F)$ generated by $\exp(i\delta\boldsymbol{\omega}^T\mathbf{j})$ for $\mathbf{j} \in \mathbb{Z}^d$. Since $\exp(i\delta\boldsymbol{\omega}^T\mathbf{j})$ has period $2\pi/\delta$ in all coordinates for any $\mathbf{j} \in \mathbb{Z}^d$, it is apparent that all elements of $\mathcal{L}_\delta(F)$ can be taken to have period $2\pi/\delta$ in each coordinate. Thus, if $H \in \mathcal{L}(F)$ is the function corresponding to the random variable $h \in \mathcal{H}(F)$ we wish to predict, then the function $\hat{H}_\delta$ corresponding to the BLP can be characterized as the best periodic approximant to H.

More specifically, by Exercise 6 in 1.3, we seek a periodic function $\hat{H}_\delta$ such that

$$\int_{\mathbb{R}^d} \left\{H(\boldsymbol{\omega}) - \hat{H}_\delta(\boldsymbol{\omega})\right\} \exp(-i\delta\boldsymbol{\omega}^T\mathbf{j})F(d\boldsymbol{\omega}) = 0 \text{ for all } \mathbf{j} \in \mathbb{Z}^d \tag{47}$$

so that $H - \hat{H}_\delta$ is orthogonal to $\mathcal{L}_\delta(F)$. Defining $A_d(r) = (-\pi r, \pi r]^d$, (47) is equivalent to

$$\sum_{\mathbf{k}\in\mathbb{Z}^d} \int_{A_d(\delta^{-1})} \left\{H(\boldsymbol{\omega} + 2\pi\delta^{-1}\mathbf{k}) - \hat{H}_\delta(\boldsymbol{\omega})\right\} \times \exp(-i\delta\boldsymbol{\omega}^T\mathbf{j})F(d\boldsymbol{\omega} + 2\pi\delta^{-1}\mathbf{k}) = 0$$

for all $\mathbf{j} \in \mathbb{Z}^d$, where, for a set B and a point $\mathbf{x}$, $B + \mathbf{x}$ is defined as $\{\mathbf{y} : \mathbf{y} - \mathbf{x} \in B\}$. Defining the measure F_δ on $A_d(\delta^{-1})$ by $F_\delta(B) = \sum_{\mathbf{k}\in\mathbb{Z}^d} F(B + 2\pi\delta^{-1}\mathbf{k})$ for Borel sets $B \subset A_d(\delta^{-1})$, it is obvious that for all $\mathbf{k} \in \mathbb{Z}^d$, $F(\cdot + 2\pi\delta^{-1}\mathbf{k})$ is absolutely continuous with respect to F_δ on $A_d(\delta^{-1})$. Let $\tau(\cdot;\mathbf{k})$ be a Radon–Nikodym derivative of $F(\cdot + 2\pi\delta^{-1}\mathbf{k})$ with respect to F_δ. Then (47) is equivalent to

$$\int_{A_d(\delta^{-1})} \exp(-i\delta\boldsymbol{\omega}^T\mathbf{j})\left\{\sum_{\mathbf{k}\in\mathbb{Z}^d} H(\boldsymbol{\omega} + 2\pi\delta^{-1}\mathbf{k})\tau(\boldsymbol{\omega};\mathbf{k}) - \hat{H}_\delta(\boldsymbol{\omega})\right\}F_\delta(d\boldsymbol{\omega}) = 0$$

for all $\mathbf{j} \in \mathbb{Z}^d$. By a basic theorem in multiple Fourier series (Stein and Weiss 1971, p. 248),

$$\left\{ \sum_{\mathbf{k} \in \mathbb{Z}^d} H(\boldsymbol{\omega} + 2\pi\delta^{-1}\mathbf{k})\tau(\boldsymbol{\omega}; \mathbf{k}) - \hat{H}_\delta(\boldsymbol{\omega}) \right\} F_\delta(\mathbf{d}\boldsymbol{\omega})$$

must be the 0 measure, so that

$$\hat{H}_\delta(\boldsymbol{\omega}) = \sum_{\mathbf{k} \in \mathbb{Z}^d} H(\boldsymbol{\omega} + 2\pi\delta^{-1}\mathbf{k})\tau(\boldsymbol{\omega}; \mathbf{k}) \tag{48}$$

almost everywhere with respect to F_δ. Since there necessarily exists $\hat{H}_\delta \in \mathcal{L}_\delta(F)$ satisfying (47) (see Exercises 5 and 6 in 1.3), for any Radon–Nikodym derivative τ, $\hat{H}_\delta(\boldsymbol{\omega})$ as defined in (48) is in $\mathcal{L}_\delta(F)$ and hence gives the function corresponding to the BLP of h.

Bound on fraction of mse of BLP attributable to a set of frequencies

Suppose for the rest of this section that F has density f with respect to Lebesgue measure, in which case, we can take $\tau(\boldsymbol{\omega}; \mathbf{k}) = f(\boldsymbol{\omega} + 2\pi\delta^{-1}\mathbf{k}) / \sum_{\mathbf{j} \in \mathbb{Z}^d} f(\boldsymbol{\omega} + 2\pi\delta^{-1}\mathbf{j})$. For a symmetric Borel set $B \subset A_d(\delta^{-1})$, define

$$M_\delta(F, B) = \operatorname*{ess\,sup}_{\omega \in B} \frac{\sum'_{\mathbf{j}} f(\boldsymbol{\omega} + 2\pi\delta^{-1}\mathbf{j})}{\sum_{\mathbf{j} \in \mathbb{Z}^d} f(\boldsymbol{\omega} + 2\pi\delta^{-1}\mathbf{j})},$$

where ess sup is essential supremum, $\sum'_{\mathbf{j}}$ indicates summation over all $\mathbf{j} \in \mathbb{Z}^d$ other than $\mathbf{0}$ and $0/0$ is defined to be 0. Corresponding to the prediction error of any BLP is a $V \in \mathcal{L}(F)$ orthogonal to $\mathcal{L}_\delta(F)$. The mse of this BLP is then given by $\|V\|_F^2$. The following result, given in Stein (1999), bounds the fraction of the mse of the BLP attributable to some range of frequencies contained in $A_d(\delta^{-1})$.

Theorem 9. *Suppose $V \in \mathcal{L}(F)$ is orthogonal to $\mathcal{L}_\delta(F)$. Then for symmetric Borel sets $B \subset A_d(\delta^{-1})$,*

$$\int_B |V(\boldsymbol{\omega})|^2 f(\boldsymbol{\omega}) \mathbf{d}\boldsymbol{\omega} \le M_\delta(F, B) \|V\|_F^2. \tag{49}$$

Furthermore, if

$$\sum_{\mathbf{j}}{}' F(\cdot + 2\pi\delta^{-1}\mathbf{j}) \quad \text{is absolutely continuous with respect to } F(\cdot) \text{ on } B, \tag{50}$$

then

$$\sup \frac{\int_B |V(\boldsymbol{\omega})|^2 f(\boldsymbol{\omega})\, \mathbf{d}\boldsymbol{\omega}}{\|V\|_F^2} = M_\delta(F, B), \tag{51}$$

where the supremum is over all $V \in \mathcal{L}(F)$ orthogonal to $\mathcal{L}_\delta(F)$ satisfying $\|V\|_F^2 > 0$. If there are no such V, define this supremum as 0.

PROOF. Equation (49) holds trivially if $\|V\|_F^2 = 0$ or $M_\delta(F,B) = 1$, so assume otherwise. Define a function $V_{\delta,B}(\boldsymbol{\omega})$ with period $2\pi\delta^{-1}$ in each coordinate, $V_{\delta,B}(\boldsymbol{\omega}) = V(\boldsymbol{\omega})$ for $\boldsymbol{\omega} \in B$ and $V_{\delta,B}(\boldsymbol{\omega}) = 0$ for $\boldsymbol{\omega} \in A_d(\delta^{-1})\backslash B$. Then $M_\delta(F,B) < 1$ and B symmetric imply $V_{\delta,B} \in \mathcal{L}_\delta(F)$ (Exercise 33). Thus,

$$\begin{aligned}
0 &= \int_{\mathbb{R}^d} V(\boldsymbol{\omega})\overline{V_{\delta,B}(\boldsymbol{\omega})} f(\boldsymbol{\omega})\, d\boldsymbol{\omega} \\
&= \int_B |V(\boldsymbol{\omega})|^2 f(\boldsymbol{\omega})\, d\boldsymbol{\omega} \\
&\quad + \sum_{\mathbf{j}}{}' \int_B V(\boldsymbol{\omega} + 2\pi\delta^{-1}\mathbf{j})\overline{V_{\delta,B}(\boldsymbol{\omega})} f(\boldsymbol{\omega} + 2\pi\delta^{-1}\mathbf{j})\, d\boldsymbol{\omega},
\end{aligned}$$

so that

$$\begin{aligned}
&\int_B |V(\boldsymbol{\omega})|^2 f(\boldsymbol{\omega})\, d\boldsymbol{\omega} \\
&\quad \le \sum_{\mathbf{j}}{}' \int_B \left|V(\boldsymbol{\omega} + 2\pi\delta^{-1}\mathbf{j}) V_{\delta,B}(\boldsymbol{\omega})\right| f(\boldsymbol{\omega} + 2\pi\delta^{-1}\mathbf{j})\, d\boldsymbol{\omega} \\
&\quad \le \left\{ \sum_{\mathbf{j}}{}' \int_B \left|V(\boldsymbol{\omega} + 2\pi\delta^{-1}\mathbf{j})\right|^2 f(\boldsymbol{\omega} + 2\pi\delta^{-1}\mathbf{j})\, d\boldsymbol{\omega} \right. \\
&\qquad \left. \times \sum_{\mathbf{j}}{}' \int_B |V_{\delta,B}(\boldsymbol{\omega})|^2 f(\boldsymbol{\omega} + 2\pi\delta^{-1}\mathbf{j})\, d\boldsymbol{\omega} \right\}^{1/2} \\
&\quad \le \left[\left\{ \|V\|_F^2 - \int_B |V(\boldsymbol{\omega})|^2 f(\boldsymbol{\omega})\, d\boldsymbol{\omega} \right\} \right. \\
&\qquad \left. \times \frac{M_\delta(F,B)}{1 - M_\delta(F,B)} \int_B |V(\boldsymbol{\omega})|^2 f(\boldsymbol{\omega})\, d\boldsymbol{\omega} \right]^{1/2},
\end{aligned}$$

where the second inequality uses the Cauchy–Schwarz inequality. Straightforward calculation yields (49).

To prove (51), note that it is trivial if $M_\delta(F,B) = 0$, so assume otherwise. Define $\tau(\boldsymbol{\omega}) = \sum_{\mathbf{j}}' \tau(\boldsymbol{\omega};\mathbf{j})$. Given $\epsilon \in (0, M_\delta(F,B))$, let B_ϵ be the subset of B on which $\tau(\boldsymbol{\omega}) > M_\delta(F,B) - \epsilon$. By the definition of $M_\delta(F,B)$, $\sum_{\mathbf{j}}' F(B_\epsilon + 2\pi\delta^{-1}\mathbf{j}) > 0$, so that by assumption, $F(B_\epsilon) > 0$. Next, define a function U by

$$U(\boldsymbol{\omega} + 2\pi\delta^{-1}\mathbf{j}) = \begin{cases} 1, & \text{for } \mathbf{j} = \mathbf{0},\ \boldsymbol{\omega} \in B_\epsilon; \\ 1 - 1/\tau(\boldsymbol{\omega}), & \text{for } \mathbf{j} \neq \mathbf{0},\ \boldsymbol{\omega} \in B_\epsilon \end{cases}$$

and 0 otherwise. Then $U \in \mathcal{L}(F)$, U is orthogonal to $\mathcal{L}_\delta(F)$ and $\|U\|_F^2 > 0$ (Exercise 34). Furthermore, $\int_B |U(\boldsymbol{\omega})|^2 f(\boldsymbol{\omega})\, \mathbf{d}\boldsymbol{\omega} = F(B_\epsilon)$ and $\sum_{\mathbf{j}}' f(\boldsymbol{\omega} + 2\pi\delta^{-1}\mathbf{j})/f(\boldsymbol{\omega}) = \tau(\boldsymbol{\omega})/\{1 - \tau(\boldsymbol{\omega})\}$ almost everywhere on B_ϵ, so that

$$\begin{aligned} \|U\|_F^2 &= F(B_\epsilon) + \int_{B_\epsilon} \left\{\frac{1}{\tau(\boldsymbol{\omega})} - 1\right\}^2 \sum_{\mathbf{j}}{}' f(\boldsymbol{\omega} + 2\pi\delta^{-1}\mathbf{j})\, \mathbf{d}\boldsymbol{\omega} \\ &= F(B_\epsilon) + \int_{B_\epsilon} \left\{\frac{1}{\tau(\boldsymbol{\omega})} - 1\right\} f(\boldsymbol{\omega})\, \mathbf{d}\boldsymbol{\omega} \\ &\leq \frac{F(B_\epsilon)}{M_\delta(F, B) - \epsilon}\,. \end{aligned}$$

Thus, $\int_B |U(\boldsymbol{\omega})|^2 f(\boldsymbol{\omega})\, \mathbf{d}\boldsymbol{\omega}/\|U\|_F^2 \geq M_\delta(F, B) - \epsilon$, which implies (51) since ϵ can be taken arbitrarily small. □

The condition in (50) always holds whenever $M_\delta(F, B) < 1$. If $M_\delta(F, B) = 1$, then (50) can be false, in which case, (51) can also be false. In particular, when $M_\delta(F, B) = 1$ and the support of F does not intersect B, then $\int_B |V(\boldsymbol{\omega})|^2 f(\boldsymbol{\omega})\, \mathbf{d}\boldsymbol{\omega}$ is trivially 0 for any V so that the left side of (51) equals 0, not 1.

As a specific example of Theorem 9, suppose there exist $\alpha > d$ and positive C_0 and C_1 such that

$$C_0\, (1 + |\boldsymbol{\omega}|)^{-\alpha} \leq f(\boldsymbol{\omega}) \leq C_1\, (1 + |\boldsymbol{\omega}|)^{-\alpha}\,. \tag{52}$$

Then for $b_d(r)$ the d-dimensional ball of radius r centered at the origin,

$$M_\delta(F, b_d(r)) \leq \frac{C_1}{C_0}(1 + r)^\alpha \left(\frac{\delta}{\pi}\right)^\alpha \xi_d(\alpha), \tag{53}$$

where $\xi_d(\alpha) = \sum_{\mathbf{j}}' |\mathbf{j}|^{-\alpha}$ (Exercise 35). We see that for r fixed and $\delta \downarrow 0$, $M_\delta(F, b_d(r))$ tends to 0 more quickly when f tends to 0 more quickly at high frequencies. As an extreme example, suppose f has bounded support B, in which case, the process is said to be bandlimited. If B is contained in $A_d(\delta^{-1})$ then $M_\delta(F, B) = 0$, which implies $\int_B |V(\boldsymbol{\omega})|^2 f(\boldsymbol{\omega})\, \mathbf{d}\boldsymbol{\omega} = 0$. Thus, $\|V\|_F^2 = 0$ since f has 0 mass outside B so that Z may be recovered without error at all $\mathbf{x}$. We have just proven a simple version of what is known as the sampling theorem for random fields (Jerri 1977).

Asymptotic optimality of pseudo-BLPs

Theorem 9 provides a useful tool for proving results on the asymptotic properties of pseudo-BLPs as $\delta \downarrow 0$ when $f_1(\boldsymbol{\omega})/f_0(\boldsymbol{\omega})$ tends to a positive finite constant as $|\boldsymbol{\omega}| \to \infty$. It does not appear to yield useful results if $f_1(\boldsymbol{\omega})/f_0(\boldsymbol{\omega})$ tends to 0 or ∞ as $|\boldsymbol{\omega}| \to \infty$, so I do not have any results analogous to Theorems 3 and 4 for this setting. The next result basically says that spectral densities with similar high frequency behavior produce uniformly similar linear predictions for small δ. Let $\mathcal{H}_{-\delta}(F_i)$ be the set of those h in $\mathcal{H}(F_i)$ for which $E_i e_i(h,\delta)^2 > 0$ and $\mathcal{L}_{-\delta}(F_i)$ is the corresponding set of functions.

Theorem 10. *For some $c > 0$, suppose $f_1(\boldsymbol{\omega})/f_0(\boldsymbol{\omega}) \to c$ as $|\boldsymbol{\omega}| \to \infty$, $f_0 \asymp f_1$, f_1 is bounded away from 0 on any bounded set and $f_1(\boldsymbol{\omega}) \to 0$ as $|\boldsymbol{\omega}| \to \infty$. Then*

$$\lim_{\delta \downarrow 0} \sup_{h \in \mathcal{H}_{-\delta}(F_1)} \left| \frac{E_1 e_1(h,\delta)^2}{E_0 e_1(h,\delta)^2} - c \right| = 0, \tag{54}$$

$$\lim_{\delta \downarrow 0} \sup_{h \in \mathcal{H}_{-\delta}(F_0)} \left| \frac{E_0 e_0(h,\delta)^2}{E_1 e_0(h,\delta)^2} - \frac{1}{c} \right| = 0 \tag{55}$$

and

$$\lim_{\delta \downarrow 0} \sup_{h \in \mathcal{H}_{-\delta}(F_0)} \frac{E_0 \{e_0(h,\delta) - e_1(h,\delta)\}^2}{E_0 e_0(h,\delta)^2} = 0. \tag{56}$$

PROOF. The result for general c is a trivial consequence of the result for $c = 1$, so assume $c = 1$. Since $f_0 \asymp f_1$, $\mathcal{H}(F_1) = \mathcal{H}(F_0)$ as sets, so the left sides of (54)–(56) are well defined. For $H \in \mathcal{L}(F_i)$, define

$$H_{\delta,F_i}(\boldsymbol{\omega}) = \frac{\sum_{\mathbf{j}} f_i(\boldsymbol{\omega} + 2\pi\delta^{-1}\mathbf{j}) H(\boldsymbol{\omega} + 2\pi\delta^{-1}\mathbf{j})}{\sum_{\mathbf{j}} f_i(\boldsymbol{\omega} + 2\pi\delta^{-1}\mathbf{j})} - H(\boldsymbol{\omega}),$$

so that if H corresponds to $h \in \mathcal{H}(F_i)$, H_{δ,F_i} corresponds to $e_i(h,\delta)$. Setting $\psi(\boldsymbol{\omega}) = \{f_1(\boldsymbol{\omega}) - f_0(\boldsymbol{\omega})\}/f_1(\boldsymbol{\omega})$,

$$\begin{aligned}
&\left| \|H_{\delta,F_1}\|_{F_1}^2 - \|H_{\delta,F_1}\|_{F_0}^2 \right| \\
&= \left| \int_{\mathbb{R}^d} f_1(\boldsymbol{\omega})\psi(\boldsymbol{\omega}) \left|H_{\delta,F_1}(\boldsymbol{\omega})\right|^2 d\boldsymbol{\omega} \right| \\
&\le \int_{A_d(\delta^{-1})} f_1(\boldsymbol{\omega})|\psi(\boldsymbol{\omega})||H_{\delta,F_1}(\boldsymbol{\omega})|^2 d\boldsymbol{\omega} + m(\delta^{-1})\|H_{\delta,F_1}\|_{F_1}^2,
\end{aligned} \tag{57}$$

where $m(r) = \sup_{\boldsymbol{\omega} \in A_d(r)^c} |\psi(\boldsymbol{\omega})|$. Note that $f_1(\boldsymbol{\omega})/f_0(\boldsymbol{\omega}) \to 1$ as $|\boldsymbol{\omega}| \to \infty$ implies $m(r) \to 0$ as $r \to \infty$.

Under the stated conditions, $|\psi|$ is bounded by a finite constant ψ_0 and $M_\delta(F_1, B) \to 0$ as $\delta \downarrow 0$ for any fixed bounded set B. Given $\epsilon > 0$, we can choose r_ϵ such that $|\psi(\boldsymbol{\omega})| < \epsilon$ on $A_d(r_\epsilon)^c$. Thus, for all δ sufficiently small,

$$\begin{aligned}
&\int_{A_d(\delta^{-1})} f_1(\boldsymbol{\omega})|\psi(\boldsymbol{\omega})| \left|H_{\delta,F_1}(\boldsymbol{\omega})\right|^2 d\boldsymbol{\omega} \\
&\le \psi_0 \int_{A_d(r_\epsilon)} f_1(\boldsymbol{\omega}) \left|H_{\delta,F_1}(\boldsymbol{\omega})\right|^2 d\boldsymbol{\omega} \\
&\qquad + \epsilon \int_{A_d(\delta^{-1}) \setminus A_d(r_\epsilon)} f_1(\boldsymbol{\omega}) \left|H_{\delta,F_1}(\boldsymbol{\omega})\right|^2 d\boldsymbol{\omega} \\
&\le \{\psi_0 M_\delta(F_1, A_d(r_\epsilon)) + \epsilon\} \|H_{\delta,F_1}\|_{F_1}^2 .
\end{aligned}$$

Now ϵ is arbitrary, $M_\delta(F_1, A_d(r_\epsilon)) \to 0$ as $\delta \downarrow 0$ and $\psi_0 M_\delta(F_1, A_d(r_\epsilon)) + \epsilon$ is independent of H, so

$$\lim_{\delta \downarrow 0} \sup_{H \in \mathcal{L}_{-\delta}(f_1)} \frac{\int_{A_d(\delta^{-1})} |\psi(\boldsymbol{\omega})| f_1(\boldsymbol{\omega}) |H_{\delta,F_1}(\boldsymbol{\omega})|^2 \, d\boldsymbol{\omega}}{\|H_{\delta,F_1}\|_{F_1}^2} = 0,$$

which together with (57) and $m(r) \to 0$ as $r \to \infty$ implies (54). Equation (55) follows from (54) by switching the roles of f_0 and f_1.

To prove (56), consider bounding $\|H_{\delta,F_1} - H_{\delta,F_0}\|_{F_0}^2$. Define f_δ by $f_\delta(\boldsymbol{\omega}) = f_0(\boldsymbol{\omega})$ for $\boldsymbol{\omega} \in A_d(\delta^{-1})$, $f_\delta(\boldsymbol{\omega}) = f_1(\boldsymbol{\omega})$ elsewhere and let F_δ be the spectral measure with density f_δ. Then

$$\|H_{\delta,F_1} - H_{\delta,F_0}\|_{F_0}^2 \le 2\|H_{\delta,F_0} - H_{\delta,F_\delta}\|_{F_0}^2 + 2\|H_{\delta,F_\delta} - H_{\delta,F_1}\|_{F_0}^2. \tag{58}$$

Define $u(r) = \sup_{\boldsymbol{\omega} \in A_d(r)^c} \psi(\boldsymbol{\omega})$ and $\ell(r) = \inf_{\boldsymbol{\omega} \in A_d(r)^c} \psi(\boldsymbol{\omega})$. Applying Theorem 1 in 3.2 with $b = 1 + u(\delta^{-1})$ and $a = 1 + \ell(\delta^{-1})$ yields

$$\|H_{\delta,F_0} - H_{\delta,F_\delta}\|_{F_0}^2 \le \frac{\{u(\delta^{-1}) - \ell(\delta^{-1})\}^2}{4\{1 + u(\delta^{-1})\}\{1 + \ell(\delta^{-1})\}} \|H_{\delta,F_0}\|_{F_0}^2. \tag{59}$$

The function $\Gamma(\boldsymbol{\omega}) = H_{\delta,F_\delta}(\boldsymbol{\omega}) - H_{\delta,F_1}(\boldsymbol{\omega})$ has period $2\pi\delta^{-1}$ in each coordinate and for $\boldsymbol{\omega} \in A_d(\delta^{-1})$,

$$\Gamma(\boldsymbol{\omega}) = \frac{f_1(\boldsymbol{\omega})\psi(\boldsymbol{\omega})H_{\delta,F_1}(\boldsymbol{\omega})}{f_0(\boldsymbol{\omega}) + \sum_{\mathbf{j}}' f_1(\boldsymbol{\omega} + 2\pi\delta^{-1}\mathbf{j})}.$$

Since $f_0 \asymp f_1$, there exist positive finite constants a and b such that $a \le f_0(\boldsymbol{\omega})/f_1(\boldsymbol{\omega}) \le b$ for all $\boldsymbol{\omega}$. Thus,

$$\begin{aligned}
&\|H_{\delta,F_\delta} - H_{\delta,F_1}\|_{F_0}^2 \\
&\quad = \int_{\mathbb{R}^d} f_0(\boldsymbol{\omega}) |\Gamma(\boldsymbol{\omega})|^2 d\boldsymbol{\omega} \\
&\quad = \int_{A_d(\delta^{-1})} \sum_{\mathbf{j}} f_0(\boldsymbol{\omega} + 2\pi\delta^{-1}\mathbf{j}) \left| \frac{f_1(\boldsymbol{\omega})\psi(\boldsymbol{\omega})H_{\delta,F_1}(\boldsymbol{\omega})}{f_0(\boldsymbol{\omega}) + \sum_{\mathbf{j}}' f_1(\boldsymbol{\omega} + 2\pi\delta^{-1}\mathbf{j})} \right|^2 d\boldsymbol{\omega} \\
&\quad \le \frac{\max(1,b)}{a} \int_{A_d(\delta^{-1})} f_1(\boldsymbol{\omega})\psi(\boldsymbol{\omega})^2 |H_{\delta,F_1}(\boldsymbol{\omega})|^2 \, d\boldsymbol{\omega},
\end{aligned} \tag{60}$$

so that for $H \in \mathcal{L}_{-\delta}(F_1)$,

$$\begin{aligned}
&\|H_{\delta,F_1} - H_{\delta,F_0}\|_{F_0}^2 \\
&\qquad \le \frac{\{u(\delta^{-1}) - \ell(\delta^{-1})\}^2}{2\{1 + u(\delta^{-1})\}\{1 + \ell(\delta^{-1})\}} \|H_{\delta,F_0}\|_{F_0}^2 \\
&\qquad\quad + \frac{2\max(1,b)}{a} \int_{A_d(\delta^{-1})} f_1(\boldsymbol{\omega})\psi(\boldsymbol{\omega})^2 |H_{\delta,F_1}(\boldsymbol{\omega})|^2 \, d\boldsymbol{\omega}.
\end{aligned} \tag{61}$$

Using (58)–(61), (56) follows by an argument similar to the one leading to (54). It is possible to give a simpler argument showing that (56) follows from (54) and (55) (see the proof of the asymptotic efficiency of the pseudo-BLP in Theorem 8 of Chapter 4). The bound in (61) is needed to obtain the rates of convergence in Theorem 12. □

Rates of convergence to optimality

Theorem 10 used Theorem 9 to show that when $f_1(\boldsymbol{\omega})/f_0(\boldsymbol{\omega}) \to 1$ as $|\boldsymbol{\omega}| \to \infty$, all linear predictors under the incorrect f_1 are uniformly asymptotically optimal and the presumed mses are uniformly asymptotically correct. Under additional assumptions on f_0 and f_1, Theorem 9 can also be used to bound the right sides of (57) and (61) and hence obtain rates of convergence for the effect of misspecifying the spectral density. We need the following lemma.

Lemma 11. *For a nonnegative function σ on $\mathbb{R}^d$, $\alpha > d$ and C_0, C_1, D and β positive, suppose that $C_0 \le f_1(\boldsymbol{\omega})(1+|\boldsymbol{\omega}|)^\alpha \le C_1$ and $\sigma(\boldsymbol{\omega}) \le D\,(1+|\boldsymbol{\omega}|)^{-\beta}$ for all $\boldsymbol{\omega}$. Then for any $H \in \mathcal{L}(F_1)$,*

$$\int_{A_d(\delta^{-1})} f_1(\boldsymbol{\omega})\sigma(\boldsymbol{\omega})\,|H_{\delta,F_1}(\boldsymbol{\omega})|^2\,\mathbf{d}\boldsymbol{\omega}$$

$$\le D\|H_{\delta,F_1}\|^2_{F_1}\left[2\left(\frac{\delta}{\pi}\right)^\beta + \frac{\beta C_1}{C_0}\left(\frac{\delta}{\pi}\right)^\alpha \xi_d(\alpha)\int_0^{\pi\delta^{-1}}(1+r)^{\alpha-\beta-1}dr\right].$$

PROOF. Using the bound on σ,

$$\begin{aligned}
&\int_{A_d(\delta^{-1})} f_1(\boldsymbol{\omega})\sigma(\boldsymbol{\omega})\,|H_{\delta,F_1}(\boldsymbol{\omega})|^2\,\mathbf{d}\boldsymbol{\omega}\\
&\qquad\le \int_{b_d(\pi\delta^{-1})} f_1(\boldsymbol{\omega})\sigma(\boldsymbol{\omega})\,|H_{\delta,F_1}(\boldsymbol{\omega})|^2\,\mathbf{d}\boldsymbol{\omega}\\
&\qquad\quad + D\left(\frac{\delta}{\pi}\right)^\beta \int_{A_d(\delta^{-1})\backslash b_d(\pi\delta^{-1})} f_1(\boldsymbol{\omega})\,|H_{\delta,F_1}(\boldsymbol{\omega})|^2\,\mathbf{d}\boldsymbol{\omega}\\
&\qquad\le D\int_0^{\pi\delta^{-1}}(1+r)^{-\beta}p(r)\,dr + D\left(\frac{\delta}{\pi}\right)^\beta \|H_{\delta,F_1}\|^2_{F_1},
\end{aligned}$$

where

$$p(r) = \int_{\partial b_d(r)} f_1(\boldsymbol{\nu})\,|H_{\delta,F_1}(\boldsymbol{\nu})|^2\,\mu(\mathbf{d}\boldsymbol{\nu})$$

and $\mu(\mathbf{d}\boldsymbol{\nu})$ indicates the surface measure on $\partial b_d(r)$. Defining $P(r) = \int_0^r p(s)\,ds$, Theorem 9 implies $P(r) \le M_\delta(F_1, b_d(r))\|H_{\delta,F_1}\|^2_{F_1}$. By definition, $P(\pi\delta^{-1}) \le \|H_{\delta,F_1}\|^2_{F_1}$. Integrating by parts,

$$\int_0^{\pi\delta^{-1}}(1+r)^{-\beta}p(r)\,dr$$

$$= \left(1 + \frac{\pi}{\delta}\right)^{-\beta} P(\pi\delta^{-1}) + \beta \int_0^{\pi\delta^{-1}} (1+r)^{-\beta-1} P(r)\,dr$$

$$\leq \left\{ \left(\frac{\delta}{\pi}\right)^{\beta} + \beta \int_0^{\pi\delta^{-1}} (1+r)^{-\beta-1} M_\delta(F_1, b_d(r))\,dr \right\} \|H_{\delta,F_1}\|_{F_1}^2.$$

Lemma 11 then follows from (53). □

Lemma 11 yields the following result, taken from Stein (1999), on the effects on linear prediction of misspecifying the spectral density.

Theorem 12. *If $f_0(\omega) \asymp f_1(\omega) \asymp (1+|\omega|)^{-\alpha}$ and for some $\gamma > 0$, $|\psi(\omega)| \ll (1+|\omega|)^{-\gamma}$, then*

$$\sup_{h \in \mathcal{H}_{-\delta}(F_1)} \frac{\left|E_1 e_1(h,\delta)^2 - E_0 e_1(h,\delta)^2\right|}{E_0 e_1(h,\delta)^2} \ll \begin{cases} \delta^{\min(\alpha,\gamma)}, & \alpha \neq \gamma \\ \delta^{\alpha} \log(\delta^{-1}), & \alpha = \gamma \end{cases} \tag{62}$$

and

$$\sup_{h \in \mathcal{H}_{-\delta}(F_1)} \frac{E_0 \left\{e_0(h,\delta) - e_1(h,\delta)\right\}^2}{E_0 e_0(h,\delta)^2} \ll \begin{cases} \delta^{\min(\alpha,2\gamma)}, & \alpha \neq 2\gamma \\ \delta^{\alpha} \log(\delta^{-1}), & \alpha = 2\gamma. \end{cases} \tag{63}$$

PROOF. To prove (62), just apply Lemma 11 and the bound on $|\psi|$ to (57). To prove (63), apply Lemma 11 and the bound on ψ^2 to (61). □

Except possibly in the case $\alpha = \gamma$ in (62) and $\alpha = 2\gamma$ in (63), these bounds are sharp in the sense that there exist f_0 and f_1 satisfying the stated conditions for which both conclusions are false if $O(\cdot)$ is replaced by $o(\cdot)$ (Stein 1999). Stein (1999) also gives some analogous results for a process observed unevenly on a bounded interval, but the arguments are much more difficult and the conditions on the spectral densities are somewhat restrictive. The general approach is similar to that taken here in that the key result is a bound of the type given by Theorem 9 on the fraction of the mse of a BLP attributable to a given range of frequencies.

Pseudo-BLPs with a misspecified mean function

Suppose that both the mean and covariance functions of Z are possibly misspecified. Take E_{ij} to be expectation under (m_i, K_j) and $e_{ij}(h,\delta)$ to be the error of the BLP of h under (m_i, K_j). Then (Exercise 36)

$$E_{00}(e_{11} - e_{00})^2 = E_{00}(e_{01} - e_{00})^2 + E_{00}(e_{10} - e_{00})^2 \tag{64}$$

and

$$E_{00}e_{11}^2 - E_{11}e_{11}^2 = (E_{01}e_{11}^2 - E_{11}e_{11}^2) + (E_{10}e_{11}^2 - E_{11}e_{11}^2). \tag{65}$$

Thus, the effect of misspecifying both the mean and covariance functions on either the actual mse of prediction or the evaluation of the mse can be decomposed into a term giving the effect of just misspecifying the covariance function and a term giving the effect of just misspecifying the mean function. In light of these decompositions, let us next consider the effect of misspecifying just the mean function on linear prediction.

Theorem 13. *Suppose* $(m_0, K_0) = (m_0, K)$ *and* $(m_1, K_1) = (m_1, K)$, *where* K *is an autocovariance function with spectrum* F *possessing spectral density* f *bounded away from 0 on any bounded set and* $f(\boldsymbol{\omega}) \to 0$ *as* $|\boldsymbol{\omega}| \to \infty$. *If* $m = m_1 - m_0$ *is square-integrable and of the form* $m(\mathbf{x}) = \int_{\mathbb{R}^d} \exp(-i\boldsymbol{\omega}^T\mathbf{x})\xi(\boldsymbol{\omega})\,\mathbf{d}\boldsymbol{\omega}$, *where*

$$\int_{\mathbb{R}^d} \frac{|\xi(\boldsymbol{\omega})|^2}{f(\boldsymbol{\omega})}\,\mathbf{d}\boldsymbol{\omega} < \infty, \tag{66}$$

then

$$\lim_{\delta \downarrow 0} \sup_{h \in \mathcal{H}_{-\delta}(F_0)} \frac{E_0\{e_0(h,\delta) - e_1(h,\delta)\}^2}{E_0 e_0(h,\delta)^2} = 0.$$

PROOF. It is a simple matter to show that $E_0 e_1(h,\delta)^2$ is unchanged by subtracting the same fixed function from m_0 and m_1, so there is no loss in generality in taking m_0 identically 0. Next, for any $V \in \mathcal{L}(F)$, the mean of the corresponding random variable in $\mathcal{H}(F)$ is $\int_{\mathbb{R}^d} \xi(\boldsymbol{\omega})\overline{V(\boldsymbol{\omega})}\,\mathbf{d}\boldsymbol{\omega}$ (Exercise 37). By (66), for $\epsilon > 0$, we can choose r_ϵ so that $\int_{b_d(r_\epsilon)^c} |\xi(\boldsymbol{\omega})|^2 f(\boldsymbol{\omega})^{-1}\mathbf{d}\boldsymbol{\omega} < \epsilon$. Using Theorem 9, $V \in \mathcal{L}(F)$ and orthogonal to $\mathcal{L}_\delta(F)$ imply

$$\begin{aligned}
&\left|\int_{\mathbb{R}^d} \xi(\boldsymbol{\omega})\overline{V(\boldsymbol{\omega})}\,\mathbf{d}\boldsymbol{\omega}\right|^2 \\
&\le 2\left|\int_{b_d(r_\epsilon)} \xi(\boldsymbol{\omega})\overline{V(\boldsymbol{\omega})}\,\mathbf{d}\boldsymbol{\omega}\right|^2 + 2\left|\int_{b_d(r_\epsilon)^c} \xi(\boldsymbol{\omega})\overline{V(\boldsymbol{\omega})}\,\mathbf{d}\boldsymbol{\omega}\right|^2 \\
&\le 2\int_{\mathbb{R}^d} \frac{|\xi(\boldsymbol{\omega})|^2}{f(\boldsymbol{\omega})}\,\mathbf{d}\boldsymbol{\omega} \int_{b_d(r_\epsilon)} f(\boldsymbol{\omega})|V(\boldsymbol{\omega})|^2\mathbf{d}\boldsymbol{\omega} \\
&\quad + 2\int_{b_d(r_\epsilon)^c} \frac{|\xi(\boldsymbol{\omega})|^2}{f(\boldsymbol{\omega})}\,\mathbf{d}\boldsymbol{\omega} \int_{\mathbb{R}^d} f(\boldsymbol{\omega})|V(\boldsymbol{\omega})|^2\mathbf{d}\boldsymbol{\omega} \\
&\le 2\int_{\mathbb{R}^d} \frac{|\xi(\boldsymbol{\omega})|^2}{f(\boldsymbol{\omega})}\,\mathbf{d}\boldsymbol{\omega}\, M_\delta(F, b_d(r_\epsilon))\|V\|_F^2 + 2\epsilon\|V\|_F^2
\end{aligned}$$

and Theorem 13 follows since $M_\delta(F, b_d(r_\epsilon)) \to 0$ as $\delta \downarrow 0$ and ϵ is arbitrary. □

There is no need to prove a separate result for $E_1 e_1(h,n)^2/E_0 e_1(h,n)^2$ since $E_0 e_1(h,n)^2 = E_1 e_0(h,n)^2$ and $E_0 e_0(h,n)^2 = E_1 e_1(h,n)^2$ when only the mean function is misspecified (see Exercise 8). As an example of when (66) holds, suppose $d = 1$ and $f(\omega) \asymp (1+\omega^2)^{-p}$ for some positive integer p. Then (66) holds if and only if m is square integrable, $m^{(p-1)}$ exists and is absolutely continuous and has almost everywhere derivative $m^{(p)}$ satisfying

$$\int_{-\infty}^{\infty}\left\{m^{(p)}(t)\right\}^2 dt < \infty$$

(see III.4.1 of Ibragimov and Rozanov (1978)). The fact that m must be square integrable eliminates such obvious candidates for a mean function as a nonzero constant. Results in 4.3 show that if the observations and predictands are restricted to a bounded region, it is generally possible to obtain asymptotically optimal predictions if the mean is misspecified by a nonzero constant. Thus, for assessing the effect of a misspecified mean, the infinite lattice setting is somewhat misleading as to what happens in the fixed-domain setting.

Under stronger assumptions on ξ we can obtain rates of convergence in Theorem 13. Specifically, suppose $f(\boldsymbol{\omega}) \asymp (1+|\boldsymbol{\omega}|)^{-\alpha}$ for some $\alpha > d$ and $|\xi(\boldsymbol{\omega})|^2/f(\boldsymbol{\omega}) \ll (1+|\boldsymbol{\omega}|)^{-d-\gamma}$ for some $\gamma > 0$, so that (66) holds. Larger values of γ correspond to smoother mean functions m.

Theorem 14. *If, in addition to the conditions of Theorem 13,* $f(\boldsymbol{\omega}) \asymp (1+|\boldsymbol{\omega}|)^{-\alpha}$ *and* $|\xi(\boldsymbol{\omega})|^2/f(\boldsymbol{\omega}) \ll (1+|\boldsymbol{\omega}|)^{-d-\gamma}$ *for some* $\gamma > 0$,

$$\sup_{h\in\mathcal{H}_{-\delta}(F)} \frac{E_0 e_1^2 - E_0 e_0^2}{E_0 e_0^2} \ll \begin{cases} \delta^{\min(\alpha,\gamma)}, & \alpha \neq \gamma \\ \delta^{\alpha}\{\log|\delta|\}^2, & \alpha = \gamma. \end{cases}$$

Proof. For $V \in \mathcal{L}(F)$ and orthogonal to $\mathcal{L}_\delta(F)$,

$$\left|\int_{\mathbb{R}^d} \xi(\boldsymbol{\omega})\overline{V(\boldsymbol{\omega})}\, \mathbf{d}\boldsymbol{\omega}\right|^2 \le 2\left|\int_{b_d(\pi\delta^{-1})} \xi(\boldsymbol{\omega})\overline{V(\boldsymbol{\omega})}\, \mathbf{d}\boldsymbol{\omega}\right|^2 + 2\left|\int_{b_d(\pi\delta^{-1})^c} \xi(\boldsymbol{\omega})\overline{V(\boldsymbol{\omega})}\, \mathbf{d}\boldsymbol{\omega}\right|^2. \tag{67}$$

By the Cauchy–Schwarz inequality,

$$\begin{aligned}\left|\int_{b_d(\pi\delta^{-1})^c} \xi(\boldsymbol{\omega})\overline{V(\boldsymbol{\omega})}\, \mathbf{d}\boldsymbol{\omega}\right|^2 &\le \int_{b_d(\pi\delta^{-1})^c} \frac{|\xi(\boldsymbol{\omega})|^2}{f(\boldsymbol{\omega})}\,\mathbf{d}\boldsymbol{\omega} \int_{b_d(\pi\delta^{-1})^c} f(\boldsymbol{\omega})|V(\boldsymbol{\omega})|^2 \mathbf{d}\boldsymbol{\omega} \\ &\ll \int_{\delta^{-1}}^{\infty} r^{-\gamma-1} dr \|V\|_F^2 \\ &\ll \delta^{\gamma}\|V\|_F^2. \end{aligned} \tag{68}$$

Similar to the proof of Lemma 11, define $p(r) = \int_{\partial b_d(r)} f(\boldsymbol{\omega})^{1/2} V(\boldsymbol{\omega})\mu(\mathbf{d}\boldsymbol{\omega})$ and $P(r) = \int_0^r p(s)ds$. Then

$$P(r) \ll \left\{r^d \int_{b_d(r)} f(\boldsymbol{\omega})|V(\boldsymbol{\omega})|^2 \mathbf{d}\boldsymbol{\omega}\right\}^{1/2} \ll \left\{r^d M_\delta(F, b_d(r))\right\}^{1/2} \|V\|_F,$$

so that using integration by parts as in the proof of Lemma 11,

$$\left\{\int_{b_d(\pi\delta^{-1})} \xi(\boldsymbol{\omega})\overline{V(\boldsymbol{\omega})}\, d\boldsymbol{\omega}\right\}^2$$

$$\ll \left\{\int_0^{\pi\delta^{-1}} (1+r)^{-(\gamma+d)/2} p(r)\, dr\right\}^2$$

$$\ll P(\pi\delta^{-1})^2 \delta^{\gamma+d}$$

$$+ \|V\|_F^2 \left[\int_0^{\pi\delta^{-1}} (1+r)^{-1-(\gamma+d)/2} \left\{r^d M_\delta(F, b_d(r))\right\}^{1/2} dr\right]^2$$

$$\ll \delta^\gamma \|V\|_F^2 + \delta^\alpha \|V\|_F^2 \left[\int_0^{\pi\delta^{-1}} (1+r)^{\{(\alpha-\gamma)/2\}-1} dr\right]^2$$

$$\ll \begin{cases} \delta^{\min(\alpha,\gamma)} \|V\|_F^2, & \alpha \neq \gamma \\ \delta^\alpha (\log \delta)^2 \|V\|_F^2, & \alpha = \gamma. \end{cases} \tag{69}$$

Theorem 14 follows from (67)–(69). □

Exercises

33 In the proof of Theorem 9, verify that $M_\delta(F, B) < 1$ and B symmetric imply $V_{\delta,B} \in \mathcal{L}_\delta(F)$.

34 In the proof of Theorem 9, verify that $U \in \mathcal{L}(F)$, U is orthogonal to $\mathcal{L}_\delta(F)$ and $\|U\|_F^2 > 0$.

35 Prove (53).

36 Prove (64) and (65).

37 Using the definitions in Theorem 13, show that for any $V \in \mathcal{L}(F)$, the mean of the corresponding random variable in $\mathcal{H}(F)$ is $\int_{\mathbb{R}^d} \xi(\boldsymbol{\omega})\overline{V(\boldsymbol{\omega})}\, d\boldsymbol{\omega}$. In addition, show that $\left|\int_{\mathbb{R}^d} \xi(\boldsymbol{\omega})\overline{V(\boldsymbol{\omega})}\, d\boldsymbol{\omega}\right|$ is finite.

38 Suppose Z is a mean 0 weakly stationary process on $\mathbb{R}$ with spectral density $f(\omega) = 1\{|\omega| < \pi\}$. Show that if Z is observed on $\mathbb{Z}$ then the observations are all uncorrelated and yet perfect interpolation at all $t \in \mathbb{R}$ is possible.

4

Equivalence of Gaussian Measures and Prediction

4.1 Introduction

The basic message of the results of 3.8 is that for interpolating a mean 0 weakly stationary random field based on observations on an infinite square lattice, the smaller the distance between neighboring observations in the lattice, the less the low frequency behavior of the spectrum matters. This suggests that if our goal is to interpolate our observations and we need to estimate the spectral density from these same observations, we should focus on getting the high frequency behavior of the spectral density as accurately as possible while not worrying so much about the low frequency behavior. Supposing that our observations and predictions will all take place in some bounded region R, a useful first question to ask is what can be done if we observe the process everywhere in R. Answering this question will put an upper bound on what one can hope to learn from some finite number of observations in R.

One simple way to formulate the question of what can be learned from observations on R is to suppose that there are only two possible probability measures for the process on R and to determine when one can tell which measure is correct and which is not. For example, consider a mean 0 Gaussian process on $\mathbb{R}$ with two possible autocovariance functions: $K_0(t) = e^{-|t|}$ and $K_1(t) = \frac{1}{2}e^{-2|t|}$. If we observe this process for all $t \in [0, T]$ with $T < \infty$, then it turns out that it is not possible to know for sure which autocovariance function is correct, despite the fact that we have an infinite number of observations. Fortunately, as demonstrated in 3.5, these models can give very similar interpolations. Indeed, under quite general conditions, measures that cannot be correctly distinguished with high probability based on a large number of available observations yield very similar predictions (see 4.3).

Let us first introduce some terminology. For two probability measures P_0 and P_1 on a measurable space $(\Omega, \mathcal{F})$, say that P_0 is absolutely continuous with respect to P_1 if for all $A \in \mathcal{F}$, $P_1(A) = 0$ implies $P_0(A) = 0$. Define P_0 and P_1 to be equivalent, written $P_0 \equiv P_1$, if P_0 is absolutely continuous with respect to P_1 and P_1 is absolutely continuous with respect to P_0. Thus, $P_0 \equiv P_1$ means that for all $A \in \mathcal{F}$, $P_0(A) = 0$ if and only if $P_1(A) = 0$. Define P_0 and P_1 to be orthogonal, written $P_0 \perp P_1$, if there exists $A \in \mathcal{F}$ such that $P_0(A) = 1$ and $P_1(A) = 0$. In this case, we also have $P_0(A^c) = 0$ and $P_1(A^c) = 1$. Thus, suppose we know that either P_0 or P_1 is the correct probability measure. If $P_0 \perp P_1$, then based on observing $\omega \in \Omega$, it is possible to determine which measure is correct with probability 1. On the other hand, if $P_0 \equiv P_1$, then no matter what is observed, it is not possible to determine which measure is correct with probability 1. More specifically, consider a decision rule of the following form. For some $A \in \mathcal{F}$, choose P_0 if A occurs and choose P_1 otherwise. If $P_0 \equiv P_1$, then for any $B \in \mathcal{F}$ such that $P_0(B) > 0$ (so that $P_1(B) > 0$), we cannot have both $P_0(A \mid B) = 1$ and $P_1(A^c \mid B) = 1$. Indeed, if $P_0(A \mid B) = 1$ then $P_1(A^c \mid B) = 0$. Thus, there is no event B receiving positive probability under either measure such that, conditionally on B, perfect discrimination between the measures is possible. Of course, measures may be neither equivalent nor orthogonal; a trivial example is to take $\Omega = \{0, 1, 2\}$, have P_0 assign probability $\frac{1}{2}$ to $\{0\}$ and $\{1\}$ and probability 0 to $\{2\}$ and P_1 assign probability $\frac{1}{2}$ to $\{1\}$ and $\{2\}$ and 0 to $\{0\}$. In this case, we would know which measure were correct if $\omega = 0$ or 2, but we would not know if $\omega = 1$. An interesting property about Gaussian measures is that in great generality they are either equivalent or orthogonal.

Section 4.2 looks at the problem of determining equivalence and orthogonality of measures for Gaussian random fields observed on a bounded region. There is a great deal known about this problem for Gaussian random fields possessing an autocovariance function. The treatment in 4.2 largely follows that of Ibragimov and Rozanov (1978, Chapter III). Other references include Yadrenko (1983), Gihman and Skorohod (1974) and Kuo (1975).

Two critical weaknesses of the theoretical results in 3.6–3.8 on the behavior of pseudo-BLPs are that they require observations over an unbounded domain and that they require regularly spaced observations. Using results on equivalence of Gaussian measures, Section 4.3 proves that if the presumed spectral density has similar high frequency behavior as the actual spectral density, pseudo-BLPs are asymptotically optimal under fixed-domain asymptotics even when the observations are not regularly spaced. It may seem curious that properties of Gaussian measures are helpful in obtaining results on linear predictors, which only depend on the first two moments of the random field. However, characterizations of equivalent Gaussian measures provide a convenient means for showing that the low frequency behavior of the spectral density has little impact on the behavior of linear predictions.

Section 4.4 provides a first attempt to consider the effect of estimating the law of a random field on subsequent predictions. In particular, 4.4 gives a quantitative Bayesian formulation of Jeffreys's law (Dawid 1984), which states that aspects of a probability law that cannot be determined from a large amount of data cannot have much impact on prediction. This law is of particular interest and value when employing fixed-domain asymptotics, since there will naturally be parameters of models that cannot be consistently estimated based on an increasing number of observations in a fixed domain.

4.2 Equivalence and orthogonality of Gaussian measures

This section develops the basic theory for determining the equivalence or orthogonality of Gaussian measures for random fields. For finite-dimensional random vectors, it is trivial to determine the equivalence or orthogonality of two possible Gaussian distributions: Gaussian random vectors on $\mathbb{R}^d$ have equivalent measures if their distributions are both nonsingular or if they are both singular and the hyperplanes that form their respective supports are the same; otherwise, they are orthogonal. It is in the infinite-dimensional setting that determining the equivalence or orthogonality of Gaussian measures becomes difficult.

Suppose Z is a random field on $\mathbb{R}^d$ with mean function m and covariance function K. For a closed set $R \subset \mathbb{R}^d$, let $G_R(m, K)$ be the Gaussian measure for the random field on R with second-order structure (m, K). When there is no chance for confusion, I write $G(m, K)$ for $G_R(m, K)$. Furthermore, I use P_j as shorthand for the Gaussian measure $G(m_j, K_j)$.

Conditions for orthogonality

Reasonably elementary arguments can be used to establish orthogonality in many cases. Note that to establish $P_0 \perp P_1$, we only have to find a set A such that $P_0(A) = 1$ and $P_1(A) = 0$. To establish equivalence, we have to show something about a whole class of sets. The following result is helpful in establishing orthogonality. $P_0 \perp P_1$ if there exists $A_1, A_2, \ldots \in \mathcal{F}$ such that

$$\lim_{n\to\infty} P_0(A_n) = 0 \text{ and } \lim_{n\to\infty} P_1(A_n) = 1 \tag{1}$$

(Exercise 1).

Since Gaussian random fields are determined by their first two moments, all statements about their equivalence or orthogonality can also be written in terms of the first two moments. For example, as a simple application of (1), consider a random field Z on some set R with two possible Gaussian measures P_0 and P_1. If there exists a sequence of linear combinations $Y_n = \sum_{j=1}^{r_n} \lambda_{jn} Z(\mathbf{x}_{jn})$, $\mathbf{x}_{1n}, \ldots, \mathbf{x}_{r_n n} \in R$ such that

$$\lim_{n\to\infty} \frac{\mathrm{var}_1(Y_n)}{\mathrm{var}_0(Y_n)} = 0 \quad (\text{or } \infty), \tag{2}$$

or

$$\lim_{n\to\infty} \frac{(E_1 Y_n - E_0 Y_n)^2}{\mathrm{var}_0(Y_n)} = \infty, \tag{3}$$

then $P_0 \perp P_1$ follows from (1) (Exercise 2). For example, suppose Z is a mean 0 stationary Gaussian process on $\mathbb{R}$, $R = [0,1]$, $K_0(t) = e^{-|t|}$ and $K_1(t) = e^{-|t|}(1+|t|)$ so that Z is mean square differentiable under P_1 but not P_0. Let $Y_n = Z(n^{-1}) - Z(0)$. Then the limit in (2) is 0 and $P_0 \perp P_1$.

More generally, suppose Z is a mean 0 stationary Gaussian process on $\mathbb{R}$ with spectral density f_j under P_j and $R = [0,1]$. If Z is not infinitely mean square differentiable under f_1 and $f_0(\omega)/f_1(\omega) \to 0$ as $\omega \to \infty$, then $P_0 \perp P_1$. To prove this, note that the condition on f_1 implies there exists a positive integer p such that $\int_{-\infty}^{\infty} \omega^{2p} f_1(\omega) d\omega = \infty$ (see Section 2.6). Define the linear operator Δ_ϵ by $\Delta_\epsilon Z(t) = \epsilon^{-1}\{Z(t+\epsilon) - Z(t)\}$. Then

$$\begin{aligned}
E_j \{(\Delta_\epsilon)^p Z(0)\}^2 &= E_j \left\{ \frac{1}{\epsilon^p} \sum_{k=0}^{p} \binom{p}{k} (-1)^{p-k} Z(k\epsilon) \right\}^2 \\
&= \int_{-\infty}^{\infty} \left| \frac{1}{\epsilon^p} \sum_{k=0}^{p} \binom{p}{k} (-e^{i\omega\epsilon})^k \right|^2 f_j(\omega)\, d\omega \\
&= \int_{-\infty}^{\infty} \left\{ \frac{2}{\epsilon} \sin\left(\frac{\omega\epsilon}{2}\right) \right\}^{2p} f_j(\omega)\, d\omega.
\end{aligned}$$

Given $\delta > 0$, we can choose T such that $f_0(\omega)/f_1(\omega) < \delta$ for $|\omega| > T$. As $\epsilon \downarrow 0$,

$$\int_{|\omega|<T} \left\{ \frac{2}{\epsilon} \sin\left(\frac{\omega\epsilon}{2}\right) \right\}^{2p} f_j(\omega)\, d\omega \to \int_{|\omega|<T} \omega^{2p} f_j(\omega)\, d\omega < \infty$$

for $j = 0, 1$ and

$$\begin{aligned}
&\int_{|\omega|>T} \left\{ \frac{2}{\epsilon} \sin\left(\frac{\omega\epsilon}{2}\right) \right\}^{2p} f_1(\omega)\, d\omega \\
&\qquad \geq \int_{T<|\omega|<\pi/\epsilon} \left\{ \frac{2}{\epsilon} \left(\frac{2}{\pi} \cdot \frac{\omega\epsilon}{2} \right) \right\}^{2p} f_1(\omega)\, d\omega \to \infty.
\end{aligned}$$

Thus,

$$\overline{\lim_{\epsilon\downarrow 0}} \frac{E_0\{(\Delta_\epsilon)^p Z(0)\}^2}{E_1\{(\Delta_\epsilon)^p Z(0)\}^2} \leq \frac{\int_{|\omega|>T}\left\{\frac{2}{\epsilon}\sin\left(\frac{\omega\epsilon}{2}\right)\right\}^{2p} f_0(\omega)\,d\omega}{\int_{|\omega|>T}\left\{\frac{2}{\epsilon}\sin\left(\frac{\omega\epsilon}{2}\right)\right\}^{2p} f_1(\omega)\,d\omega} \leq \delta,$$

since $f_0(\omega)/f_1(\omega) < \delta$ for $|\omega| > T$, so $P_0 \perp P_1$ follows by the arbitrariness of δ and (2).

If $m_0 = m_1$ and $f_0 \asymp f_1$, then neither (2) nor (3) can happen. However, we may still be able to prove orthogonality of Gaussian measures by considering sequences of sums of squares of linear combinations of Z. Suppose Z is a stationary Gaussian process on $[0, 1]$ with autocovariance function K satisfying $K(t) = C - D|t| + o(|t|)$ as $t \to 0$ for some $D > 0$. Define

$$U_n = \frac{1}{2}\sum_{j=1}^{n}\left\{Z\left(\frac{j}{n}\right) - Z\left(\frac{j-1}{n}\right)\right\}^2. \tag{4}$$

Then

$$EU_n = n\left\{K(0) - K\left(\frac{1}{n}\right)\right\} \to D$$

as $n \to \infty$. Now X_1, X_2, X_3, X_4 jointly Gaussian with mean 0 and $\operatorname{cov}(X_i, X_j) = \sigma_{ij}$ implies $\operatorname{cov}(X_1X_2, X_3X_4) = \sigma_{13}\sigma_{24} + \sigma_{14}\sigma_{23}$ (see Appendix A), so

$$\begin{aligned}
&\operatorname{var} U_n \\
&= \frac{n}{4}\operatorname{var}\left[\left\{Z\left(\frac{1}{n}\right) - Z(0)\right\}^2\right] \\
&\quad + \frac{1}{2}\sum_{j=1}^{n-1}(n-j)\operatorname{cov}\left[\left\{Z\left(\frac{1}{n}\right) - Z(0)\right\}^2, \left\{Z\left(\frac{j+1}{n}\right) - Z\left(\frac{j}{n}\right)\right\}^2\right] \\
&= 2n\left\{K(0) - K\left(\frac{1}{n}\right)\right\}^2 \\
&\quad + \sum_{j=1}^{n-1}(n-j)\left\{2K\left(\frac{j}{n}\right) - K\left(\frac{j+1}{n}\right) - K\left(\frac{j-1}{n}\right)\right\}^2.
\end{aligned} \tag{5}$$

If K has a bounded second derivative on $(0, 1]$ then $EU_n = D + O(n^{-1})$ and $\operatorname{var} U_n = 2n^{-1}D^2 + O(n^{-2})$ (Exercise 3). That is, we can estimate D with asymptotically the same mse as when the first differences, $Z(j/n) - Z((j-1)/n)$ for $j = 1, \ldots, n$, are independent and identically distributed $N(0, 2Dn^{-1})$. Note that if Z were Brownian motion with $\operatorname{var}\{Z(t) - Z(0)\} = 2D|t|$, then the first differences would be independent and identically distributed $N(0, 2Dn^{-1})$.

So, for $j = 0, 1$, suppose K_j is an autocovariance function on $\mathbb{R}$ with $K_j(t) = C_j - D_j|t| + o(|t|)$ as $t \to 0$ for some $D_j > 0$ and K_j has a bounded second derivative on $(0, 1]$. If $D_0 \neq D_1$ and $R = [0, 1]$, then $G_R(0, K_0) \perp$

$G_R(0, K_1)$, since under $G_R(0, K_0)$, $U_n \xrightarrow{L^2} D_0$ and under $G_R(0, K_1)$, $U_n \xrightarrow{L^2} D_1$.

Although assuming K_j has a bounded second derivative on $(0, 1]$ simplifies matters considerably, the following result implies that $K_j(t) = C_j - D_j|t| + o(|t|)$ as $t \to 0$ for $j = 0, 1$ with D_0 and D_1 unequal and positive is sufficient to conclude $G_R(0, K_0) \perp G_R(0, K_1)$ for R any interval of positive length (Exercise 4).

Theorem 1. *For a mean 0 stationary Gaussian process on $\mathbb{R}$ with autocovariance function satisfying $K(t) = C - D|t| + o(|t|)$ as $t \to 0$ for some $D > 0$ and U_n as defined in (4),* $\text{var}(U_n) \to 0$ *as $n \to \infty$.*

A proof using spectral methods is given at the end of this section. Exercise 25 outlines a proof in the time domain.

I conjecture that the conditions of Theorem 1 imply the stronger result $\text{var}(U_n) = O(n^{-1})$ as $n \to \infty$. For example, if $K(t) = \left(\frac{1}{2} - |t|\right)^+$, then since K is not even once differentiable on $(0, 1]$, one might imagine this case would violate my conjecture. However, direct calculation shows that we still have $\text{var}(U_n) = O(n^{-1})$ (Exercise 8).

Before attempting to develop a general theory of equivalence and orthogonality of Gaussian measures, let us consider one more example: $R = [0, 2]$, $m_0 = m_1 = 0$, $K_0(t) = (1 - |t|)^+$ and $K_1(t) = e^{-|t|}$. Define

$$W_n = \sum_{j=1}^{n} \left\{ Z\left(1 + \frac{j}{n}\right) - Z\left(1 + \frac{j-1}{n}\right) \right\} \left\{ Z\left(\frac{j}{n}\right) - Z\left(\frac{j-1}{n}\right) \right\}. \quad (6)$$

Then as $n \to \infty$ (Exercise 9), $E_0 W_n \to -1$, $E_1 W_n \to 0$, $\text{var}_0 W_n \to 0$ and $\text{var}_1 W_n \to 0$, so that $G_R(0, K_0) \perp G_R(0, K_1)$, despite the fact that the autocovariance functions behave similarly at the origin. Recall that the triangular autocovariance function $K_0(t) = (1 - |t|)^+$ produced strange linear predictions (Section 3.5), so it is encouraging that we should be able to distinguish between K_0 and K_1 based on observations on $[0, 2]$. It is important that we chose $R = [0, 2]$; it is possible to show that if $R = [0, T]$, then $G_R(0, K_0) \perp G_R(0, K_1)$ if $T > 1$ and $G_R(0, K_0) \equiv G_R(0, K_1)$ if $T \leq 1$ (see Exercise 19). However, it is also true that if $T \leq 1$, K_0 produces no unusual predictors on $[0, T]$.

Gaussian measures are equivalent or orthogonal

Suppose m_0 and m_1 are continuous functions on $\mathbb{R}^d$, K_0 and K_1 are continuous and p.d. on $\mathbb{R}^d \times \mathbb{R}^d$ and R is a closed subset of $\mathbb{R}^d$. We now demonstrate that the Gaussian measures $P_0 = G_R(m_0, K_0)$ and $P_1 = G_R(m_1, K_1)$ are always either equivalent or orthogonal. We follow the approach of Ibragimov and Rozanov (1978, pp. 74–77). The following notation and assumptions are used throughout the rest of this section and in 4.3. For a random field Z on $\mathbb{R}^d$, let $\mathcal{H}_R^0$ be the real linear manifold of $Z(\mathbf{x})$ for $x \in R$ and let

$\mathcal{H}_R(m,K)$ be the closure of $\mathcal{H}_R^0$ with respect to the norm given by second moments under (m,K). The continuity assumptions about m_j and K_j imply that $\mathcal{H}_R(m_j,K_j)$ is separable for $j=0,1$ (Exercise 10). If there is a basis for $\mathcal{H}_R(m_0,K_0)$ and $\mathcal{H}_R(m_1,K_1)$ that is linearly independent under one of the covariance functions but not the other, then trivially $P_0 \perp P_1$, so assume there exists $h_1, h_2, \ldots$ in $\mathcal{H}_R^0$ forming a linearly independent basis for both $\mathcal{H}_R(m_0,K_0)$ and $\mathcal{H}_R(m_1,K_1)$. For example, taking $h_i = Z(\mathbf{x}_i)$ with $\mathbf{x}_1, \mathbf{x}_2, \ldots$ dense in R yields a basis for $\mathcal{H}_R(m_j,K_j)$ and if the $\mathbf{x}_i$s are distinct, the h_is will commonly be linearly independent (although see Exercise 16 in 2.7). Ibragimov and Rozanov (1978, Lemma 1 of Chapter 3) show that two Gaussian measures on the σ-algebra generated by $h_1, h_2, \ldots$ are equivalent if and only if they are equivalent on the σ-algebra generated by $Z(\mathbf{x})$ for $\mathbf{x} \in R$, so I do not explicitly consider the distinction between these two σ-algebras subsequently. To determine when P_0 and P_1 are equivalent or orthogonal (or neither), it makes no difference if we subtract a sequence of constants from the h_is, so without loss of generality, assume $E_0 h_i = 0$ for all i. Now we can linearly transform $h_1, \ldots, h_n$ to $h_{1n}, \ldots, h_{nn}$ such that for $j,k = 1, \ldots, n$,

$$K_0(h_{kn}, h_{jn}) = \delta_{kj} \quad \text{and} \quad K_1(h_{kn}, h_{jn}) = \sigma_{kn}^2 \delta_{kj}, \tag{7}$$

where $\delta_{kj} = 1$ if $k=j$ and is 0 otherwise. Set $m_{kn} = m_1(h_{kn})$. Note that here we are considering m and K_i to be operators on spaces of random variables rather than functions on regions of Euclidean space.

Likelihood ratios play a critical role in statistical theory. In particular, the famous Neyman–Pearson Lemma (Casella and Berger 1990, p. 366) shows that for testing one simple hypothesis (a hypothesis containing only one probability law for the observations) against another simple hypothesis, tests based on the likelihood ratio are optimal in a well-defined sense. Thus, it should not be surprising that likelihood ratios can be used to determine equivalence and orthogonality of Gaussian measures. Based on observations $h_1, \ldots, h_n$, the likelihood ratio of P_1 to P_0, denoted by p_n, is just the joint density of $h_1, \ldots, h_n$ under P_1 divided by their joint density under P_0. Direct calculation yields (Exercise 11)

$$\log p_n = -\sum_{k=1}^{n} \log \sigma_{kn} - \frac{1}{2} \sum_{k=1}^{n} \left\{ \frac{(h_{kn} - m_{kn})^2}{\sigma_{kn}^2} - h_{kn}^2 \right\}. \tag{8}$$

Using the definitions in (7), we have (Exercise 11)

$$E_0 \log p_n = \frac{1}{2} \sum_{k=1}^{n} \left\{ -\log \sigma_{kn}^2 - \frac{1}{\sigma_{kn}^2} + 1 - \left(\frac{m_{kn}}{\sigma_{kn}} \right)^2 \right\},$$

$$\text{var}_0(\log p_n) = \frac{1}{2} \sum_{k=1}^{n} \frac{(1-\sigma_{kn}^2)^2 + 2m_{kn}^2}{\sigma_{kn}^4},$$

$$E_1 \log p_n = \frac{1}{2} \sum_{k=1}^{n} (-\log \sigma_{kn}^2 + \sigma_{kn}^2 - 1 + m_{kn}^2)$$

and

$$\operatorname{var}_1(\log p_n) = \frac{1}{2} \sum_{k=1}^{n} \{(1 - \sigma_{kn}^2)^2 + 2\sigma_{kn}^2 m_{kn}^2\}. \tag{9}$$

Define the entropy distance between the measures P_0 and P_1 based on $h_1, \ldots, h_n$ by

$$\begin{aligned} r_n &= -E_0 \log p_n + E_1 \log p_n \\ &= \frac{1}{2} \sum_{k=1}^{n} \left(\sigma_{kn}^2 + \frac{1}{\sigma_{kn}^2} - 2 + m_{kn}^2 + \frac{m_{kn}^2}{\sigma_{kn}^2} \right). \end{aligned} \tag{10}$$

The quantity $E_1 \log p_n$ is known as the Kullback divergence of P_0 from P_1 based on $h_1, \ldots, h_n$, so that r_n is a symmetrized Kullback divergence. Section 4.4 discusses an interesting connection between Kullback divergence and prediction. Now $E_1 \log p_n$ is monotonically increasing in n (Exercise 12) and hence so is $-E_0 \log p_n$. Thus, r_n is monotonically increasing in n, so it tends to a limit, possibly infinite.

Lemma 2. *If $r_n \to \infty$, then $P_0 \perp P_1$.*

Proof. (Ibragimov and Rozanov 1978, p. 76). From (10) and the monotonicity of r_n, either $\inf_{k,n} \sigma_{kn}^2 = 0$ or $\sup_{k,n} \sigma_{kn}^2 = \infty$ implies both that $r_n \to \infty$ and, from (2), $P_0 \perp P_1$. Thus, from now on, suppose

$$\sigma_{kn}^2 \asymp 1, \tag{11}$$

so that

$$\log \sigma_{kn}^2 + \sigma_{kn}^{-2} - 1 \asymp -\log \sigma_{kn}^2 + \sigma_{kn}^2 - 1 \asymp (1 - \sigma_{kn}^2)^2$$

and

$$\begin{aligned} -E_0 \log p_n &\asymp E_1 \log p_n \asymp \operatorname{var}_0(\log p_n) \asymp \operatorname{var}_1(\log p_n) \\ &\asymp r_n \asymp \sum_{k=1}^{n} \{(1 - \sigma_{kn}^2)^2 + m_{kn}^2\}. \end{aligned} \tag{12}$$

Define the event

$$A_n = \left\{ \log p_n - E_0 \log p_n \geq \tfrac{1}{2} r_n \right\}.$$

By Chebyshev's inequality,

$$P_0(A_n) \leq \frac{4 \operatorname{var}_0(\log p_n)}{r_n^2} \asymp r_n^{-1} \to 0$$

and

$$\begin{aligned}P_1(A_n) &= 1 - P_1\left(-\log p_n + E_0 \log p_n \geq -\tfrac{1}{2} r_n\right) \\ &= 1 - P_1\left(-\log p_n + E_1 \log p_n \geq \tfrac{1}{2} r_n\right) \\ &\geq 1 - \frac{4\,\mathrm{var}_1(\log p_n)}{r_n^2} \rightarrow 1.\end{aligned}$$

Thus $P_0 \perp P_1$. □

Lemma 3. *If* $r_n \rightarrow r < \infty$ *then* $P_0 \equiv P_1$.

PROOF. (Ibragimov and Rozanov 1978, p. 77). Suppose there exists $A \in \mathcal{F}$ such that $P_0(A) = 0$ and $P_1(A) > 0$. Let $P_2 = P_0 + P_1$. Then there exists a sequence of events $A_1, A_2, \ldots$ such that A_n is measurable with respect to the σ-field generated by $h_1, \ldots, h_n$ and $P_2(A \circ A_n) \rightarrow 0$ as $n \rightarrow \infty$, where $\circ$ indicates symmetric difference (Exercise 13). Thus,

$$P_0(A \circ A_n) \rightarrow 0 \quad \text{and} \quad P_1(A \circ A_n) \rightarrow 0 \quad \text{as} \quad n \rightarrow \infty. \tag{13}$$

Consider $\mathcal{F}_n' = \{\varnothing, \Omega, A_n, A_n^c\}$. For $\omega \in \Omega$, define

$$X_n(\omega) = E_1\left\{p_n(\omega)^{-1} \mid \mathcal{F}_n'\right\} = \begin{cases} \dfrac{P_0(A_n)}{P_1(A_n)} & \text{for } \omega \in A_n \\[2ex] \dfrac{1 - P_0(A_n)}{1 - P_1(A_n)} & \text{for } \omega \in A_n^c. \end{cases}$$

Then

$$E_1 \log X_n = P_1(A_n) \log \frac{P_0(A_n)}{P_1(A_n)} + \{1 - P_1(A_n)\} \log \frac{1 - P_0(A_n)}{1 - P_1(A_n)}.$$

By (13), $P_1(A_n) \rightarrow P_1(A) > 0$ and $P_0(A_n) \rightarrow P_0(A) = 0$ so $-E_1 \log X_n \rightarrow \infty$. By Exercise 12, $-E_1 \log X_n \leq -E_1 \log p_n^{-1} \leq r_n$ so that $r_n \rightarrow \infty$, which yields a contradiction. Hence, we cannot have $P_0(A) = 0$ and $P_1(A) > 0$. Similarly, there cannot exist $A \in \mathcal{F}$ with $P_1(A) = 0$ and $P_0(A) > 0$. The lemma follows. □

Combining these two lemmas yields the following.

Theorem 4. P_0 *and* P_1 *are either equivalent or orthogonal and are orthogonal if and only if* $r_n \rightarrow \infty$.

As the following corollary notes, we can determine the equivalence or orthogonality of P_0 and P_1 by first considering the equivalence or orthogonality of $G(0, K_0)$ and $G(m_1, K_0)$ and then that of $G(0, K_0)$ and $G(0, K_1)$.

Corollary 5. $G(0, K_0) \equiv G(m_1, K_1)$ *if and only if* $G(0, K_0) \equiv G(m_1, K_0)$ *and* $G(0, K_0) \equiv G(0, K_1)$.

PROOF. If (11) is false, then $G(0, K_0) \perp G(m_1, K_1)$ and $G(0, K_0) \perp G(0, K_1)$, so the corollary holds in this case. If (11) is true, then from

(12), r_n tends to a finite limit if and only if both $\sum_{k=1}^n \left(1-\sigma_{kn}^2\right)^2$ and $\sum_{k=1}^n m_{kn}^2$ are bounded in n. □

Determining equivalence or orthogonality for periodic random fields

Quite a bit is known about how to establish equivalence of Gaussian measures for processes possessing an autocovariance function; see Ibragimov and Rozanov (1978, Chapter III) for processes on $\mathbb{R}$ and Yadrenko (1983) for random fields. The proofs in these works are rather technical. By restricting attention to periodic processes, the problem of establishing equivalence or orthogonality of Gaussian measures becomes straightforward. We prove results in this simple case and then just state the corresponding results for nonperiodic processes.

A process with period 2π in each coordinate will have a spectrum with support on $\mathbb{Z}^d$. For $\mathbf{j} \in \mathbb{Z}^d$, take $\mathbf{j} > \mathbf{0}$ to mean $\mathbf{j} \neq \mathbf{0}$ and the first nonzero component of $\mathbf{j}$ is positive. Consider the Gaussian random field on $\mathbb{R}^d$ with spectral representation

$$Z(\mathbf{x}) = X(\mathbf{0}) + \sum_{\mathbf{j}>\mathbf{0}} \left\{X(\mathbf{j})\cos(\mathbf{j}^T\mathbf{x}) + Y(\mathbf{j})\sin(\mathbf{j}^T\mathbf{x})\right\}, \tag{14}$$

where the $X(\mathbf{j})$s and $Y(\mathbf{j})$s are independent Gaussian random variables with $EX(\mathbf{j}) = \mu(\mathbf{j})$, $EY(\mathbf{j}) = \nu(\mathbf{j})$, $\mathrm{var}\{X(\mathbf{0})\} = f(\mathbf{0})$ and $\mathrm{var}\{X(\mathbf{j})\} = \mathrm{var}\{Y(\mathbf{j})\} = 2f(\mathbf{j})$ for $\mathbf{j} > \mathbf{0}$. Then $EZ(\mathbf{x}) = \mu(\mathbf{0}) + \sum_{\mathbf{j}>\mathbf{0}}\{\mu(\mathbf{j})\cos(\mathbf{j}^T\mathbf{x}) + \nu(\mathbf{j})\sin(\mathbf{j}^T\mathbf{x})\}$ and

$$K(\mathbf{x}-\mathbf{y}) = \mathrm{cov}\left\{Z(\mathbf{x}), Z(\mathbf{y})\right\} = \sum_{\mathbf{j}\in\mathbb{Z}^d} f(\mathbf{j})\cos\left\{\mathbf{j}^T(\mathbf{x}-\mathbf{y})\right\} \tag{15}$$

if we set $f(-\mathbf{j}) = f(\mathbf{j})$. Under these conditions, for the sum (14) to exist as an L^2 limit of finite sums for all $\mathbf{x}$, it is necessary and sufficient that $\sum f(\mathbf{j}) < \infty$ and $\sum_{\mathbf{j}>\mathbf{0}}\{\mu(\mathbf{j})^2 + \nu(\mathbf{j})^2\} < \infty$. Indeed, by a relatively simple version of Bochner's Theorem, it is not difficult to show that a function K from $\mathbb{R}^d$ to the reals is a continuous positive definite function on $\mathbb{R}^d$ with period 2π in each coordinate if and only if it can be written as in (15) with all $f(\mathbf{j})$s nonnegative, f even and $\sum f(\mathbf{j}) < \infty$. We see that f is the spectral density of the process with respect to counting measure on the integer lattice.

The explicit representation in (14) of a periodic random field as a sum of independent random variables makes it relatively easy to study. In particular, for two such Gaussian measures with mean functions m_0 and m_1 having period 2π in each coordinate and autocovariance functions defined by spectral densities f_0 and f_1 on $\mathbb{Z}^d$, it is a simple matter to determine their equivalence or orthogonality. First, as previously noted, we can assume without loss of generality that the mean under P_0 is 0 and let $\mu(\mathbf{j})$

and $\nu(\mathbf{j})$ be the Fourier coefficients for the mean under P_1. Letting $\mathbf{j}_1, \mathbf{j}_2, \ldots$ be some listing of those $\mathbf{j}$ in $\mathbb{Z}^d$ for which $\mathbf{j} > \mathbf{0}$, the sequence of random variables $X(\mathbf{0}), Y(\mathbf{j}_1), X(\mathbf{j}_1), Y(\mathbf{j}_2), X(\mathbf{j}_2), \ldots$ forms a basis for the Hilbert space of random variables generated by $Z(\mathbf{x})$ for $\mathbf{x} \in (0, 2\pi]^d$ under the inner product defined by f_0. Defining r_n as in (10) and $r = \lim_{n\to\infty} r_n$, we get

$$r = \sum_{\mathbf{j}\in\mathbb{Z}^d} \left[\frac{\{f_1(\mathbf{j}) - f_0(\mathbf{j})\}^2}{f_0(\mathbf{j})f_1(\mathbf{j})} + \{\mu(\mathbf{j})^2 + \nu(\mathbf{j})^2\} \left\{ \frac{1}{f_0(\mathbf{j})} + \frac{1}{f_1(\mathbf{j})} \right\} \right],$$

where μ and ν are taken to be even functions and $\nu(\mathbf{0}) = 0$. This definition is appropriate even when some $f_i(\mathbf{j})$s are 0 as long as 0/0 is defined to be 0 and a positive number over 0 is defined to be ∞. By Theorem 4, P_0 and P_1 are equivalent if r is finite and are otherwise orthogonal. If $f_0(\mathbf{j}) \asymp f_1(\mathbf{j})$, which is necessary for equivalence, then $r < \infty$ if and only if

$$\sum_{\mathbf{j}\in\mathbb{Z}^d} \left[\left\{ \frac{f_1(\mathbf{j}) - f_0(\mathbf{j})}{f_0(\mathbf{j})} \right\}^2 + \frac{\mu(\mathbf{j})^2 + \nu(\mathbf{j})^2}{f_0(\mathbf{j})} \right] < \infty. \tag{16}$$

In one dimension, we can rewrite (16) in terms of the mean and autocovariance functions when $f_0(j) \asymp f_1(j) \asymp (1 + j^2)^{-p}$ for $j \in \mathbb{Z}$ and some positive integer p. First consider the case where the autocovariance functions are equal. Set $P_0 = G_{(0,2\pi]}(0, K)$ and $P_1 = G_{(0,2\pi]}(m_1, K)$ where $K(t) = \sum_{j=-\infty}^{\infty} f(j)\cos(jt)$ and $m_1(t) = \mu_0 + \sum_{j=1}^{\infty}\{\mu(j)\cos(jt) + \nu(j)\sin(jt)\}$. Then $\sum_{j=0}^{\infty}\{\mu(j)^2 + \nu(j)^2\}/f(j)$ is finite if and only if $\sum_{j=0}^{\infty}\{\mu(j)^2 + \nu(j)^2\}j^{2p}$ is, which in turn is equivalent to $m_1^{(p-1)}$ existing and being absolutely continuous on $\mathbb{R}$ with almost everywhere derivative $m_1^{(p)}$ satisfying

$$\int_0^{2\pi} \left\{m_1^{(p)}(t)\right\}^2 dt < \infty \tag{17}$$

(Exercise 14). Now consider the case where the means are both 0 and define $k(t) = K_0(t) - K_1(t)$. Then (16) holds if and only if $\sum_{j\in\mathbb{Z}}\{f_1(j) - f_0(j)\}^2 j^{4p} < \infty$, which in turn holds if and only if $k^{(2p-1)}$ exists and is absolutely continuous on $\mathbb{R}$ with almost everywhere derivative $k^{(2p)}$ satisfying (Exercise 14)

$$\int_0^{2\pi} \left\{k^{(2p)}(t)\right\}^2 dt < \infty. \tag{18}$$

Determining equivalence or orthogonality for nonperiodic random fields

The results for nonperiodic random fields possessing a spectral density with respect to Lebesgue measure look quite similar to those for periodic random fields but are considerably harder to prove. Let us first consider results in the spectral domain. Define $\mathcal{Q}^d$ to be those functions $f : \mathbb{R}^d \to \mathbb{R}$ such that

$$f(\boldsymbol{\omega}) \asymp |\phi(\boldsymbol{\omega})|^2 \text{ as } |\boldsymbol{\omega}| \to \infty \tag{19}$$

for some function ϕ that is the Fourier transform of a square-integrable function with bounded support, where $a(\boldsymbol{\omega}) \asymp b(\boldsymbol{\omega})$ as $|\boldsymbol{\omega}| \to \infty$ means there exists a finite constant A such that $a(\boldsymbol{\omega}) \asymp b(\boldsymbol{\omega})$ on $|\boldsymbol{\omega}| > A$. Then for $f_0 \in \mathcal{Q}^d$ and any bounded region $R \subset \mathbb{R}^d$, $G_R(0, K_0) \equiv G_R(0, K_1)$ if for some $C < \infty$,

$$\int_{|\boldsymbol{\omega}|>C} \left\{ \frac{f_1(\boldsymbol{\omega}) - f_0(\boldsymbol{\omega})}{f_0(\boldsymbol{\omega})} \right\}^2 d\boldsymbol{\omega} < \infty. \tag{20}$$

This result is a minor extension of Theorem 4 in Yadrenko (1983, p. 156). Moreover, Yadrenko (1983) gives nontrivial sufficient conditions for orthogonality of Gaussian measures with different spectral densities, but the results are rather messy. Ibragimov and Rozanov (1978, Theorem 17 of Chapter III) state that for a process on $\mathbb{R}$, (20) is necessary and sufficient for equivalence of the Gaussian measures on any finite interval. However, the claim of necessity is false (Exercise 15) and furthermore, appears to be unintended by the authors in light of the discussion on page 107 on conditions for orthogonality. A reasonable conjecture is that if (20) does not hold then there exists a bounded region on which the corresponding Gaussian measures are orthogonal.

The condition $f_0 \in \mathcal{Q}^d$ does not have an analogue for periodic random fields, so it is worth further scrutiny. For simplicity, I only consider processes on $\mathbb{R}$ here. Exercise 16 asks you to show that if f is a spectral density on $\mathbb{R}$ and $f(\omega) \asymp \omega^{-\alpha}$ as $\omega \to \infty$ for some $\alpha > 1$, then f satisfies (19). Now let us consider what kinds of spectral densities $\mathcal{Q}^d$ excludes. Suppose $\phi(\omega) = \int_{-\infty}^{\infty} c(t)e^{i\omega t}dt$, where c has bounded support and is not 0 on a set of positive measure. Denoting by c_2 the convolution of c with itself, we have that $|\phi|^2$ is the Fourier transform of c_2. Now c_2 has bounded support and is not 0 on a set of positive measure, so c_2 cannot be analytic on the real line. This lack of smoothness in c_2 implies that $|\phi|^2$ cannot be arbitrarily small at high frequencies. To see this, note that c square integrable implies $|\phi|^2$ is integrable, so that $c_2(t) = (2\pi)^{-1} \int_{-\infty}^{\infty} |\phi(\omega)|^2 e^{-i\omega t} d\omega$. Thus, for example, $|\phi|^2$ cannot possess a Laplace transform in a neighborhood of the origin, since that would imply that c_2 is analytic on the real line (Exercise 17). Therefore, $\mathcal{Q}^1$ excludes spectral densities such as $e^{-|\omega|}$ and $e^{-\omega^2}$.

If f_0 possesses a Laplace transform in a neighborhood of the origin, then (20) no longer implies the equivalence of the corresponding Gaussian measures. More specifically, two nonidentical stationary Gaussian measures are orthogonal on any interval of positive length if either of them has a spectral density possessing a Laplace transform in a neighborhood of the origin. This can be proven by first recalling (Exercise 16 in 2.7) that if a stationary Gaussian process Z with analytic autocovariance function is observed on $[-T, 0]$ for $T > 0$, then for any $t \in \mathbb{R}$, $Z(t)$ can be predicted without error with probability one. Next, Yaglom (1987a, p. 234) shows that the autocovariance function K of a Gaussian process observed on all of $\mathbb{R}$ can be obtained at any given $t \geq 0$ with probability one if

$$\lim_{T \to \infty} T^{-1} \int_0^T K(t)^2 dt = 0. \tag{21}$$

If the spectral density exists, then (21) follows from the Riemann–Lebesgue Lemma, which says that the Fourier transform of an integrable function tends to 0 as its argument tends to $\pm\infty$ (Stein and Weiss 1971, p. 2). Now suppose that K_0 and K_1 are two autocovariance functions, and K_0 is analytic and has a spectral density. Then for any $T > 0$, $G_T(0, K_0) \equiv G_T(0, K_1)$ if and only if $K_0 = K_1$, where $G_T(m, K) = G_{[0,T]}(m, K)$ (Ibragimov and Rozanov (1978, p. 95) or Exercise 18). The reason analytic autocovariance functions are not excluded from (16) for periodic processes is that the perfect extrapolation of the process from $[0, 2\pi]$ to $\mathbb{R}$, although possible, does not provide any new information about the autocovariance function K, and hence K cannot be reconstructed with probability one. Of course, (21) is not satisfied for a periodic process with continuous autocovariance K unless K is identically 0.

Let us next consider spectral conditions for equivalence if only the mean function is misspecified. For a closed region $R \subset \mathbb{R}^d$ and f bounded, Yadrenko (1983, p. 138) shows $G_R(0, K) \equiv G_R(m_1, K)$ if and only if m_1 can be extended to a square-integrable function on all of $\mathbb{R}^d$ whose Fourier transform $\tilde{m}_1$ satisfies

$$\int_{\mathbb{R}^d} \frac{|\tilde{m}_1(\boldsymbol{\omega})|^2}{f(\boldsymbol{\omega})} \, d\boldsymbol{\omega} < \infty.$$

If $R = \mathbb{R}^d$, then there is no need to extend m_1. Comparing this result to Theorem 13 of 3.8, if f is bounded away from 0 and ∞ on bounded sets, we see that $G_{\mathbb{R}^d}(0, K) \equiv G_{\mathbb{R}^d}(m_1, K)$ implies the uniform asymptotic optimality of pseudo-BLPs using the wrong mean function based on observations at $\delta\mathbf{j}$ for all $\mathbf{j} \in \mathbb{Z}^d$ as $\delta \downarrow 0$.

In one dimension, analogous to (17) and (18) in the periodic setting, there are results in the time domain on the equivalence of Gaussian measures. Consider Gaussian measures $G_T(0, K)$ and $G_T(m_1, K)$ for $T > 0$, where K has spectral density f. If for some positive integer p,

$$f(\omega)\omega^{2p} \asymp 1 \quad \text{as} \quad \omega \to \infty, \tag{22}$$

then $G_T(0, K) \equiv G_T(m_1, K)$ if and only if $m_1^{(p-1)}$ exists and is absolutely continuous on $[0, T]$ with almost everywhere derivative $m_1^{(p)}$ satisfying

$$\int_0^T \left\{ m_1^{(p)}(t) \right\}^2 dt < \infty \tag{23}$$

(Ibragimov and Rozanov, 1978, p. 92), which can be compared with (17). Since (22) means Z is exactly $p-1$ times mean square differentiable under $G(0, K)$, loosely speaking, we see that (23) says that if the difference of mean functions has one more derivative than the stochastic part of Z, the two models are equivalent.

Next consider Gaussian measures $G_T(0, K_0)$ and $G_T(0, K_1)$ for a stationary process Z on $\mathbb{R}$. Define $k(t) = K_0(t) - K_1(t)$. For f_0 satisfying (22), $G_T(0, K_0) \equiv G_T(0, K_1)$ if and only if $k^{(2p-1)}$ exists and is absolutely continuous on $(-T, T)$ with almost everywhere derivative $k^{(2p)}$ satisfying

$$\int_0^T \left\{k^{(2p)}(t)\right\}^2 (T-t)\, dt < \infty, \tag{24}$$

which follows from Theorems 13 and 14 of Ibragimov and Rozanov (1978, Chapter III). Note that (22) implies $K_0^{(2p-2)}$ is not differentiable at 0, so for equivalence we require that the difference between the autocovariance functions be smoother than either of them separately. If, say, $f_1(\omega) \geq f_0(\omega)$ for all ω, then we can define independent 0 mean Gaussian processes X and Y with spectral densities f_0 and $f_1 - f_0$, respectively, so that $G_T(0, K_0)$ is the law of X on $[0, T]$ and $G_T(0, K_1)$ is the law of $X + Y$ on $[0, T]$. In this case, (24) has the loose interpretation that we cannot distinguish between X and $X+Y$ if Y has one more derivative than X. Note that (24) allows us to verify the claim of the preceding section that if $K_0(t) = e^{-|t|}$ and $K_1(t) = (1-|t|)^+$, then $G_T(0, K_0) \equiv G_T(0, K_1)$ if and only if $T \leq 1$ (Exercise 19).

Measurement errors and equivalence and orthogonality

This subsection examines the equivalence and orthogonality of Gaussian measures on a sequence of observations of a random field with measurement error. Let us suppose that the observation locations are all contained in a region R and that the sequence is dense in R. The basic message is: if the variance of the measurement error is different under the two measures, then the measures are orthogonal and, if the variances are equal, then the measures are equivalent if and only if the Gaussian measures for the random field on R are equivalent.

To be more specific, for $i = 0, 1$, let $G_R(m_i, K_i)$ be two Gaussian measures for a random field Z on a region R and let $\mathcal{X} = \{\mathbf{x}_1, \mathbf{x}_2, \ldots\}$ be a sequence of points in R. For $j = 1, 2, \ldots$, define $Y_j = Z(\mathbf{x}_j) + \epsilon_j$, where under model i, $\epsilon_1, \epsilon_2, \ldots$ are independent of Z and independent and identically distributed $N(0, \sigma_i^2)$. Thus, under model i, the distribution of $(Y_1, Y_2, \ldots, Y_n)^T$ is $N(E_i \mathbf{Z}_n, \operatorname{cov}_i(\mathbf{Z}_n, \mathbf{Z}_n^T) + \sigma_i^2 \mathbf{I})$, where $\mathbf{Z}_n = (Z(\mathbf{x}_1), \ldots, Z(\mathbf{x}_n))^T$. Write $G_{\mathcal{X}}(m_i, K_i, \sigma_i^2)$ for the probability measure of $Y_1, Y_2, \ldots$ under model i.

Theorem 6. *Suppose all points in R are limit points of R, Z is mean square continuous on R under $G_R(m_0, K_0)$ and that $\mathcal{X}$ is a dense sequence of points in R. If $\sigma_0^2 \neq \sigma_1^2$, then $G_{\mathcal{X}}(m_0, K_0, \sigma_0^2) \perp G_{\mathcal{X}}(m_1, K_1, \sigma_1^2)$. If $\sigma_0^2 = \sigma_1^2$, then $G_{\mathcal{X}}(m_0, K_0, \sigma_0^2) \equiv G_{\mathcal{X}}(m_1, K_1, \sigma_1^2)$ if and only if $G_R(m_0, K_0) \equiv G_R(m_1, K_1)$.*

PROOF. I just provide an outline of a proof, leaving the details as a series of exercises. The key point to the proof is to show that $\mathcal{H}_R(m_0, K_0)$ is contained in the Hilbert space generated by $Y_1, Y_2, \ldots$ (Exercise 21). This result can be proven by noting that for any $\mathbf{x} \in R$, we can find $a_n \downarrow 0$ such that $d(\mathbf{x}, n) = \sum_{j=1}^{n} 1\{|\mathbf{x} - \mathbf{x}_j| \leq a_n\} \to \infty$ as $n \to \infty$, so that

$$\frac{1}{d(\mathbf{x}, n)} \sum_{j=1}^{n} 1\{|\mathbf{x} - \mathbf{x}_j| \leq a_n\} Y_j \xrightarrow{L^2} Z(\mathbf{x}) \tag{25}$$

as $n \to \infty$ under $G_R(m_0, K_0)$. It follows that if Z is not mean square continuous on R under $G_R(m_1, K_1)$, then $G_R(m_0, K_0) \perp G_R(m_1, K_1)$ and $G_{\mathcal{X}}(m_0, K_0, \sigma_0^2) \perp G_{\mathcal{X}}(m_1, K_1, \sigma_1^2)$, whether or not $\sigma_0^2 = \sigma_1^2$ (Exercise 22), so let us now suppose Z is mean square continuous on R under either measure. We can then correctly recover the measurement error variance with probability 1 under either model (Exercise 23), so that $G_{\mathcal{X}}(m_0, K_0, \sigma_0^2) \perp G_{\mathcal{X}}(m_1, K_1, \sigma_1^2)$ if $\sigma_0^2 \neq \sigma_1^2$. If $\sigma_0^2 = \sigma_1^2$ and $G_R(m_0, K_0) \equiv G_R(m_1, K_1)$, then the two Gaussian measures on the σ-algebra generated by $\{Z(\mathbf{x}) : \mathbf{x} \in R\}$ and $\epsilon_1, \epsilon_2, \ldots$ are equivalent. In addition, $Y_1, Y_2, \ldots$ are measurable on this σ-algebra, so that $G_{\mathcal{X}}(m_0, K_0, \sigma_0^2) \equiv G_{\mathcal{X}}(m_1, K_1, \sigma_1^2)$ (Exercise 24). Finally, if $\sigma_0^2 = \sigma_1^2$ but $G_R(m_0, K_0) \perp G_R(m_1, K_1)$, then the fact that $\mathcal{H}_R(m_i, K_i)$ is contained in the Hilbert space generated by $Y_1, Y_2, \ldots$ under $G_{\mathcal{X}}(m_i, K_i, \sigma_i^2)$ implies $G_{\mathcal{X}}(m_0, K_0, \sigma_0^2) \perp G_{\mathcal{X}}(m_1, K_1, \sigma_1^2)$ (Exercise 24). □

Proof of Theorem 1

We now return to proving Theorem 1, stated earlier in this section. Let $\tilde{F}$ be the spectral distribution function for the process, so that if F is the spectral measure, $\tilde{F}(\omega) = F((-\infty, \omega])$. Recalling that $A_d(r) = (-\pi r, \pi r]^d$, then

$$\begin{aligned} 2K\left(\frac{j}{n}\right) &- K\left(\frac{j+1}{n}\right) - K\left(\frac{j-1}{n}\right) \\ &= 2\int_{-\infty}^{\infty} e^{i\omega j/n}\left(1 - \cos\frac{\omega}{n}\right) d\tilde{F}(\omega) \\ &= 2\int_{A_1(n)} e^{i\omega j/n}\left(1 - \cos\frac{\omega}{n}\right) d\tilde{F}_n(\omega), \end{aligned}$$

where $\tilde{F}_n(\omega) = \sum_{j=-\infty}^{\infty} F\left((\pi n(2j-1), 2\pi jn + \omega]\right)$ for $\omega \in A_1(n)$ and the integrals are interpreted in the Riemann–Stieltjes sense. Using F symmetric about 0, it follows that

$$\sum_{j=1}^{n-1}(n-j)\left\{2K\left(\frac{j}{n}\right)-K\left(\frac{j+1}{n}\right)-K\left(\frac{j-1}{n}\right)\right\}^2 \tag{26}$$
$$= 4\int_{A_2(n)}\left(1-\cos\frac{\omega}{n}\right)\left(1-\cos\frac{\nu}{n}\right)\sum_{j=1}^{n-1}(n-j)e^{i(\omega-\nu)j/n}d\tilde{F}_n(\omega)\,d\tilde{F}_n(\nu).$$

From (5), it suffices to show (26) tends to 0 as $n \to \infty$. First,

$$\left|\sum_{j=1}^{n-1}(n-j)e^{i(\omega-\nu)j/n}\right|$$
$$= \left|\frac{ne^{i(\omega-\nu)/n}\{1-e^{i(\omega-\nu)/n}\}-e^{i(\omega-\nu)/n}\{1-e^{i(\omega-\nu)}\}}{\{1-e^{i(\omega-\nu)/n}\}^2}\right|$$
$$\ll \frac{n^2}{1+n\sin\left|\frac{\omega-\nu}{2n}\right|} \tag{27}$$

for $\nu, \omega \in A_1(n)$ (Exercise 5). Thus,

$$\sum_{j=1}^{n-1}(n-j)\left\{2K\left(\frac{j}{n}\right)-K\left(\frac{j+1}{n}\right)-K\left(\frac{j-1}{n}\right)\right\}^2$$
$$\ll \frac{1}{n^2}\int_{A_2(n)}\frac{\omega^2\nu^2}{1+n\sin\left|\frac{\omega-\nu}{2n}\right|}\,d\tilde{F}_n(\omega)\,d\tilde{F}_n(\nu).$$

Suppose $\{c_n\}$ and $\{d_n\}$ are positive sequences such that $c_n \to \infty$, $c_n = o(n^{1/2})$, $d_n \to \infty$ and $d_n/c_n \to 0$ as $n \to \infty$. Divide $A_2(n)$ into three regions: $R_1 = A_2(c_n)$; R_2, the part of $A_2(n)\backslash R_1$ for which $n\sin|(\omega-\nu)/(2n)| < d_n$; and R_3, the rest of $A_2(n)$ (see Figure 1). Now, $c_n = o(n^{1/2})$ implies

$$\frac{1}{n^2}\int_{R_1}\frac{\omega^2\nu^2}{1+n\sin\left|\frac{\omega-\nu}{2n}\right|}\,d\tilde{F}_n(\omega)\,d\tilde{F}_n(\nu) \to 0$$

as $n \to \infty$. By Pitman's Tauberian theorem (Theorem 4 of 2.8), $\tilde{F}(-t) \sim C/t$ as $t \to \infty$ for a positive constant C. Thus, defining $H_n(\omega) = \tilde{F}_n(\pi n) - \tilde{F}_n(\omega)$, we have $H_n(\omega) \ll 1/\omega$ for $0 < \omega \le \pi n$ (Exercise 6). For $\alpha > 1$,

$$\int_{(0,\pi n]}\omega^\alpha d\tilde{F}_n(\omega) = -\omega^\alpha H_n(\omega)\Big|_0^{\pi n} + \alpha\int_0^{\pi n}\omega^{\alpha-1}H_n(\omega)\,d\omega$$
$$\ll n^{\alpha-1}, \tag{28}$$

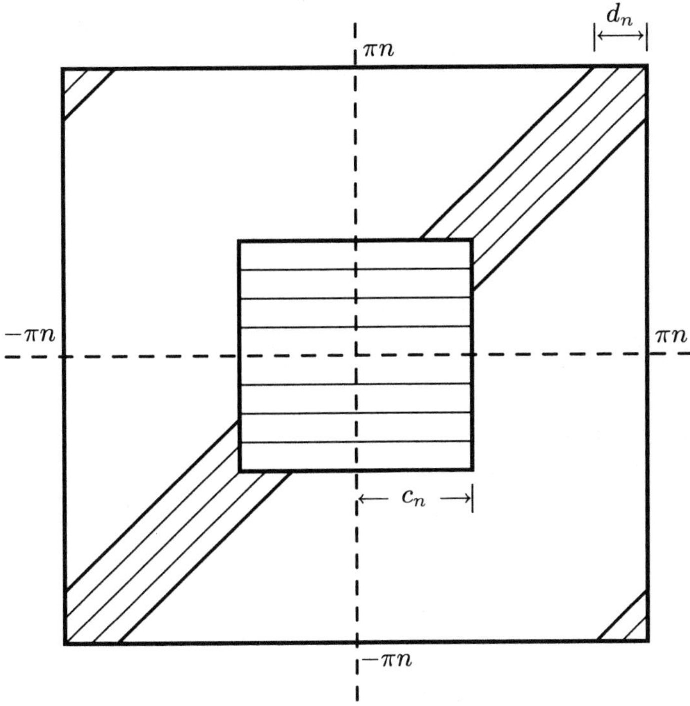

FIGURE 1. Regions of integration in proof of Theorem 1. Horizontal stripes indicate R_1, diagonal stripes indicate R_2 and unstriped area is R_3.

so that

$$\begin{aligned}\frac{1}{n^2}\int_{R_3}&\frac{\omega^2\nu^2}{1+n\sin\left|\frac{\omega-\nu}{2n}\right|}\,d\tilde{F}_n(\omega)\,d\tilde{F}_n(\nu)\\&\ll\frac{1}{n^2 d_n}\int_{R_3}\omega^2\nu^2 d\tilde{F}_n(\omega)\,d\tilde{F}_n(\nu)\\&\ll\frac{1}{n^2 d_n}\left\{\int_{(0,\pi n]}\omega^2 d\tilde{F}_n(\omega)\right\}^2\to 0\end{aligned}$$

as $n\to\infty$, since $d_n\to\infty$. Finally,

$$\begin{aligned}\frac{1}{n^2}\int_{R_2}&\frac{\omega^2\nu^2}{1+n\sin\left|\frac{\omega-\nu}{2n}\right|}\,d\tilde{F}_n(\omega)\,d\tilde{F}_n(\nu)\\&\ll\frac{1}{n^2}\int_{R_2}\nu^4 d\tilde{F}_n(\omega)\,d\tilde{F}_n(\nu)\\&\le\frac{2}{n^2}\int_{(c_n-d_n,\pi n]}\nu^4\left\{\tilde{F}_n(\nu+d_n)-\tilde{F}_n(\nu-d_n)\right\}d\tilde{F}_n(\nu),\end{aligned}$$

where if $\nu + d_n \geq \pi n$, $\tilde{F}_n(\nu + d_n) = \tilde{F}_n(\pi n) + \tilde{F}_n(\nu + d_n - 2\pi n)$. Now, for any positive integer j_0 and all $\nu \in (c_n - d_n, \pi n]$,

$$\nu\{\tilde{F}_n(\nu + d_n) - \tilde{F}_n(\nu - d_n)\} \tag{29}$$
$$\leq \nu \sum_{j=-j_0}^{j_0} \{\tilde{F}_n(2\pi jn + \nu + d_n) - \tilde{F}_n(2\pi jn + \nu - d_n)\} + 2\pi n\tilde{F}(-2\pi j_0 n).$$

Since $\tilde{F}(-t) \sim C/t$ as $t \to \infty$, given $\epsilon > 0$, we can choose j_0 such that $2\pi n\tilde{F}(-2\pi j_0 n) < \epsilon$ for all n sufficiently large. Again using Pitman's Tauberian theorem, it is possible to show that the first term on the right side of (29) tends to 0 uniformly for all $\nu \in (c_n - d_n, \pi n]$ (Exercise 7). Since ϵ is arbitrary, it follows that $\nu\{\tilde{F}_n(\nu + d_n) - \tilde{F}_n(\nu - d_n)\} \to 0$ uniformly for all $\nu \in (c_n - d_n, \pi n]$ (Exercise 7), so by (28), the contribution to (26) from R_2 tends to 0 as $n \to \infty$. Since the first term on the right side of (5) is $O(n^{-1})$, the theorem follows. Exercise 25 outlines a time-domain proof of Theorem 1.

Exercises

1 Prove (1).

2 Use (1) to show that (2) or (3) imply $P_0 \perp P_1$.

3 Suppose U_n is defined as in (4) and $K(t) = C - D|t| + o(|t|)$ as $t \to 0$ for some $D > 0$. If K has a bounded second derivative on $(0, 1]$ then $EU_n = D + O(n^{-1})$ and $\operatorname{var} U_n = 2D^2n^{-1} + O(n^{-2})$. If, in addition, K'' is continuous on $(0, 1]$, show that $EU_n = D + \alpha n^{-1} + o(n^{-1})$ and $\operatorname{var} U_n = 2D^2n^{-1} + \beta n^{-2} + o(n^{-2})$ as $n \to \infty$ and give explicit expressions for α and β.

4 Use Theorem 1 to show that if $K_j(t) = C_j - D_j|t| + o(|t|)$ as $t \to 0$ for $D_j > 0$ for $j = 0, 1$ with $D_0 \neq D_1$, then $G_R(0, K_0) \perp G_R(0, K_1)$ on any interval of positive length.

5 Verify (27).

6 Verify the claim in the proof of Theorem 1 that $H_n(\omega) \ll \omega^{-1}$ for $0 < \omega \leq \pi n$.

7 In the proof of Theorem 1, show that the first term on the right side of (29) tends to 0 uniformly for all $\nu \in (c_n - d_n, \pi n]$. Show that it follows that $\nu\{\tilde{F}_n(\nu + d_n) - \tilde{F}_n(\nu - d_n)\} \to 0$ as $n \to \infty$ uniformly for all $\nu \in (c_n - d_n, \pi n]$.

8 For $K(t) = \left(\frac{1}{2} - |t|\right)^+$, show that for U_n as given in (4), $\operatorname{var} U_{2n} \sim \alpha n^{-1}$ and $\operatorname{var} U_{2n+1} \sim \beta n^{-1}$ as $n \to \infty$. Find α and β.

9 For W_n as defined in (6) and Z a Gaussian process, show that as $n \to \infty$, $E_0 W_n \to -1$, $E_1 W_n \to 0$, $\operatorname{var}_0 W_n \to 0$ and $\operatorname{var}_1 W_n \to 0$ for $R = [0, 2]$, $m_0 = m_1 = 0$, $K_0(t) = (1 - |t|)^+$ and $K_1(t) = e^{-|t|}$.

10 Show that for a mean square continuous random field Z on $\mathbb{R}^d$, the closed real linear manifold of $Z(\mathbf{x})$ for $\mathbf{x} \in \mathbb{R}^d$ is a separable Hilbert space.

11 Verify the expressions for $\log p_n$ in (8) and for the mean and variance of $\log p_n$ under P_0 and P_1 in (9).

12 Consider probability spaces $(\Omega, \mathcal{F}, P_0)$ and $(\Omega, \mathcal{F}, P_1)$ with P_1 absolutely continuous with respect to P_0 and let p be the Radon–Nikodym derivative of P_1 with respect to P_0. If $\mathcal{F}' \subset \mathcal{F}$ is a σ-algebra on Ω, show that

$$E_0[\log\{E_0(p^{-1}|\mathcal{F}')\}] \leq \log\{E_0(p^{-1})\}.$$

Note that to apply this result in the proof of Lemma 3, we should take $\Omega = \mathbb{R}^n$, $\mathcal{F}$ the Borel sets on $\mathbb{R}^n$ and $\mathcal{F}'$ to be $\mathcal{F}'_n$ as defined in Lemma 3.

13 In the proof of Lemma 3, verify that there exists a sequence of events $A_1, A_2, \ldots$ where A_n is measurable with respect to the σ-field generated by $h_1, \ldots, h_n$ such that $P_2(A \circ A_n) \to 0$ as $n \to \infty$.

14 Consider a Gaussian process on $\mathbb{R}$ with period 2π. For a positive integer p, suppose $f_0(j) \asymp f_1(j) \asymp (1 + j^2)^{-p}$ for $j \in \mathbb{Z}$. Show that $G_{\mathbb{R}}(0, K_0) = G_{\mathbb{R}}(m, K_0)$ if and only if $m^{(p-1)}$ exists and is absolutely continuous and has almost everywhere derivative satisfying (17). In addition, for $k = K_0 - K_1$, show that $G_{\mathbb{R}}(0, K_0) = G_{\mathbb{R}}(0, K_1)$ if and only if $k^{(2p-1)}$ exists and is absolutely continuous and has almost everywhere derivative satisfying (18).

15 Show by example that for $d = 1$ the integral in (20) can be infinite and yet $G_T(0, K_0) \equiv G_T(0, K_1)$ for some positive T.

16 Show that if f is a spectral density on $\mathbb{R}$ and $f(\omega) \asymp \omega^{-\alpha}$ as $\omega \to \infty$ for some $\alpha > 1$, then f satisfies (19). This result is (4.31) in Chapter III of Ibragimov and Rozanov (1978).

17 Show that if a function on $\mathbb{R}$ possesses a Laplace transform in a neighborhood of the origin, then its Fourier transform is analytic on the real line.

18 Applying (21), provide the details showing that if K_0 and K_1 are two autocovariance functions on $\mathbb{R}$, K_0 is analytic and has a spectral density, then $G_T(0, K_0) \equiv G_T(0, K_1)$ for $T > 0$ if and only if $K_0 = K_1$.

19 If $K_0(t) = e^{-|t|}$ and $K_1(t) = (1 - |t|)^+$, show that (24) implies that $G_T(0, K_0) \equiv G_T(0, K_1)$ if and only if $T \leq 1$.

20 If $T > 0$, $K_0(t) = e^{-|t|}$ and in a neighborhood of the origin, $K_1(t) = K_1(0) - |t| + D|t|^\gamma + o(|t|^\gamma)$ as $t \to 0$ for $\gamma \in (1, \frac{3}{2}]$ and some $D \neq 0$, show that $G_T(0, K_0) \perp G_T(0, K_1)$.

In Exercises 21–24, assume that the conditions of Theorem 6 hold.

21 Show that (25) holds and hence that $\mathcal{H}_R(m_0, K_0)$ is contained in the Hilbert space generated by $Y_1, Y_2, \ldots$.

22 If Z is not mean square continuous on R under $G_R(m_1, K_1)$, show that $G_R(m_0, K_0) \perp G_R(m_1, K_1)$ and $G_{\mathcal{X}}(m_0, K_0, \sigma_0^2) \perp G_{\mathcal{X}}(m_1, K_1, \sigma_1^2)$.

23 Show that it is possible to recover σ_0^2 with probability 1 from $Y_1, Y_2, \ldots$ under $G_{\mathcal{X}}(m_0, K_0, \sigma_0^2)$. Hence, show $G_{\mathcal{X}}(m_0, K_0, \sigma_0^2) \perp G_{\mathcal{X}}(m_1, K_1, \sigma_1^2)$ if $\sigma_0^2 \neq \sigma_1^2$.

24 Suppose $G_R(m_0, K_0) \equiv G_R(m_1, K_1)$. Prove that $G_{\mathcal{X}}(m_0, K_0, \sigma^2) \equiv G_{\mathcal{X}}(m_1, K_1, \sigma^2)$. Suppose $G_R(m_0, K_0) \perp G_R(m_1, K_1)$ and Z is mean square continuous on R under both models. Prove that $G_{\mathcal{X}}(m_0, K_0, \sigma^2)$ is orthogonal to $G_{\mathcal{X}}(m_1, K_1, \sigma^2)$.

25 Prove Theorem 1 by completing the following argument. Suppose for $n = 1, 2, \ldots$ that $X_{1n}, \ldots, X_{nn}$ is a sequence of random variables satisfying $\operatorname{var} X_{in} = 1$ for all i and n, $\operatorname{cov}(X_{in}, X_{jn})$ depends only on $i - j$ and n for all $i, j \leq n$ and all n, and for any fixed $j > 1$, $\operatorname{cov}(X_{1n}, X_{jn}) \to 0$ as $n \to \infty$. Let $\overline{X}_n = n^{-1} \sum_{j=1}^{n} X_{jn}$. The plan is to prove $\operatorname{var} \overline{X}_n \to 0$ as $n \to \infty$ and then to show that this result implies Theorem 1.

i. Given $\epsilon > 0$ and j finite, choose $n_0 > j$ such that for all $n \geq n_0$ and $1 < k \leq j$, $\operatorname{cov}(X_{1n}, X_{kn}) < \epsilon/j$. Let $\mathbf{\Sigma}_n$ be the covariance matrix of $(X_{1n}, \ldots, X_{jn})$. Using Theorem 8.1.3 of Golub and Van Loan (1996), show that for all $n \geq n_0$, the eigenvalues of $\mathbf{\Sigma}_n$ are all in $[1 - \epsilon, 1 + \epsilon]$.

ii. Let $\boldsymbol{\sigma}_{kn} = \operatorname{cov}\left((X_{kn}, \ldots, X_{k+j-1,n})^T, X_{1n}\right)$. Prove $\boldsymbol{\sigma}_{kn}^T \mathbf{\Sigma}_n^{-1} \boldsymbol{\sigma}_{kn} \leq 1$ for all $1 \leq k \leq n - j$ and all $n \geq j$.

iii. Show that $\boldsymbol{\sigma}_{kn}^T \boldsymbol{\sigma}_{kn} \leq 1/(1-\epsilon)$ for all $n \geq n_0$ and all $1 \leq k \leq n - j$.

iv. Show that

$$\sum_{j=1}^{n} \operatorname{cov}(X_{1n}, X_{jn}) \leq n^{1/2} \left\{ \sum_{j=1}^{n} \operatorname{cov}(X_{1n}, X_{jn})^2 \right\}^{1/2}.$$

v. Show that

$$\sum_{j=1}^{n} \operatorname{cov}(X_{1n}, X_{jn})^2 \leq \left(\frac{n}{j} + 1 \right) \frac{1}{1 - \epsilon}.$$

vi. Show that $\operatorname{var} \overline{X}_n \to 0$ as $n \to \infty$.

vii. Prove Theorem 1.

4.3 Applications of equivalence of Gaussian measures to linear prediction

Suppose P_0 and P_1 are equivalent Gaussian measures for a random field Z on a closed set R and let $\mathcal{H}$ be the Hilbert space generated by $Z(\mathbf{x})$ for $\mathbf{x} \in R$. Then there is only a finite amount of information in $\mathcal{H}$ for distinguishing between P_0 and P_1. Furthermore, if $\mathcal{H}$ is separable with basis $h_1, h_2, \ldots$, then for n large, most of the information for distinguishing between these measures is contained in $\mathcal{H}_n$, the Hilbert space generated by $h_1, \ldots, h_n$. Under P_0, $e_0(h,n)$ is independent of $\mathcal{H}_n$, by which I mean that $e_0(h,n)$ is independent of all elements of $\mathcal{H}_n$. If $e_0(h,n)$ were not nearly independent of $\mathcal{H}_n$ under P_1, then that would mean that h contained a nonnegligible amount of information for distinguishing P_1 from P_0 not contained in $\mathcal{H}_n$. This suggests $e_0(h,n) \approx e_1(h,n)$ when $P_0 \equiv P_1$. Furthermore, if $E_1 e_1(h,n)^2$ is not approximately $E_0 e_1(h,n)^2$, then again h would contain a nonnegligible amount of information not contained in $\mathcal{H}_n$ for distinguishing the measures, suggesting $E_1 e_1(h,n)^2 \approx E_0 e_1(h,n)^2$ when $P_0 \equiv P_1$. The goal of this section is to formalize this argument for arbitrary separable Hilbert spaces of Gaussian random variables and then apply it to obtain results on the asymptotic optimality of pseudo-BLPs under fixed-domain asymptotics.

I use the following notation throughout this section. Let $h_1, h_2, \ldots$ be a sequence of random variables that are linearly independent under $(0, K_0)$, let $\mathcal{H}^0$ be the real linear manifold of this sequence and define $\mathcal{H}(m,K)$ to be the closure of $\mathcal{H}^0$ with respect to the inner product given by (m,K). As noted in 4.2, we can take the mean under P_0 to be 0 without loss of generality and I generally do so throughout this section. Let m_1 be a real linear functional on $\mathcal{H}^0$. Take $\psi_1, \psi_2, \ldots$ to be the Gram–Schmidt orthogonalization of $h_1, h_2, \ldots$ under $(0, K_0)$ so that $K_0(\psi_j, \psi_k) = \delta_{jk}$. Define

$$b_{jk} = K_1(\psi_j, \psi_k) - \delta_{jk} \quad \text{and} \quad \mu_j = E_1 \psi_j. \tag{30}$$

As in the previous section (see (7)), let $h_{1n}, \ldots, h_{nn}$ be a linear transformation of $h_1, \ldots, h_n$ such that for $j,k = 1, \ldots, n$, $K_0(h_{kn}, h_{jn}) = \delta_{kj}$ and $K_1(h_{kn}, h_{jn}) = \sigma_{jn}^2 \delta_{kj}$ and set $m_{kn} = m_1(h_{kn})$.

The next theorem shows how to determine the equivalence or orthogonality of Gaussian measures in terms of the b_{jk}s and μ_js. It combines (2.20) and the last equation on page 78 in Ibragimov and Rozanov (1978).

Theorem 7. *Suppose* $\operatorname{var}_0 h \asymp \operatorname{var}_1 h$ *for* $h \in \mathcal{H}^0$. *Then* $G(0, K_0) \equiv G(m_1, K_1)$ *if and only if*

$$\sum_{j,k=1}^{\infty} b_{jk}^2 < \infty \tag{31}$$

and

$$\sum_{j=1}^{\infty} \mu_j^2 < \infty. \tag{32}$$

PROOF. There exists an $n \times n$ orthogonal matrix $\mathbf{A}$ such that $\mathbf{A}(\psi_1 \ldots \psi_n)^T = (h_{1n}, \ldots, h_{nn})^T$. Letting $\mathbf{K}$ be the covariance matrix under K_1 of $\psi_1, \ldots, \psi_n$, it follows that $\sigma_{1n}^2, \ldots, \sigma_{nn}^2$ are the eigenvalues of $\mathbf{K}$ and hence that $(1 - \sigma_{jn}^2)^2$ for $j = 1, \ldots, n$ are the eigenvalues of $(\mathbf{I} - \mathbf{K})^2$. Since the trace of a matrix equals the sum of the eigenvalues, we have

$$\sum_{j=1}^{n} (1 - \sigma_{jn}^2)^2 = \operatorname{tr}\{(\mathbf{I} - \mathbf{K})^2\} = \sum_{j,k=1}^{n} b_{jk}^2. \tag{33}$$

Letting $\mathbf{m} = (\mu_1, \ldots, \mu_n)^T$, then $\mathbf{A}$ orthogonal implies $\|\mathbf{Am}\|^2 = \|\mathbf{m}\|^2$, or

$$\sum_{j=1}^{n} m_{jn}^2 = \sum_{j=1}^{n} \mu_j^2. \tag{34}$$

Define $\alpha = \sum_{j,k=1}^{\infty} b_{jk}^2$ and $\beta = \sum_{j=1}^{\infty} \mu_j^2$, so that by (33) and (34), $\sum_{j=1}^{n} (1 - \sigma_{jn}^2)^2 \to \alpha$ and $\sum_{j=1}^{n} m_{jn}^2 \to \beta$ as $n \to \infty$. Theorem 4 and (12) imply that α and β are finite if and only if $G(0, K_0) \equiv G(m_1, K_1)$. □

Asymptotically optimal pseudo-BLPs

If $(0, K_0)$ is the correct second-order structure, we can apply Theorem 7 to show that pseudo-BLPs under a second-order structure (m_1, K_1) satisfying $G(0, K_0) \equiv G(m_1, K_1)$ are asymptotically optimal and the presumed mses of the pseudo-BLPs are asymptotically correct. The following result combines Theorem 3.1 and Corollary 3.1 of Stein (1990a).

Theorem 8. *Suppose* $(0, K_0)$ *and* (m_1, K_1) *are two possible second-order structures on* $\mathcal{H}^0$ *with* $G(0, K_0) \equiv G(m_1, K_1)$. *Let* $\mathcal{H}_{-n}$ *be made up of elements* h *of* $\mathcal{H}(0, K_0)$ *for which* $E_0 e_0(h, n)^2 > 0$. *Then*

$$\lim_{n \to \infty} \sup_{\psi \in \mathcal{H}_{-n}} \left| \frac{E_1 e_0(\psi, n)^2 - E_0 e_0(\psi, n)^2}{E_0 e_0(\psi, n)^2} \right| = 0, \tag{35}$$

$$\lim_{n \to \infty} \sup_{\psi \in \mathcal{H}_{-n}} \left| \frac{E_0 e_1(\psi, n)^2 - E_1 e_1(\psi, n)^2}{E_1 e_1(\psi, n)^2} \right| = 0, \tag{36}$$

$$\lim_{n \to \infty} \sup_{\psi \in \mathcal{H}_{-n}} \frac{E_0 e_1(\psi, n)^2 - E_0 e_0(\psi, n)^2}{E_0 e_0(\psi, n)^2} = 0 \tag{37}$$

and

$$\lim_{n \to \infty} \sup_{\psi \in \mathcal{H}_{-n}} \frac{E_1 e_0(\psi, n)^2 - E_1 e_1(\psi, n)^2}{E_1 e_1(\psi, n)^2} = 0. \tag{38}$$

Before proving this result, a few comments on the assumed conditions are in order. The restriction to $\mathcal{H}_{-n}$ is just to avoid dividing by 0; if we defined 0/0 to be 0, the suprema in (35)–(38) could be over $\mathcal{H}(0, K_0)$ independent of n. A more consequential condition is that the sequence of observations is also the basis for all possible predictands. This condition generally excludes "distant" extrapolations such as $h_1, h_2, \ldots$ being a dense sequence of observations of a process Z on, say, $[0, 1]$ and predicting $Z(2)$.

PROOF OF THEOREM 8. For $\psi \in \mathcal{H}(0, K_0)$, we can write $\psi = \sum_{j=1}^{\infty} c_j \psi_j$, the limit existing in L^2 under either $(0, K_0)$ or (m_1, K_1), where $\sum c_j^2 < \infty$. Then the error of the BLP for ψ given $\mathcal{H}_n$ under $(0, K_0)$ is

$$e_0(\psi, n) = \sum_{j=n+1}^{\infty} c_j \psi_j.$$

Define b_{jk} and μ_j as in (30). If $E_0 e_0(\psi, n)^2 > 0$, then as $n \to \infty$,

$$\begin{aligned}\left| \frac{E_1 e_0(\psi, n)^2 - E_0 e_0(\psi, n)^2}{E_0 e_0(\psi, n)^2} \right| &\leq \frac{\left| \sum_{j,k=n+1}^{\infty} c_j c_k b_{jk} \right| + \left(\sum_{j=n+1}^{\infty} c_j \mu_j \right)^2}{\sum_{j=n+1}^{\infty} c_j^2} \\ &\leq \left\{ \sum_{j,k=n+1}^{\infty} b_{jk}^2 \right\}^{1/2} + \sum_{j=n+1}^{\infty} \mu_j^2, \end{aligned} \tag{39}$$

by twice applying the Cauchy–Schwarz inequality. The right side of (39) does not depend on ψ and, by Theorem 7, tends to 0 as $n \to \infty$, so (35) follows. Switching the roles of K_0 and K_1 yields (36). Next, since $E_1 e_1^2 \leq E_1 e_0^2$,

$$\begin{aligned}\frac{E_0 e_1(\psi, n)^2}{E_0 e_0(\psi, n)^2} &= \frac{E_0 e_1(\psi, n)^2}{E_1 e_1(\psi, n)^2} \cdot \frac{E_1 e_1(\psi, n)^2}{E_1 e_0(\psi, n)^2} \cdot \frac{E_1 e_0(\psi, n)^2}{E_0 e_0(\psi, n)^2} \\ &\leq \frac{E_0 e_1(\psi, n)^2}{E_1 e_1(\psi, n)^2} \cdot \frac{E_1 e_0(\psi, n)^2}{E_0 e_0(\psi, n)^2},\end{aligned}$$

so (37) follows from (35) and (36). Again switching the roles of K_0 and K_1 yields (38).

Note that we have only used the properties of Gaussian measures through the result that equivalence of Gaussian measures implies (31) and (32). Thus, (35)–(38) follow from (31) and (32), whether or not the elements of $\mathcal{H}^0$ are jointly Gaussian. □

We can combine Theorems 6 and 8 to prove that if a random field Z is observed with measurement error whose variance is the same under either model, the conclusions of Theorem 8 on the behavior of pseudo-BLPs still apply.

Corollary 9. *Suppose Z is a random field on $\mathbb{R}^d$, $\mathbf{x}_1, \mathbf{x}_2, \ldots$ is a sequence of points in $R \subseteq \mathbb{R}^d$ and $Y_j = Z(\mathbf{x}_j) + \epsilon_j$ for $j = 1, 2, \ldots$ where the ϵ_js have mean 0, are uncorrelated with Z and each other and have variance σ^2 under either (m_0, K_0) or (m_1, K_1). For $\psi \in H_R(0, K_0)$, define $e_i(\psi, n)$ to be the error of the BLP of ψ based on $Y_1, \ldots, Y_n$ when Z has second-order structure (m_i, K_i). Assume all points in R are limit points of R, Z is mean square continuous under $(0, K_0)$, $G_R(0, K_0) \equiv G_R(m_1, K_1)$ and $\mathbf{x}_1, \mathbf{x}_2, \ldots$ are dense in R. Then (35)–(38) in Theorem 8 all hold where, independent of n, $\mathcal{H}_{-n}$ is defined as all nonzero elements of $\mathcal{H}_R(0, K_0)$.*

The proof is left to the reader (Exercise 26).

Observations not part of a sequence

Taking the observations to be a sequence forming a basis for the Hilbert space of possible predictands is convenient mathematically but excludes some settings of interest in fixed-domain asymptotics. For example, for $R = [0, 1]$, it excludes taking observations at j/n for $j = 0, \ldots, n$ and letting $n \to \infty$, since the observations are not nested as n increases. Furthermore, if $R = [0, T]$, it excludes taking $\mathcal{H}_n$ to be the Hilbert space generated by $Z(t)$ for $t \in [0, T - \epsilon_n]$ with $\epsilon_n \downarrow 0$ as $n \to \infty$, which was considered in Stein (1990d). The following result covers both of these settings.

Theorem 10. *Suppose $(0, K_0)$ and (m_1, K_1) are two possible second-order structures on $\mathcal{H}^0$ with $G(0, K_0) \equiv G(m_1, K_1)$. For $n = 1, 2, \ldots,$ let $\mathcal{H}_n$ be a sequence of subspaces of $\mathcal{H}(0, K_0)$ such that for any given $h \in \mathcal{H}(0, K_0)$, $E_0 e_0(h, n)^2 \to 0$ as $n \to \infty$. Then (35)–(38) hold.*

In contrast to Theorem 8, we have the additional assumption that $E_0 e_0(h, n)^2 \to 0$ as $n \to \infty$ for any $h \in \mathcal{H}(0, K_0)$. This condition is an immediate consequence of the formulation of Theorem 8 (Exercise 27), so it is unnecessary to include it as an assumption in that result. The role of this assumption in Theorem 10 is to ensure that $\mathcal{H}_n$ "approximates" $\mathcal{H}(0, K_0)$ well when n is large, so that most of the information in $\mathcal{H}(0, K_0)$ for distinguishing between $G(0, K_0)$ and $G(m_1, K_1)$ is contained in $\mathcal{H}_n$ when Z is Gaussian.

PROOF OF THEOREM 10. I provide an outline of a proof; the details are left as a series of exercises. Under the inner product defined by $(0, K_0)$, let $u_1^n, u_2^n, \ldots$ be an orthonormal basis for $\mathcal{H}_n$ of length θ_n and $v_1^n, v_2^n, \ldots$ an orthonormal basis for its orthogonal complement of length γ_n (both θ_n and γ_n may be ∞) so that the two sequences together form an orthonormal basis for $\mathcal{H}(0, K_0)$. Define $a_{jk}^n = K_1(u_j^n, u_k^n) - \delta_{jk}$, $b_{jk}^n = K_1(v_j^n, v_k^n) - \delta_{jk}$, $c_{jk}^n = K_1(u_j^n, v_k^n)$, $\nu_j^n = E_1 u_j^n$ and $\mu_j^n = E_1 v_j^n$. Let $\mathcal{H}_{-n}$ be the subset of

$\mathcal{H}(0, K_0)$ for which $E_0 e_0(\psi, n)^2 > 0$. For $\psi \in \mathcal{H}_{-n}$, similar to (39),

$$\left|\frac{E_1 e_0(\psi, n)^2 - E_0 e_0(\psi, n)^2}{E_0 e_0(\psi, n)^2}\right| \leq \left\{\sum_{j,k=1}^{\gamma_n} (b_{jk}^n)^2\right\}^{1/2} + \sum_{j=1}^{\gamma_n} (\mu_j^n)^2.$$

Defining α and β as in the proof of Theorem 7, then similar to (33) and (34),

$$\sum_{j,k=1}^{\theta_n} (a_{jk}^n)^2 + 2\sum_{j=1}^{\theta_n}\sum_{k=1}^{\gamma_n} (c_{jk}^n)^2 + \sum_{j,k=1}^{\gamma_n} (b_{jk}^n)^2 = \alpha$$

and

$$\sum_{j=1}^{\theta_n} (\nu_j^n)^2 + \sum_{j=1}^{\gamma_n} (\mu_j^n)^2 = \beta$$

for all n. Thus, (35) holds if

$$\lim_{n\to\infty} \sum_{j,k=1}^{\theta_n} (a_{jk}^n)^2 = \alpha \tag{40}$$

and

$$\lim_{n\to\infty} \sum_{j=1}^{\theta_n} (\nu_j^n)^2 = \beta. \tag{41}$$

Consider (40); the proof of (41) is left as an exercise (Exercise 28). As in (7), let $h_1, h_2, \ldots$ be a linearly independent basis for $\mathcal{H}(0, K_0)$ and $h_{1p}, \ldots, h_{pp}$ a linear transformation of $h_1, \ldots, h_p$ such that $K_0(h_{jp}, h_{kp}) = \delta_{jk}$ and $K_1(h_{jp}, h_{kp}) = \sigma_{jp}^2 \delta_{jk}$ for $j, k \leq p$. From Theorem 7, $\sum_{j=1}^p (1 - \sigma_{jp}^2)^2$ converges to a finite limit as $p \to \infty$; call this limit α. Thus, given $\epsilon > 0$, we can find p such that $\sum_{j=1}^p (1 - \sigma_{jp}^2)^2 > \alpha - \epsilon$. Define q_{jp}^n to be the BLP of h_{jp} based on $\mathcal{H}_n$ and $\mathbf{Q}_{ip}^n$ the covariance matrix of $(q_{1p}^n, \ldots, q_{pp}^n)$ under K_i. Then $\mathbf{Q}_{0p}^n \to \mathbf{I}$ as $n \to \infty$ and $\mathbf{Q}_{1p}^n$ converges to the diagonal matrix with diagonal elements $\sigma_{1p}^2, \ldots, \sigma_{pp}^2$ (Exercise 29). Using (33) it can be shown that

$$\lim_{n\to\infty} \operatorname{tr}\left[\left\{\mathbf{I} - (\mathbf{Q}_{0p}^n)^{-1}\mathbf{Q}_{1p}^n\right\}^2\right] = \sum_{j=1}^{p} (1 - \sigma_{jp}^2)^2 \tag{42}$$

(Exercise 30). Furthermore, for all n,

$$\operatorname{tr}\left[\left\{\mathbf{I} - (\mathbf{Q}_{0p}^n)^{-1}\mathbf{Q}_{1p}^n\right\}^2\right] \leq \sum_{j,k=1}^{\theta_n} (a_{jk}^n)^2 \tag{43}$$

(Exercise 31), and (40) follows since ϵ is arbitrary. □

As we show in Theorem 12, the condition $G(0, K_0) \equiv G(0, K_1)$ is stronger than necessary to obtain uniformly asymptotically optimal predictions. We do have a converse of sorts to Theorem 10 when only the mean is misspecified. Let $\mathcal{S}$ be the subspace of $\mathcal{H}(0, K)$ generated by the observations. Note that $\mathcal{S}$ is fixed so that the following result is not asymptotic. Furthermore, if Z is in fact Gaussian, taking $G_{\mathcal{S}}(0, K) \equiv G_{\mathcal{S}}(m, K)$ is not restrictive in practice, since it would be silly to use a model that is demonstrably wrong with probability 1 based on the available observations.

Theorem 11. *Let* $(0, K_0) = (0, K)$ *and* $(m_1, K_1) = (m, K)$. *If* $G_{\mathcal{S}}(0, K) \equiv G_{\mathcal{S}}(m, K)$ *and* $G_{\mathcal{H}}(0, K) \perp G_{\mathcal{H}}(m, K)$, *then*

$$\sup_{h \notin \mathcal{S}(0,K)} \left| \frac{E_1 e_0(h, \mathcal{S})^2 - E_0 e_0(h, \mathcal{S})^2}{E_0 e_0(h, \mathcal{S})^2} \right| = \infty.$$

PROOF. Let $\mathcal{S}^\perp$ be the orthogonal complement to $\mathcal{S}$ under the inner product defined by K; that is, $\mathcal{S}^\perp$ is made up of those $h \in \mathcal{H}(0, K)$ for which $E_0(h\psi) = 0$ for all $\psi \in \mathcal{S}$. It is possible to construct an orthonormal basis $\{r_j\}_{j=1}^\infty$ for $\mathcal{S}^\perp$ such that $\mu_j = m(r_j)$ is finite for all j (Exercise 32). Then $G_{\mathcal{S}}(0, K) \equiv G_{\mathcal{S}}(m, K)$ and $G_{\mathcal{H}}(0, K) \perp G_{\mathcal{H}}(m, K)$ imply $G_{\mathcal{S}^\perp}(0, K) \perp G_{\mathcal{S}^\perp}(m, K)$ so that $\sum_{j=1}^\infty \mu_j^2 = \infty$ (Exercise 33). Defining $\tau_p = \sum_{j=1}^p \mu_j r_j$, the theorem follows since $e_0(\tau_p, \mathcal{S}) = \tau_p$ and

$$\frac{E_1 e_0(\tau_p, \mathcal{S})^2 - E_0 e_0(\tau_p, \mathcal{S})^2}{E_0 e_0(\tau_p, \mathcal{S})^2} = \sum_{j=1}^p \mu_j^2 \to \infty$$

as $p \to \infty$. □

A theorem of Blackwell and Dubins

Theorem 8, which says that second-order structures corresponding to equivalent Gaussian measures yield asymptotically similar linear predictions, is essentially a special case of a much more general result on comparing conditional distributions for equivalent measures due to Blackwell and Dubins (1962). Let P_0 and P_1 be two probability measures on a sequence of random variables $X_1, X_2, \ldots$ and let P_0^n and P_1^n be the corresponding conditional measures given $\mathcal{F}_n = \sigma(X_1, \ldots, X_n)$, the σ-algebra generated by $X_1, \ldots, X_n$. Under a mild technical condition, the Main Theorem of Blackwell and Dubins (1962) says that if P_1 is absolutely continuous with respect to P_0, then the variation distance between P_0^n and P_1^n tends to 0 with P_1-probability 1. (The variation distance between two measures $(\Omega, \mathcal{F}, P)$ and $(\Omega, \mathcal{F}, Q)$ is the supremum over $A \in \mathcal{F}$ of $|P(A) - Q(A)|$.) Theorem 8 is a straightforward consequence of this result. Specifically, for Gaussian measures P_0 and P_1 on $X_1, X_2, \ldots$, if P_1 is absolutely continuous with respect to P_0, then Theorem 4 implies $P_0 \equiv P_1$. For h in the Hilbert space $\mathcal{H}$ generated by $X_1, X_2, \ldots$, we have as a special case of the Main Theorem of

Blackwell and Dubins (1962),

$$P_1\left\{\lim_{n\to\infty}\sup_{h\in\mathcal{H},t\in\mathbb{R}}|P_0^n(h\le t)-P_1^n(h\le t)|=0\right\}=1. \tag{44}$$

Using the fact that the conditional distribution of h given $\mathcal{F}_n$ is Gaussian under P_0 or P_1, it is possible to show (44) holds if and only if

$$\lim_{n\to\infty}\sup_{h\in\mathcal{H}_{-n}}\frac{\operatorname{var}_0(h\mid\mathcal{F}_n)}{\operatorname{var}_1(h\mid\mathcal{F}_n)}=1 \quad \text{and} \tag{45}$$

$$\lim_{n\to\infty}\sup_{h\in\mathcal{H}_{-n}}\frac{E_1\left\{E_0(h\mid\mathcal{F}_n)-E_1(h\mid\mathcal{F}_n)\right\}^2}{\operatorname{var}_1(h\mid\mathcal{F}_n)}=0. \tag{46}$$

Taking h to be $e_0(h,n)$ in (46) yields

$$\lim_{n\to\infty}\sup_{h\in\mathcal{H}_{-n}}\frac{E_1\left\{e_0(h,n)-e_1(h,n)\right\}^2}{E_1e_1(h,n)^2}=0, \tag{47}$$

which is the same as (38). Combining (47) and (45) yields

$$\lim_{n\to\infty}\sup_{h\in\mathcal{H}_{-n}}\left|\frac{E_0e_0(h,n)^2-E_1e_0(h,n)^2}{E_1e_0(h,n)^2}\right|=0 \tag{48}$$

and Theorem 8 follows, since (47) and (48) are only statements about the first two moments of the process and do not require Gaussianity. That is, for a random field on R, Theorem 8 only requires the equivalence of the Gaussian measures defined by the second-order structures $(0,K_0)$ and (m_1,K_1), not that the random field actually be Gaussian. However, as I have already discussed (see, for example, 1.4), focusing on linear predictors and their unconditional mses can be a serious mistake if the random field is not Gaussian.

The result of Blackwell and Dubins also yields conclusions about nonlinear predictions of Gaussian processes. Thus, for example, consider a Gaussian process Z on $R=[0,1]$ with $G_R(m_0,K_0)\equiv G_R(m_1,K_1)$ and $t_1,t_2,\ldots$ a dense sequence of observations on $[0,1]$. Then the conditional distribution of, say, $\int_0^1 e^{Z(t)}dt$ given $Z(t_1),\ldots,Z(t_n)$ is very nearly the same under $G_R(m_0,K_0)$ and $G_R(m_1,K_1)$ for n large.

Weaker conditions for asymptotic optimality of pseudo-BLPs

Theorem 10 of Chapter 3 showed that pseudo-BLPs based on observations at $\delta\mathbf{j}$ for $\mathbf{j}\in\mathbb{Z}^d$ are asymptotically optimal and the evaluations of mse are asymptotically correct as $\delta\downarrow 0$ if $f_1(\boldsymbol{\omega})/f_0(\boldsymbol{\omega})\to 1$ as $|\boldsymbol{\omega}|\to\infty$. Theorem 12 shows that a similar result holds for the fixed-domain setting when $f_0\in\mathcal{Q}^d$. The condition $f_1(\boldsymbol{\omega})/f_0(\boldsymbol{\omega})\to 1$ as $|\boldsymbol{\omega}|\to\infty$ is in practice substantially weaker than required to obtain equivalence of the corresponding

Gaussian measures. For example, in one dimension, if $f_0(\omega) \asymp \omega^{-\alpha}$ for some $\alpha > 1$ as $\omega \to \infty$, then Ibragimov and Rozanov (1978, p. 107) note that $\omega^{1/2}\{f_0(\omega) - f_1(\omega)\}/f_0(\omega) \to \infty$ as $\omega \to \infty$ implies the corresponding Gaussian measures are orthogonal on any interval of positive length.

Theorem 12. *Consider continuous autocovariance functions K_0 and K_1 with corresponding spectral densities f_0 and f_1 for a mean 0 process Z on $\mathbb{R}^d$. For R a bounded subset of $\mathbb{R}^d$, suppose $\mathcal{H}_n$, $n = 1, 2, \ldots$ is a sequence of subspaces of $\mathcal{H}_R(0, K_0)$ satisfying $E_0 e_0(h,n)^2 \to 0$ as $n \to \infty$ for all $h \in \mathcal{H}_R(0, K_0)$. If $f_0 \in \mathcal{Q}^d$ and $f_1(\boldsymbol{\omega})/f_0(\boldsymbol{\omega}) \to c$ as $|\boldsymbol{\omega}| \to \infty$ for some positive finite c, then*

$$\lim_{n\to\infty} \sup_{h\in\mathcal{H}_{-n}} \frac{E_0 e_1(h,n)^2}{E_0 e_0(h,n)^2} = 1 \tag{49}$$

and

$$\lim_{n\to\infty} \sup_{h\in\mathcal{H}_{-n}} \left| \frac{E_1 e_1(h,n)^2}{E_0 e_1(h,n)^2} - c \right| = 0. \tag{50}$$

PROOF. Given $\epsilon > 0$, there exists C_ϵ finite such that

$$\sup_{|\boldsymbol{\omega}|>C_\epsilon} \left| \frac{f_1(\boldsymbol{\omega})}{c f_0(\boldsymbol{\omega})} - 1 \right| < \epsilon.$$

Define

$$g_\epsilon(\boldsymbol{\omega}) = \begin{cases} c^{-1} f_1(\boldsymbol{\omega}) & \text{for } |\boldsymbol{\omega}| \le C_\epsilon, \\ f_0(\boldsymbol{\omega}) & \text{for } |\boldsymbol{\omega}| > C_\epsilon. \end{cases}$$

Using the subscript ϵ to indicate a calculation done assuming g_ϵ is the spectral density,

$$\begin{aligned} &E_0 e_1(h,n)^2 \\ &\quad = E_0 \left[\{e_1(h,n) - e_\epsilon(h,n)\} + e_\epsilon(h,n)\right]^2 \\ &\quad \le E_0 \{e_1(h,n) - e_\epsilon(h,n)\}^2 \\ &\qquad + 2\left[E_0 \{e_1(h,n) - e_\epsilon(h,n)\}^2 E_0 e_\epsilon(h,n)^2\right]^{1/2} + E_0 e_\epsilon(h,n)^2 . \end{aligned} \tag{51}$$

By (20), $G_R(0, f_0) \equiv G_R(0, g_\epsilon)$, so by Theorem 8,

$$\lim_{n\to\infty} \sup_{h\in\mathcal{H}_{-n}} \frac{E_0 e_\epsilon(h,n)^2}{E_0 e_0(h,n)^2} = 1. \tag{52}$$

Set $\epsilon_0 = \frac{1}{2}$ so that $f_0(\boldsymbol{\omega}) \le 2c^{-1} f_1(\boldsymbol{\omega})$ for all $|\boldsymbol{\omega}| \ge C_{\epsilon_0}$. Using $G_R(0, f_0) \equiv G_R(0, g_{\epsilon_0})$ and Exercise 2 in 4.2, there exists $\alpha_0 < \infty$ such that

$$E_0 h^2 \le \alpha_0 E_{\epsilon_0} h^2 \quad \text{for all } h \in \mathcal{H}_R(0, K_0).$$

For all $\epsilon > 0$, let us choose $C_\epsilon \ge C_{\epsilon_0}$, which we can always do. Then $g_{\epsilon_0}(\boldsymbol{\omega}) \le 2 g_\epsilon(\boldsymbol{\omega})$, so that $E_0 h^2 \le 2\alpha_0 E_\epsilon h^2$ for all ϵ. By Theorem 10, for any particular ϵ,

$$\lim_{n\to\infty} \sup_{h\in\mathcal{H}_{-n}} \left| \frac{E_\epsilon e_\epsilon(h,n)^2}{E_0 e_0(h,n)^2} - 1 \right| = 0$$

so that for all n sufficiently large,

$$\frac{E_0 \left\{e_1(h,n) - e_\epsilon(h,n)\right\}^2}{E_0 e_0(h,n)^2} \leq 4\alpha_0 \frac{E_\epsilon \left\{e_1(h,n) - e_\epsilon(h,n)\right\}^2}{E_\epsilon e_\epsilon(h,n)^2}$$

for all $h \in \mathcal{H}_{-n}$, where α_0 is independent of ϵ. But by Theorem 1 of Chapter 3,

$$\frac{E_\epsilon \left\{e_1(h,n) - e_\epsilon(h,n)\right\}^2}{E_\epsilon e_\epsilon(h,n)^2} \leq \frac{\epsilon^2}{1-\epsilon^2}\,. \tag{53}$$

By (51)–(53),

$$\overline{\lim_{n\to\infty}} \sup_{h\in\mathcal{H}_{-n}} \frac{E_0 e_1(h,n)^2}{E_0 e_0(h,n)^2} \leq \frac{4\alpha_0\epsilon^2}{1-\epsilon^2} + 4\epsilon\left(\frac{\alpha_0}{1-\epsilon^2}\right)^{1/2} + 1$$

and (49) follows since ϵ is arbitrary. The proof of (50) is left as Exercise 34. A result similar to Theorem 12 is given in Stein (1993a), although the proof there is not valid if $f_1(\boldsymbol{\omega})/f_0(\boldsymbol{\omega})$ is unbounded. □

The conditions in Theorem 12 are still stronger than necessary. For example, if f_0 satisfies (19) and there exists a density f_2 such that

$$\int_{\mathbb{R}^d} \left\{ \frac{f_2(\boldsymbol{\omega}) - f_0(\boldsymbol{\omega})}{f_0(\boldsymbol{\omega})} \right\}^2 \mathbf{d}\boldsymbol{\omega} < \infty \tag{54}$$

and for some positive finite c,

$$\lim_{|\omega|\to\infty} \frac{f_2(\boldsymbol{\omega})}{f_1(\boldsymbol{\omega})} = c \tag{55}$$

then (49) and (50) hold (Exercise 35 or Stein (1993a)).

Whether $f_0 \in \mathcal{Q}^d$ or something like it is needed in Theorem 12 is unknown. This condition is invoked in the proof to show that the Gaussian measures corresponding to the spectral densities f_0 and g_ϵ are equivalent so that Theorem 10 can be invoked. Theorem 10 in 3.8 regarding observations on an infinite lattice does not require any assumptions analogous to $f_0 \in \mathcal{Q}^d$, which suggests that Theorem 12 may hold even if $f_0 \notin \mathcal{Q}^d$.

Similar to Corollary 9 of Theorem 8, we can obtain a corollary to Theorem 12 for a random field observed with known measurement error σ^2.

Corollary 13. *For $n = 1, 2, \ldots$, let $\mathcal{X}_n = \{\mathbf{x}_{1n}, \ldots, \mathbf{x}_{j_n n}\}$ be a finite subset of a bounded set R and set $Y_{jn} = Z(\mathbf{x}_{jn}) + \epsilon_{jn}$ for $j = 1, \ldots, j_n$, where the ϵ_{jn}s have mean 0, are uncorrelated with each other and with Z and have common variance σ^2 not depending on the model for Z. For $\psi \in H_R(0, K_0)$,* *define* $e_i(\psi, n)$ *to be the error of the BLP of* ψ *based on* $Y_{1n}, \ldots, Y_{j_n n}$ *when* Z *has second-order structure* $(0, K_i)$. *Suppose* K_0 *and* K_1 *are continuous autocovariance functions with corresponding spectral densities* f_0 *and* f_1 *for a mean 0 random field* Z *on* $\mathbb{R}^d$. *Assume* $f_0 \in \mathcal{Q}^d$ *and* $f_1(\boldsymbol{\omega})/f_0(\boldsymbol{\omega}) \to 1$ *as* $|\boldsymbol{\omega}| \to \infty$. *If* R *is a bounded subset of* $\mathbb{R}^d$ *such that all points in* R *are limit points in* R *and*

$$\lim_{n\to\infty} \sup_{\mathbf{x}\in R} \inf_{\mathbf{y}\in\mathcal{X}_n} |\mathbf{x} - \mathbf{y}| = 0, \tag{56}$$

then (49) and (50) hold with $c = 1$, *where* $\mathcal{H}_{-n}$ *is taken as all nonzero elements of* $\mathcal{H}_R(0, K_0)$.

Corollary 13 can be proven by first obtaining an analog to Theorem 8 for observations with measurement error and then essentially repeating the proof of Theorem 12. The condition (56) just says that every point in R is near a point in $\mathcal{X}_n$ when n is large.

Rates of convergence to asymptotic optimality

In 3.8, we were able to give rates of convergence to 0 in (35)–(38) under additional conditions on f_0 and f_1. Obtaining rates is much more difficult with observations confined to a bounded region, although I have obtained some limited results in one dimension (Stein 1990b, 1999). No rate results are presently available in more than one dimension for random fields possessing a spectral density with respect to Lebesgue measure.

Stein (1990a) gives some rates of convergence in one and two dimensions for the easier problem of the mean function being misspecified. The basic message is that if $m_1 - m_0$ is much smoother than the stochastic component of Z, the effect of using m_1 rather than the correct m_0 disappears rapidly as the observations get denser in the region of interest.

Asymptotic optimality of BLUPs

Theorem 8 can be used to prove that there is asymptotically little difference between BLPs and BLUPs (see 1.5) if $EZ(\mathbf{x}) = \boldsymbol{\beta}^T \mathbf{m}(\mathbf{x})$ and the components of $\mathbf{m}$ are much smoother than the stochastic component of Z, despite the fact that the BLUE of $\boldsymbol{\beta}$ will not be consistent in this case (Stein 1990a, Theorem 5.2). The basic idea of the proof is to show that the BLUP cannot do too much worse than a pseudo-BLP based on a fixed but incorrect value for $\boldsymbol{\beta}$. Thus, this result is hardly a victory for best linear unbiased prediction, but merely is a restatement of the fact that the mean often does not matter asymptotically for prediction when using fixed-domain asymptotics. Exercise 10 of Chapter 3 gives an explicit example of the asymptotic optimality of BLUPs.

Exercises

26 Prove Corollary 9.

27 Under the conditions of Theorem 8, show that for all $h \in \mathcal{H}(0, K_0)$, $E_0 e_0(h,n)^2 \to 0$ as $n \to \infty$.

28 Prove (41).

29 In the proof of Theorem 8, show that $\mathbf{Q}_0^n \to \mathbf{I}$ as $n \to \infty$ and $\mathbf{Q}_1^n$ converges to the matrix with elements $\sigma_{1p}^2, \ldots, \sigma_{pp}^2$ along the diagonal and 0 elsewhere.

30 Prove (42).

31 Prove that (40) follows from (42) and (43).

32 In the proof of Theorem 11, show that it is possible to construct an orthonormal basis $\{r_j\}_{j=1}^{\infty}$ for $\mathcal{S}^{\perp}$ such that $\mu_j = m(r_j)$ is finite for all j.

33 Under the conditions of Theorem 11, show that $G_{\mathcal{S}}(0, K) \equiv G_{\mathcal{S}}(m, K)$ and $G_{\mathcal{H}}(0, K) \perp G_{\mathcal{H}}(m, K)$ imply $G_{\mathcal{S}^{\perp}}(0, K) \perp G_{\mathcal{S}^{\perp}}(m, K)$.

34 Prove (50).

35 Show that (54) and (55) imply (49) and (50).

36 Suppose Z is a mean 0 weakly stationary random field on $\mathbb{R}^d$ with actual autocovariance function K_0 and presumed autocovariance function K_1. For a bounded set $R \subseteq \mathbb{R}^d$, define the function $\tilde{K}_j(\mathbf{x}, \mathbf{y})$ on $R \times R$ by $\tilde{K}_j(\mathbf{x}, \mathbf{y}) = K_j(\mathbf{x} - \mathbf{y})$. Suppose that for $j = 0, 1$, $\tilde{K}_j$ can be extended to a function on $\mathbb{R}^d \times \mathbb{R}^d$ such that $\tilde{K}_j(\mathbf{x}, \mathbf{y})$ depends only on $\mathbf{x} - \mathbf{y}$ and $\int_{\mathbb{R}^d} |\tilde{K}_j(\mathbf{x}, \mathbf{0})| d\mathbf{x} < \infty$. Define

$$f_j(\boldsymbol{\omega}) = \frac{1}{(2\pi)^d} \int_{\mathbb{R}^d} \exp\left\{-i\boldsymbol{\omega}^T \mathbf{x}\right\} \tilde{K}_j(\mathbf{x}, \mathbf{0})\, d\mathbf{x}.$$

If f_0 satisfies (20) and $f_1(\boldsymbol{\omega})/f_0(\boldsymbol{\omega}) \to c$ as $|\boldsymbol{\omega}| \to \infty$ for some positive, finite c, show that (47) and (48) follow for $\mathcal{H}_n$ defined as in Theorem 10.

37 Suppose Z is a mean 0 weakly stationary process on $\mathbb{R}$, $K_0(t) = e^{-|t|}$ and $K_1(t) = (1 - |t|)^+$. Suppose $R = [0, T]$ and that $\mathcal{H}_n$ is as in Theorem 10.

(i) For $T < 1$, use Pólya's criteria (Exercise 37 of Chapter 2) and the previous exercise to show that (47) and (48) hold for $\mathcal{H}_n$ defined as in Theorem 10.

Show that (47) and (48) also hold for $T = 1$ by filling in the details of the following argument.

(ii) Define the function $\phi_a(t)$ by taking it to have period $4a^{-1}$ and setting $\phi_a(t) = 1 - a|t|$ for $|t| \le 2a^{-1}$. Show that ϕ_a is p.d. for all $a > 0$.

(iii) Define $\psi(t) = \frac{1}{2}\int_0^2 \phi_a(t)\,da$. Show that ψ is p.d. and that $\psi(t) = 1 - |t|$ on $[-1, 1]$.

(iv) Show that ψ has spectral density

$$f(\omega) = \frac{4}{\pi^3}\sum_{j=0}^{\infty}\frac{1\{|\omega| \le (2j+1)\pi\}}{(2j+1)^3}.$$

(v) Prove that $\pi(1+\omega^2)f(\omega) \to 1$ as $\omega \to \infty$.

4.4 Jeffreys's law

Dawid (1984) discusses principles of forecasting, which he defines as making predictions about the $n+1$th element of an uncertain sequence based on observing the first n elements of that sequence. He is specifically concerned with this problem when there is a parametric family of models $\mathcal{P} = \{P_\theta : \theta \in \Theta\}$ on this infinite sequence of observations with θ unknown. He notes that even if θ cannot be consistently estimated as the number of observations increases, it should still be possible to obtain forecasts based on estimated values of θ that do asymptotically as well as forecasts using the true θ. Again, a more succinct statement is "things we shall never find much out about cannot be very important for prediction" (Dawid 1984, p. 285). Dawid calls this principle Jeffreys's law based on the following statement of Jeffreys (1938, p. 718): "When a law has been applied to a large body of data without any systematic discrepancy being detected, it leads to the result that the probability of a further inference from the law approaches certainty whether the law is true or not."

Dawid (1984) considers the Main Theorem of Blackwell and Dubins (1962) to be a mathematical statement of Jeffreys's law, so that Theorem 8 can be thought of as an example of this law. The Kullback divergence can be used to obtain a more quantitative connection between discrepancies between Gaussian measures and linear prediction. Consider two probability measures P_0 and P_1 on $(\Omega, \mathcal{F})$. Suppose $\mathbf{Y}$ is an $\mathcal{F}$-measurable random vector and, for simplicity, assume that under P_j, $\mathbf{Y}$ has density p_j with respect to Lebesgue measure. Then $I(P_0, P_1; \mathbf{Y})$, the Kullback divergence of P_1 from P_0 based on $\mathbf{Y}$, is given by $E_0 \log\{p_0(\mathbf{Y})/p_1(\mathbf{Y})\}$. The larger the value of $I(P_0, P_1; \mathbf{Y})$, the more information in $\mathbf{Y}$, on average, for determining that P_1 is the wrong measure when P_0 is correct. Note that r_n as defined in (10) is just $I(P_0, P_1; (h_1, \ldots, h_n)) + I(P_1, P_0; (h_1, \ldots, h_n))$. See Kullback (1968) for the role of the Kullback divergence in estimation problems and Christakos (1992, Chapter 2, Section 13; Chapter 9, Section 8; and

Chapter 10, Section 9) for a rather different use of information measures in modeling and prediction of random fields.

Suppose $\mathbf{Y}$ is a vector of observations, Z is the predictand and assume $(\mathbf{Y}, Z)$ has a joint density with respect to Lebesgue measure under either P_0 or P_1. Then a measure of the additional information in Z not contained in $\mathbf{Y}$ for distinguishing between P_0 and P_1 when P_0 is true is

$$E_0 \log \frac{p_0(\mathbf{Y}, Z)}{p_1(\mathbf{Y}, Z)} - E_0 \log \frac{p_0(\mathbf{Y})}{p_1(\mathbf{Y})} = E_0 \log \frac{p_0(Z \mid \mathbf{Y})}{p_1(Z \mid \mathbf{Y})},$$

where p_j generically indicates the marginal or conditional density of a random vector under P_j. If $(\mathbf{Y}, Z)$ is Gaussian under P_0 and P_1 and e_j is the error of the BLP of Z under P_j, then

$$E_0 \log \frac{p_0(Z \mid \mathbf{Y})}{p_1(Z \mid \mathbf{Y})} = \frac{1}{2}\left(\frac{E_0 e_0^2}{E_1 e_1^2} - 1 - \log \frac{E_0 e_0^2}{E_1 e_1^2}\right) + \frac{E_0(e_1 - e_0)^2}{2E_1 e_1^2} \tag{57}$$

(Exercise 38). Now consider a sequence of prediction problems in which $\mathbf{Y}_n$ is observed, Z_n is the predictand, $(\mathbf{Y}_n, Z_n)$ is jointly Gaussian under P_0 or P_1 and $e_j(n)$ is the prediction error under P_j. Suppose that as $n \to \infty$, there is asymptotically negligible harm in using P_1 rather than P_0 for predicting Z_n: $E_0\{e_1(n) - e_0(n)\}^2/E_0 e_0(n)^2 \to 0$ and $\{E_0 e_1(n)^2 - E_1 e_1(n)^2\}/E_0 e_1(n)^2 \to 0$ as $n \to \infty$, which is exactly the case in Theorem 10 when $c = 1$. Then

$$\begin{aligned} E_0 \log \frac{p_0(Z_n \mid \mathbf{Y}_n)}{p_1(Z_n \mid \mathbf{Y}_n)} &\sim \frac{1}{4}\left\{\frac{E_0 e_0(n)^2}{E_1 e_1(n)^2} - 1\right\}^2 + \frac{E_0\{e_1(n) - e_0(n)\}^2}{2E_1 e_1(n)^2} \\ &\sim \left\{\frac{E_0 e_1(n)^2 - E_1 e_1(n)^2}{2E_0 e_1(n)^2}\right\}^2 + \frac{E_0\{e_1(n) - e_0(n)\}^2}{2E_0 e_0(n)^2} \end{aligned} \tag{58}$$

(Exercise 39). Thus, the additional information in Z_n for distinguishing the measures is approximately $\frac{1}{4}$ times the square of the relative misspecification of the mse plus $\frac{1}{2}$ times the relative increase in mse due to using P_1. Note that the two terms on the right side of (58) do not necessarily tend to 0 at the same rate, although results in Chapter 3 (compare (32) to (36) or (34) to (38)) suggest that they sometimes do.

A Bayesian version

It is possible to give an exact quantification of Jeffreys's law by taking a Bayesian perspective. Let $\mathcal{P} = \{P_{\boldsymbol{\theta}} : \boldsymbol{\theta} \in \Theta\}$ be a finite-dimensional parametric family of distributions for $(\mathbf{Y}, Z)$. Suppose $(\boldsymbol{\theta}, \mathbf{Y}, Z)$ have a joint density with respect to Lebesgue measure and use p generically to denote a marginal or conditional density, so that, in particular, $p(\boldsymbol{\theta})$ is the prior density for $\boldsymbol{\theta}$. Define the predictive density for Z given $\mathbf{Y}$

$$p(Z \mid \mathbf{Y}) = \int_{\Theta} p(Z \mid \boldsymbol{\theta}, \mathbf{Y}) p(\boldsymbol{\theta} \mid \mathbf{Y})\, \mathbf{d}\boldsymbol{\theta}$$
$$= \frac{\int_{\Theta} p(Z \mid \boldsymbol{\theta}, Y) p(\mathbf{Y} \mid \boldsymbol{\theta}) p(\boldsymbol{\theta})\, \mathbf{d}\boldsymbol{\theta}}{\int_{\Theta} p(\mathbf{Y} \mid \boldsymbol{\theta}) p(\boldsymbol{\theta})\, \mathbf{d}\boldsymbol{\theta}}.$$

Then

$$E\left[I\left\{p(Z \mid \boldsymbol{\theta}, \mathbf{Y}), p(Z \mid \mathbf{Y})\right\} \mid \mathbf{Y}\right] = E\left[I\left\{p(\boldsymbol{\theta} \mid \mathbf{Y}, Z), p(\boldsymbol{\theta} \mid \mathbf{Y})\right\} \mid \mathbf{Y}\right], \quad (59)$$

which was suggested to me by Wing Wong. The proof of (59) is just to note that both sides of it equal

$$E\left\{\log \frac{p(\boldsymbol{\theta}, Z \mid \mathbf{Y})}{p(\boldsymbol{\theta} \mid \mathbf{Y}) p(Z \mid \mathbf{Y})} \mid \mathbf{Y}\right\}.$$

This expression is 0 if $\boldsymbol{\theta}$ and Z are conditionally independent given $\mathbf{Y}$, so that both sides of (59) measure the conditional dependence of $\boldsymbol{\theta}$ and Z given $\mathbf{Y}$. To see why (59) can be viewed as a quantification of Jeffreys's law, we need to take a closer look at both sides of this equality. For any particular $\boldsymbol{\theta}_0$, $I\left\{p(Z \mid \boldsymbol{\theta}_0, \mathbf{Y}), p(Z \mid \mathbf{Y})\right\}$ measures how far the predictive distribution for Z diverges from the conditional distribution for Z we would obtain if $\boldsymbol{\theta} = \boldsymbol{\theta}_0$ were known. The left side of (59) is then just the average of this divergence over all possible values of $\boldsymbol{\theta}$. Thus, the left side of (59) measures how much information $\boldsymbol{\theta}$ contains about Z that is not already contained in $\mathbf{Y}$. Similarly, the right-hand side of (59) is a measure of how much information Z contains about $\boldsymbol{\theta}$ that is not already contained in $\mathbf{Y}$. The conclusion I draw from (59) is a sharpening of Jeffreys's law: if the quantity we wish to predict tells us very little new about the parameters of our model, then our predictions will be close to those we would obtain if we knew the true values of the parameters.

Let us now reexamine (40) in 3.6 in light of this result. To review, suppose Z is a mean 0 Gaussian process on $\mathbb{R}$ observed at $Z(-j/n)$ for $j = 0, \ldots, n$, $K_0(t) = e^{-|t|}$ and $K_1(t) = \frac{1}{2}e^{-2|t|}$. For any finite interval R, $G_R(0, K_0) \equiv G_R(0, K_1)$, so that even if n is large, it will be difficult to distinguish between these measures. Now consider predicting $Z(t)$ for $t > 0$. Since prediction of $Z(t)$ does not depend on n under either model, denote the prediction error for $Z(t)$ under K_j by $e_j(t)$. Using (40) in 3.6, Figure 2 plots $E_0 e_1(t)^2 / E_0 e_0(t)^2$ and $E_1 e_1(t)^2 / E_0 e_1(t)^2$ as functions of t. Both functions are near 1 for t small, which, considering (58), implies that $Z(t)$ for t small does not provide much new information for distinguishing the measures. For larger t, $Z(t)$ does provide nonnegligible additional information for distinguishing between the measures and this is reflected particularly in $E_1 e_1(t)^2 / E_0 e_1(t)^2$, which tends to $\frac{1}{2}$ as $t \to \infty$. Thus, the statement "things we shall never find much out about cannot be very important for prediction" (Dawid 1984) is incorrect in this setting because

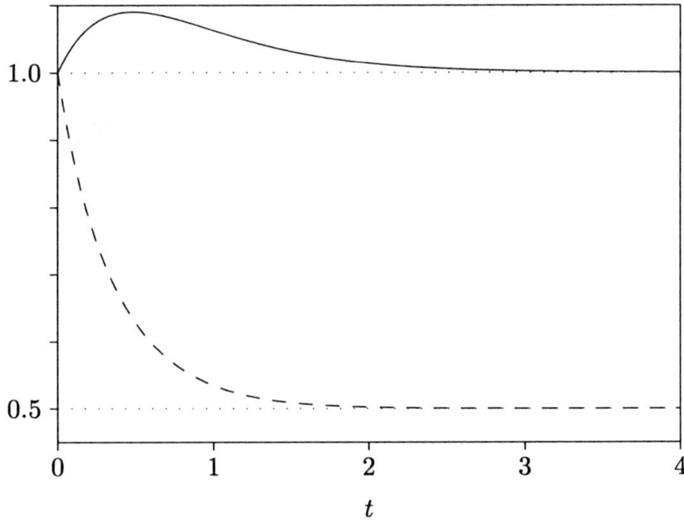

FIGURE 2. Ratios of mean squared errors for predicting $Z(t)$ from $Z(0)$ when $K_0(t) = e^{-|t|}$ and $K_1(t) = \frac{1}{2}e^{-2|t|}$. Solid curve gives $E_0 e_1^2 / E_0 e_0^2$ and dashed curve $E_1 e_1^2 / E_0 e_1^2$.

it does not anticipate the possibility that what is to be predicted provides substantial additional information for distinguishing the measures.

Exercises

38 Prove (57).

39 Prove (58).

5 Integration of Random Fields

5.1 Introduction

This chapter studies the prediction of integrals of random fields based on observations on a lattice. The goal here is not to give a full exposition of the topic (see Ritter (1995) for a more detailed treatment) but to make two specific points about properties of systematic designs. The first is that simple averages over observations from systematic designs can be very poor predictors of integrals of random fields, especially in higher dimensions. The second is that, at least for random fields that are not too anisotropic, the problem with this predictor is the simple average aspect of it, not the systematic design. These two points are of interest on their own, but they are also critical to understanding a serious flaw in an argument of Matheron (1971) purporting to demonstrate that statistical inference is "impossible" for differentiable random fields (see 6.3).

Suppose Z is a mean 0 weakly stationary random field on $\mathbb{R}^d$. Define $\mathcal{G}_m = \{1, \ldots, m\}^d$ and let $\mathbf{h}$ be the vector of length d with each component equal to $\frac{1}{2}$. Consider predicting $I(Z) = \int_{[0,1]^d} Z(\mathbf{x})\,\mathbf{dx}$ based on observing Z at $m^{-1}(\mathbf{j} - \mathbf{h})$ for $\mathbf{j} \in \mathcal{G}_m$. This set of observations is called a centered systematic sample because it places an observation at the center of each cube of the form $\times_{\alpha=1}^{d} \left[m^{-1}(j_\alpha - 1), m^{-1}j_\alpha\right]$ for $\mathbf{j} = (j_1, \ldots, j_d) \in \mathcal{G}_m$ (see Figure 1). A natural predictor of the integral is just the simple average of the observations, $\overline{Z}_m = m^{-d} \sum_{\mathbf{j} \in \mathcal{G}_m} Z\big(m^{-1}(\mathbf{j} - \mathbf{h})\big)$. Although it may be natural, it is not necessarily a good predictor. Section 5.2 looks at the asymptotic mse of $\overline{Z}_m$ as $m \to \infty$. Results in 5.3 and 5.4 show that if Z has

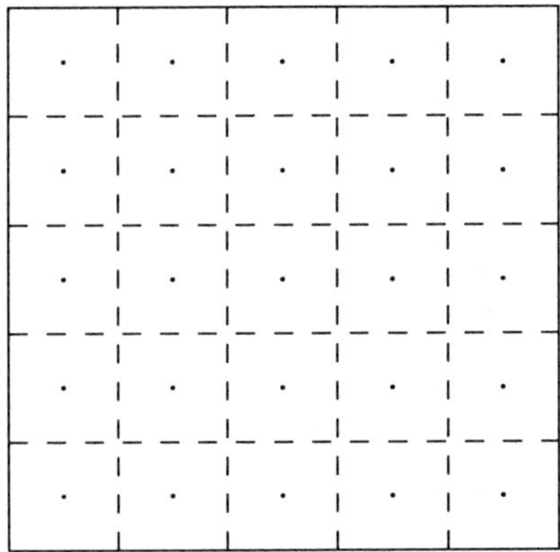

FIGURE 1. Centered systematic sample for $d = 2$ and $m = 5$. Dots indicate design points and dashed lines the squares in which the design points are centered.

spectral density f satisfying $f(\boldsymbol{\omega}) \asymp (1+|\boldsymbol{\omega}|)^{-p}$, then $\overline{Z}_m$ is asymptotically optimal for $p < 4$ and its mse tends to 0 at a slower rate than the mse of the BLP for $p > 4$. Section 5.4 also shows how to obtain an easily computed predictor that is asymptotically optimal as $m \to \infty$ for any particular p.

In principle, if the autocovariance function K is known, it is possible to find the BLP of $I(Z)$, which I denote by $\hat{Z}_m$. However, this requires calculating $\operatorname{cov}\{I(Z), Z(m^{-1}(\mathbf{j}-\mathbf{h}))\}$ for all $\mathbf{j} \in \mathcal{G}_m$, which will generally need to be done numerically and could be a formidable task for m^d large. Calculating the mse can also be quite difficult since it requires the calculation of a $2d$-dimensional integral, which again must generally be done numerically. Theorem 7 of Section 5.4 gives asymptotically valid and fairly readily computable approximations to the mse of the BLP or any asymptotically optimal predictor under certain conditions on f. Section 5.5 provides some numerical results for $d = 1$ indicating the applicability of the large sample results to finite samples.

5.2 Asymptotic properties of simple average

The even spacing of the observations in a centered systematic sample suggests the use of spectral methods for analyzing the behavior of $\overline{Z}_m$. If Z has mean 0 and spectral density f, then

$$\begin{aligned} &\operatorname{var}\{I(Z) - \overline{Z}_m\} \qquad (1) \\ &= \int_{\mathbb{R}^d} f(\boldsymbol{\omega}) \left| \int_{[0,1]^d} \exp(i\boldsymbol{\omega}^T \mathbf{x})\, d\mathbf{x} - \sum_{\mathbf{j} \in \mathcal{G}_m} \frac{1}{m^d} \exp\{im^{-1}\boldsymbol{\omega}^T(\mathbf{j}-\mathbf{h})\} \right|^2 d\boldsymbol{\omega}. \end{aligned}$$

It follows that

$$\operatorname{var}\left\{I(Z) - \overline{Z}_m\right\} = \int_{\mathbb{R}^d} g_m(\boldsymbol{\omega})\,\mathbf{d}\boldsymbol{\omega},$$

where

$$g_m(\boldsymbol{\omega}) = f(\boldsymbol{\omega}) \prod_{\alpha=1}^{d} \frac{\sin^2\left(\frac{\omega_\alpha}{2}\right)}{m^2 \sin^2\left(\frac{\omega_\alpha}{2m}\right)} \left\{1 - \prod_{\alpha=1}^{d} \operatorname{sinc}\left(\frac{\omega_\alpha}{2m}\right)\right\}^2,$$

$\boldsymbol{\omega} = (\omega_1, \ldots, \omega_d)^T$ and $\operatorname{sinc} t = t^{-1} \sin t$ (Exercise 2). Thus,

$$\begin{aligned} \operatorname{var}\left\{I(Z) - \overline{Z}_m\right\} &= \sum_{\mathbf{j} \in \mathbb{Z}^d} \int_{A_d(m)} g_m(\boldsymbol{\omega} + 2\pi m \mathbf{j})\,\mathbf{d}\boldsymbol{\omega} \\ &= \sum_{\mathbf{j} \in \mathbb{Z}^d} \int_{\mathbb{R}^d} g_m(\boldsymbol{\omega}; \mathbf{j})\,\mathbf{d}\boldsymbol{\omega}, \end{aligned} \tag{2}$$

where

$$\begin{aligned} g_m(\boldsymbol{\omega}; \mathbf{j}) = {} & f(\boldsymbol{\omega} + 2\pi m \mathbf{j}) \prod_{\alpha=1}^{d} \frac{\sin^2\left(\frac{\omega_\alpha}{2}\right)}{m^2 \sin^2\left(\frac{\omega_\alpha}{2m}\right)} \\ & \times \left\{1 - \prod_{\alpha=1}^{d} \frac{(-1)^{j_\alpha} 2m \sin\left(\frac{\omega_\alpha}{2m}\right)}{\omega_\alpha + 2\pi m j_\alpha}\right\}^2 1\{\boldsymbol{\omega} \in A_d(m)\}. \end{aligned}$$

The key to finding the asymptotic mse of $\overline{Z}_m$ is to determine whether the term $\mathbf{j} = \mathbf{0}$ or the terms $\mathbf{j} \neq \mathbf{0}$ dominate the sum on the right side of (2). We first need some preliminary approximations for $g_m(\boldsymbol{\omega}; \mathbf{j})$. For any fixed $\boldsymbol{\omega}$ and $\mathbf{j} \neq \mathbf{0}$,

$$g_m(\boldsymbol{\omega}; \mathbf{j}) \sim f(\boldsymbol{\omega} + 2\pi m \mathbf{j}) \prod_{\alpha=1}^{d} \operatorname{sinc}^2\left(\frac{\omega_\alpha}{2}\right) \tag{3}$$

as $m \to \infty$. And, for fixed $\boldsymbol{\omega}$,

$$\begin{aligned} \prod_{\alpha=1}^{d} \operatorname{sinc}\left(\frac{\omega_\alpha}{2m}\right) &= \prod_{\alpha=1}^{d} \left(1 - \frac{\omega_\alpha^2}{24m^2}\right) + O(m^{-4}) \\ &= 1 - \frac{1}{24m^2} |\boldsymbol{\omega}|^2 + O(m^{-4}) \end{aligned}$$

as $m \to \infty$, so that

$$m^4 g_m(\boldsymbol{\omega}; \mathbf{0}) \to G(\boldsymbol{\omega}) \tag{4}$$

as $m \to \infty$ for fixed $\boldsymbol{\omega}$, where

$$G(\boldsymbol{\omega}) = \frac{1}{576} |\boldsymbol{\omega}|^4 f(\boldsymbol{\omega}) \prod_{\alpha=1}^{d} \operatorname{sinc}^2\left(\frac{\omega_\alpha}{2}\right).$$

Results for sufficiently smooth random fields

Which terms in the sum over $\mathbf{j}$ on the right side of (2) dominate depends on how fast f decays at high frequencies. Equations (2)–(4) can be used to show that if $f(\boldsymbol{\omega}) = o(|\boldsymbol{\omega}|^{-4})$ as $|\boldsymbol{\omega}| \to \infty$, the term $\mathbf{j} = \mathbf{0}$ dominates the sum. The following result is taken from Stein (1993c). Tubilla (1975) obtains a similar result in terms of the autocovariance function rather than the spectral density under much stronger conditions than assumed here.

Theorem 1. *If* $f(\boldsymbol{\omega}) = o(|\boldsymbol{\omega}|^{-4})$ *as* $|\boldsymbol{\omega}| \to \infty$,

$$m^4 \operatorname{var}\left\{I(Z) - \overline{Z}_m\right\} \to \int_{\mathbb{R}^d} G(\boldsymbol{\omega})\, \mathbf{d}\boldsymbol{\omega}.$$

PROOF. For fixed $\boldsymbol{\omega}$, $m^4 g_m(\boldsymbol{\omega};\mathbf{0})$ is dominated by $G(\boldsymbol{\omega})$ and G is integrable (Exercise 3), which combined with (4) implies

$$m^4 \int_{\mathbb{R}^d} g_m(\boldsymbol{\omega};\mathbf{0})\, \mathbf{d}\boldsymbol{\omega} \to \int_{\mathbb{R}^d} G(\boldsymbol{\omega})\, \mathbf{d}\boldsymbol{\omega} \tag{5}$$

by the dominated covergence theorem. As in Chapter 3, let $\sum'_{\mathbf{j}}$ indicate summation over all $\mathbf{j} \in \mathbb{Z}^d$ other than $\mathbf{0}$. If $f(\boldsymbol{\omega}) = o(|\boldsymbol{\omega}|^{-4})$ as $|\boldsymbol{\omega}| \to \infty$, then

$$m^4 \sum_{\mathbf{j}}{}' g_m(\boldsymbol{\omega};\mathbf{j}) \prod_{\alpha=1}^{d} (1+\omega_\alpha^2) \to 0 \tag{6}$$

as $m \to \infty$ uniformly for $\boldsymbol{\omega} \in A_d(m)$ (Exercise 4), which implies

$$m^4 \int_{\mathbb{R}^d} \sum_{\mathbf{j}}{}' g_m(\boldsymbol{\omega};\mathbf{j})\, \mathbf{d}\boldsymbol{\omega} \to 0. \tag{7}$$

Combining (5) and (7) yields Theorem 1. □

Note that for $d \geq 5$, the mse of order m^{-4} is larger than the reciprocal of the number of observations, or m^{-d}. Since we can always get an mse of order m^{-d} by taking a uniform simple random sample on $[0,1]^d$ of size m^d and averaging the observations, it is tempting to conclude that the centered systematic sample is a poor design in higher dimensions. We show in 5.4 (Exercise 16) that under a quite weak condition on the spectral density, centered systematic sampling together with an appropriate and easily computed weighting of the observations yields a predictor with mse that is $o(m^{-d})$. Thus, it is not true that centered systematic sampling performs worse asymptotically than a simple average based on a simple random sample in high dimensions. It is possible to argue that these asymptotic results are misleading when d is very large, since m^d then grows so quickly with d as to make them irrelevant to practice. However, that is a rather different argument than claiming that systematic sampling is asymptotically inferior to simple random sampling.

Results for sufficiently rough random fields

Theorem 1 showed that when $|\boldsymbol{\omega}|^4 f(\boldsymbol{\omega}) \to 0$ as $|\boldsymbol{\omega}| \to \infty$, the $\mathbf{j} = \mathbf{0}$ term in (2) dominates the mse. The following result essentially shows that when $|\boldsymbol{\omega}|^4 f(\boldsymbol{\omega}) \to \infty$ as $|\boldsymbol{\omega}| \to \infty$, the terms $\mathbf{j} \neq \mathbf{0}$ dominate the mse.

Theorem 2. *Suppose for some p satisfying $d < p < 4$, $f(\boldsymbol{\omega}) \asymp |\boldsymbol{\omega}|^{-p}$ as $|\boldsymbol{\omega}| \to \infty$. Then*

$$\operatorname{var}\left\{I(Z) - \overline{Z}_m\right\} \sim \int_{A_d(m)} \sum_{\mathbf{j}}{}' f(\boldsymbol{\omega} + 2\pi m\mathbf{j}) \prod_{\alpha=1}^{d} \operatorname{sinc}^2\left(\frac{\omega_\alpha}{2}\right) \mathbf{d}\boldsymbol{\omega}.$$

PROOF. For $\boldsymbol{\omega} \in A_d(m)$ and $\mathbf{j} \neq \mathbf{0}$,

$$g_m(\boldsymbol{\omega}; \mathbf{j}) \prod_{\alpha=1}^{d} \left\{\operatorname{sinc}\left(\frac{\omega_\alpha}{2}\right)\right\}^{-2} \asymp (m|\mathbf{j}|)^{-p} \tag{8}$$

(Exercise 5) so that

$$\sum_{\mathbf{j}}{}' \int_{\mathbb{R}^d} g_m(\boldsymbol{\omega}; \mathbf{j})\, \mathbf{d}\boldsymbol{\omega} \asymp m^{-p},$$

which combined with

$$\int_{\mathbb{R}^d} g_m(\boldsymbol{\omega}; \mathbf{0})\, \mathbf{d}\boldsymbol{\omega} \ll m^{-4}(1 + \langle m \rangle^{3-p}), \tag{9}$$

where $\langle m \rangle^q = m^q$ for $q \neq 0$ and $\langle m \rangle^0 = \log m$ (Exercise 6), yields

$$\operatorname{var}\left\{I(Z) - \overline{Z}_m\right\} \sim \sum_{\mathbf{j}}{}' \int_{\mathbb{R}^d} g_m(\boldsymbol{\omega}; \mathbf{j})\, \mathbf{d}\boldsymbol{\omega} \tag{10}$$

as $m \to \infty$. To simplify this result, note that for $\boldsymbol{\omega} \in A_d(m)$ and $\mathbf{j} \neq \mathbf{0}$,

$$\begin{aligned} &\left| \frac{g_m(\boldsymbol{\omega}; \mathbf{j})}{f(\boldsymbol{\omega} + 2\pi m\mathbf{j})} - \prod_{\alpha=1}^{d} \operatorname{sinc}^2\left(\frac{\omega_\alpha}{2}\right) \right| \\ &\quad \leq \prod_{\alpha=1}^{d} \frac{\sin^2\left(\frac{\omega_\alpha}{2}\right)}{m^2 \sin^2\left(\frac{\omega_\alpha}{2m}\right)} \left[\left| \left\{1 - \prod_{\alpha=1}^{d} \frac{(-1)^{j_\alpha} 2m \sin\left(\frac{\omega_\alpha}{2m}\right)}{\omega_\alpha + 2\pi m j_\alpha}\right\}^2 - 1 \right| \right. \\ &\qquad\qquad \left. + \left| 1 - \prod_{\alpha=1}^{d} \operatorname{sinc}^2\left(\frac{\omega_\alpha}{2m}\right) \right| \right]. \end{aligned}$$

Now, for $\boldsymbol{\omega} \in A_d(m)$ and $\mathbf{j} \neq \mathbf{0}$,

$$\prod_{\alpha=1}^{d} \frac{\sin^2\left(\frac{\omega_\alpha}{2}\right)}{m^2 \sin^2\left(\frac{\omega_\alpha}{2m}\right)} \ll \prod_{\alpha=1}^{d} \operatorname{sinc}^2\left(\frac{\omega_\alpha}{2}\right),$$

$$\prod_{\alpha=1}^{d} \frac{(-1)^{j_\alpha} 2m \sin\left(\frac{\omega_\alpha}{2m}\right)}{\omega_\alpha + 2\pi m j_\alpha} \ll \prod_{\alpha: j_\alpha \neq 0} \frac{|\omega_\alpha|}{m|j_\alpha|}$$

and

$$1 - \prod_{\alpha=1}^{d} \operatorname{sinc}^2\left(\frac{\omega_\alpha}{2m}\right) \ll \left|\frac{\boldsymbol{\omega}}{m}\right|^2,$$

so that

$$\left|\frac{g_m(\boldsymbol{\omega};\mathbf{j})}{f(\boldsymbol{\omega}+2\pi m\mathbf{j})} - \prod_{\alpha=1}^{d} \operatorname{sinc}^2\left(\frac{\omega_\alpha}{2}\right)\right| \ll \left|\frac{\boldsymbol{\omega}}{m}\right|^2 \prod_{\alpha=1}^{d} \operatorname{sinc}^2\left(\frac{\omega_\alpha}{2}\right)$$

and

$$\begin{aligned}
&\left|\int_{\mathbb{R}^d} g_m(\boldsymbol{\omega};\mathbf{j})\,\mathbf{d}\boldsymbol{\omega} - \int_{A_d(m)} f(\boldsymbol{\omega}+2\pi m\mathbf{j}) \prod_{\alpha=1}^{d} \operatorname{sinc}^2\left(\frac{\omega_\alpha}{2}\right) \mathbf{d}\boldsymbol{\omega}\right| \\
&\quad \ll (m|\mathbf{j}|)^{-p} m^{-2} \int_{A_d(m)} |\boldsymbol{\omega}|^2 \prod_{\alpha=1}^{d} \operatorname{sinc}^2\left(\frac{\omega_\alpha}{2}\right) \mathbf{d}\boldsymbol{\omega} \\
&\quad \ll (m|\mathbf{j}|)^{-p} m^{-2} \int_{0<\omega_d<\dots<\omega_1<\pi m} \frac{\omega_1^2}{\prod_{\alpha=1}^{d}(1+\omega_\alpha^2)} \mathbf{d}\boldsymbol{\omega} \\
&\quad \ll m^{-p-1}|\mathbf{j}|^{-p}.
\end{aligned}$$

Theorem 2 then follows from (10). □

We can obtain a yet simpler result by making stronger assumptions about f. The following is essentially a special case of Theorem 2 of Stein (1993c); its proof is left as an exercise.

Theorem 3. *Suppose* $f(\boldsymbol{\omega}) \asymp |\boldsymbol{\omega}|^{-p}$ *as* $|\boldsymbol{\omega}| \to \infty$ *for some* $p < 4$ *and there exists a function* $\overline{f} : \mathbb{R}^d \to \mathbb{R}$ *such that for any* $\boldsymbol{\nu} \in \mathbb{R}^d$ *and* $\boldsymbol{\omega} \neq \mathbf{0}$

$$\lim_{t\to\infty} t^p f(\boldsymbol{\nu}+t\boldsymbol{\omega}) = \overline{f}(\boldsymbol{\omega}). \tag{11}$$

Then

$$m^p \operatorname{var}\left\{I(Z) - \overline{Z}_m\right\} \to (2\pi)^{d-p} \sum_{\mathbf{j}}{}' \overline{f}(\mathbf{j}).$$

Note that (11) holds if, for example, $f(\boldsymbol{\omega}) \sim |\mathbf{A}\boldsymbol{\omega}|^{-p}$ as $|\boldsymbol{\omega}| \to \infty$ for some nonsingular $d \times d$ matrix $\mathbf{A}$, in which case $\overline{f}(\boldsymbol{\omega}) = |\mathbf{A}\boldsymbol{\omega}|^{-p}$.

Exercises

1 For a weakly stationary, mean square continuous random field Z on $\mathbb{R}^d$, show that $I(Z)$ can be defined as an L^2 limit of finite sums.

2 Verify $\operatorname{var}\left\{I(Z) - \overline{Z}_m\right\} = \int_{\mathbb{R}^d} g_m(\boldsymbol{\omega})\,\mathbf{d}\boldsymbol{\omega}$.

3 Show that $m^4 g_m(\boldsymbol{\omega};\mathbf{0})/G(\boldsymbol{\omega})$ is bounded in m and $\boldsymbol{\omega}$. Show that G is integrable if $f(\boldsymbol{\omega}) \ll |\boldsymbol{\omega}|^{-3-\epsilon}$ for some $\epsilon > 0$ as $|\boldsymbol{\omega}| \to \infty$.

4 Verify (6).

5 Show that (8) holds for $\boldsymbol{\omega} \in A_d(m)$ and $\mathbf{j} \neq \mathbf{0}$.

6 Verify (9).

7 Prove Theorem 3.

8 Find a simple expression for the asymptotic mse of $\overline{Z}_m$ under (11) with $p = 4$ (Stein 1993c).

9 For $f(\boldsymbol{\omega}) \asymp |\boldsymbol{\omega}|^{-4}$ as $|\boldsymbol{\omega}| \to \infty$, show that

$$\operatorname{var}\left\{I(Z) - \overline{Z}_m\right\} \sim m^{-4} \int_{\mathbb{R}^d} G(\boldsymbol{\omega})\, d\boldsymbol{\omega} + \int_{A_d(m)} \sum_{\mathbf{j}}{}' f(\boldsymbol{\omega} + 2\pi m \mathbf{j}) \prod_{\alpha=1}^{d} \operatorname{sinc}^2\left(\frac{\omega_\alpha}{2}\right) d\boldsymbol{\omega}.$$

Show that the second term on the right side is $O(m^{-4})$ but not $o(m^{-4})$. By combining these results with Theorem 5 in 5.4, show that $\overline{Z}_m$ is not asymptotically optimal when $f(\boldsymbol{\omega}) \asymp |\boldsymbol{\omega}|^{-4}$ as $|\boldsymbol{\omega}| \to \infty$ but does have mse converging at the optimal rate.

5.3 Observations on an infinite lattice

As we saw in 3.8, it is straightforward to calculate certain properties of BLPs if the observations are on the infinite lattice $\delta\mathbf{j}$, $\mathbf{j} \in \mathbb{Z}^d$ for some $\delta > 0$. Here we consider the slightly more general setting where the mean 0 weakly stationary process Z is observed at points $\delta(\mathbf{j} - \boldsymbol{\nu})$ for $\mathbf{j} \in \mathbb{Z}^d$, where $\delta > 0$ and $\boldsymbol{\nu}$ is a fixed point in $[0, 1)^d$. The reason for including $\boldsymbol{\nu}$ is that by taking $\boldsymbol{\nu} = \mathbf{h}$ we get that a centered systematic sample of size m^d is a subset of the infinite lattice with $\delta = m^{-1}$. Therefore, the mse of the BLP based on the infinite lattice provides a lower bound for the mse of any linear predictor based on the centered systematic sample. In particular, if a sequence of integration rules based on centered systematic samples of size m^d has asymptotically the same mse as the BLP based on observing Z at $m^{-1}(\mathbf{j} - \mathbf{h})$ for all $\mathbf{j} \in \mathbb{Z}^d$, then this sequence of rules is necessarily asymptotically optimal relative to all linear predictors based on centered systematic samples.

Asymptotic mse of BLP

This section considers predicting $I(Z; v) = \int_{\mathbb{R}^d} v(\mathbf{x}) Z(\mathbf{x})\, d\mathbf{x}$, where both $\int_{\mathbb{R}^d} v(\mathbf{x})^2 d\mathbf{x}$ and $\operatorname{var}\{I(Z; v)\}$ are positive and finite. In 5.2 we took $v(\mathbf{x}) = 1\{\mathbf{x} \in [0, 1]^d\}$. Let f be the spectral density of Z and set $V(\boldsymbol{\omega}) = \int_{\mathbb{R}^d} v(\mathbf{x}) \exp(i\boldsymbol{\omega}^T \mathbf{x})\, d\mathbf{x}$, so that $\operatorname{var}\{I(Z; v)\} = \int_{\mathbb{R}^d} f(\boldsymbol{\omega}) |V(\boldsymbol{\omega})|^2 d\boldsymbol{\omega}$. Note that we have not assumed v is integrable so we have to interpret its Fourier transform V as an L^2 limit of Fourier transforms of integrable functions (Stein and Weiss 1971, Section I.2). Define

$$S(t) = \int_{A_d(t)} \sum_{\mathbf{j}}{}' f(\boldsymbol{\omega} + 2\pi t\mathbf{j})|V(\boldsymbol{\omega})|^2 \mathbf{d}\boldsymbol{\omega}$$

and let $\hat{I}_\delta(Z;v)$ be the BLP of $I(Z;v)$ based on observing Z at $\delta(\mathbf{j} - \boldsymbol{\nu})$ for all $\mathbf{j} \in \mathbb{Z}^d$ and some fixed $\boldsymbol{\nu} \in [0,1)^d$. The following is a special case of Theorem 4 of Stein (1995a).

Theorem 4. *If* $f(\boldsymbol{\omega}) \asymp (1 + |\boldsymbol{\omega}|)^{-p}$, *then* $\operatorname{var}\left\{I(Z;v) - \hat{I}_\delta(Z;v)\right\} \sim S(\delta^{-1})$ *as* $\delta \downarrow 0$.

Proof. I only consider the case $\boldsymbol{\nu} = \mathbf{0}$ here as it simplifies the notation. See Stein (1995a) for the more general case. The basic idea of the proof is to show that there is a family of linear predictors depending on δ that has $S(\delta^{-1})$ as its asymptotic mse and then to show that the BLP cannot do better asymptotically.

The following simple result is helpful. Suppose $\{c_n\}$ is a sequence of nonnegative and measurable functions on $\mathbb{R}^d$ and $\{a_n\}$ and $\{b_n\}$ are sequences of measurable complex functions on $\mathbb{R}^d$ such that $\int_{\mathbb{R}^d} \{|a_n(\boldsymbol{\omega})|^2 + |b_n(\boldsymbol{\omega})|^2\} c_n(\boldsymbol{\omega})\, \mathbf{d}\boldsymbol{\omega} < \infty$ for all n and

$$\lim_{n\to\infty} \frac{\int_{\mathbb{R}^d} |b_n(\boldsymbol{\omega})|^2 c_n(\boldsymbol{\omega})\, \mathbf{d}\boldsymbol{\omega}}{\int_{\mathbb{R}^d} |a_n(\boldsymbol{\omega})|^2 c_n(\boldsymbol{\omega})\, \mathbf{d}\boldsymbol{\omega}} = 0.$$

Then

$$\int_{\mathbb{R}^d} |a_n(\boldsymbol{\omega}) + b_n(\boldsymbol{\omega})|^2 c_n(\boldsymbol{\omega})\, \mathbf{d}\boldsymbol{\omega} \sim \int_{\mathbb{R}^d} |a_n(\boldsymbol{\omega})|^2 c_n(\boldsymbol{\omega})\, \mathbf{d}\boldsymbol{\omega} \tag{12}$$

as $n \to \infty$ (Exercise 10).

Now every linear predictor based on observing Z at $\delta\mathbf{j}$ for all $\mathbf{j} \in \mathbb{Z}^d$ corresponds to a function in $\mathcal{L}_\delta(f)$, the closed real linear manifold of the functions $\exp(i\delta\boldsymbol{\omega}^T\mathbf{j})$ for $\mathbf{j} \in \mathbb{Z}^d$ with respect to the norm defined by f. Thus, to find a family of predictors that has mse asymptotically equal to $S(\delta^{-1})$, it suffices to find $U_\delta \in \mathcal{L}_\delta(f)$ such that $\int_{\mathbb{R}^d} f(\boldsymbol{\omega})|U_\delta(\boldsymbol{\omega}) - V(\boldsymbol{\omega})|^2 \mathbf{d}\boldsymbol{\omega} \sim S(\delta^{-1})$ as $\delta \downarrow 0$. Let $U_\delta(\boldsymbol{\omega}) = V(\boldsymbol{\omega})$ for $\boldsymbol{\omega} \in A_d(\delta^{-1})$ and take U_δ to have period $2\pi\delta^{-1}$ in each coordinate, so that $U_\delta \in \mathcal{L}_\delta(f)$ (Exercise 11). Next,

$$\int_{A_d(\delta^{-1})^c} f(\boldsymbol{\omega})|V(\boldsymbol{\omega})|^2 \mathbf{d}\boldsymbol{\omega} \ll \delta^p \int_{A_d(\delta^{-1})^c} |V(\boldsymbol{\omega})|^2 \mathbf{d}\boldsymbol{\omega} = o(\delta^p), \tag{13}$$

where the last step holds because v square integrable implies V is as well. Then

$$\int_{\mathbb{R}^d} f(\boldsymbol{\omega})|U_\delta(\boldsymbol{\omega}) - V(\boldsymbol{\omega})|^2 \mathbf{d}\boldsymbol{\omega} \sim S(\delta^{-1}) \tag{14}$$

follows from (12), (13), $U_\delta(\boldsymbol{\omega}) = V(\boldsymbol{\omega})$ on $A_d(\delta^{-1})$ and $\int_{A_d(\delta^{-1})^c} f(\boldsymbol{\omega}) \times |U_\delta(\boldsymbol{\omega})|^2 \mathbf{d}\boldsymbol{\omega} = S(\delta^{-1}) \asymp \delta^p$.

We next show that the BLPs cannot do better asymptotically than the predictors corresponding to U_δ. The element of $\mathcal{L}_\delta(f)$ corresponding to the BLP $\hat{I}_\delta(Z; v)$ is (see 3.8)

$$\hat{V}_\delta(\boldsymbol{\omega}) = \frac{\sum_{\mathbf{j}} f(\boldsymbol{\omega} + 2\pi\delta^{-1}\mathbf{j}) V(\boldsymbol{\omega} + 2\pi\delta^{-1}\mathbf{j})}{\sum_{\mathbf{j}} f(\boldsymbol{\omega} + 2\pi\delta^{-1}\mathbf{j})},$$

so that

$$\begin{aligned} &\operatorname{var}\left\{I(Z; v) - \hat{I}_\delta(Z; v)\right\} \\ &\quad \geq \int_{A_d(\delta^{-1})^c} f(\boldsymbol{\omega}) \mid U_\delta(\boldsymbol{\omega}) - U_\delta(\boldsymbol{\omega}) + \hat{V}_\delta(\boldsymbol{\omega}) - V(\boldsymbol{\omega})|^2 \mathbf{d}\boldsymbol{\omega} \\ &\quad \geq \int_{A_d(\delta^{-1})^c} f(\boldsymbol{\omega}) \left\{ |U_\delta(\boldsymbol{\omega})| - |U_\delta(\boldsymbol{\omega}) - \hat{V}_\delta(\boldsymbol{\omega})| - |V(\boldsymbol{\omega})| \right\}^2 \mathbf{d}\boldsymbol{\omega} \\ &\quad \sim S(\delta^{-1}), \end{aligned} \tag{15}$$

where the last step follows from (12), (13) and

$$\begin{aligned} &\int_{A_d(\delta^{-1})^c} f(\boldsymbol{\omega}) |U_\delta(\boldsymbol{\omega}) - \hat{V}_\delta(\boldsymbol{\omega})|^2 \mathbf{d}\boldsymbol{\omega} \\ &= \int_{A_d(\delta^{-1})} |V(\boldsymbol{\omega}) - \hat{V}_\delta(\boldsymbol{\omega})|^2 \sum_{\mathbf{j}}{}' f(\boldsymbol{\omega} + 2\pi\delta^{-1}\mathbf{j})\, \mathbf{d}\boldsymbol{\omega} \\ &\ll \delta^p \int_{A_d(\delta^{-1})} f(\boldsymbol{\omega})^{-2} \left| \sum_{\mathbf{j}}{}' f(\boldsymbol{\omega} + 2\pi\delta^{-1}\mathbf{j}) \left\{ V(\boldsymbol{\omega}) - V(\boldsymbol{\omega} + 2\pi\delta^{-1}\mathbf{j}) \right\} \right|^2 \mathbf{d}\boldsymbol{\omega} \\ &\ll \delta^{3p} \int_{A_d(\delta^{-1})} (1 + |\boldsymbol{\omega}|)^{2p} |V(\boldsymbol{\omega})|^2 \mathbf{d}\boldsymbol{\omega} \\ &\quad + \int_{A_d(\delta^{-1})} \sum_{\mathbf{j}}{}' f(\boldsymbol{\omega} + 2\pi\delta^{-1}\mathbf{j}) |V(\boldsymbol{\omega} + 2\pi\delta^{-1}\mathbf{j})|^2 \mathbf{d}\boldsymbol{\omega} \\ &= o(\delta^p). \end{aligned} \tag{16}$$

Exercise 12 asks you to provide the details for (16). Theorem 4 follows from (14) and (15). □

Let us examine what makes this proof work. If $f(\boldsymbol{\omega}) \asymp (1 + |\boldsymbol{\omega}|)^{-p}$, then any family of predictors t_δ with mse tending to 0 as $\delta \downarrow 0$ must have corresponding functions in $\hat{t}_\delta \in \mathcal{L}_\delta(f)$ satisfying $\int_B |\hat{t}_\delta(\boldsymbol{\omega}) - V(\boldsymbol{\omega})|^2 \mathbf{d}\boldsymbol{\omega} \to 0$ as $\delta \downarrow 0$ for any bounded set B. Since $\hat{t}_\delta$ is periodic and V is "small" at high frequencies, for δ small, there is then no way to avoid a contribution to the mse from frequencies outside $A_d(\delta^{-1})$ of approximately $S(\delta^{-1})$.

Asymptotic optimality of simple average

Theorems 2 and 4 yield that $\overline{Z}_m$ is asymptotically optimal for $I(Z)$ if $f(\boldsymbol{\omega}) \asymp (1 + |\boldsymbol{\omega}|)^{-p}$ for $p < 4$. For $p > 4$, it turns out that $\overline{Z}_m$ has mse tending to 0 at a slower rate than the mse of the BLP, which follows from Theorem 1 in 5.2 and Theorem 5 in the next section. Stein (1995a) shows that for v sufficiently smooth and $p < 4$, var $\{I(Z; v) - \overline{Z}_m(v)\} \sim S(m)$, where

$$\overline{Z}_m(v) = m^{-d} \sum_{\mathbf{j} \in \mathcal{G}_m} v\left(\frac{\mathbf{j} - \mathbf{h}}{m}\right) Z\left(\frac{\mathbf{j} - \mathbf{h}}{m}\right).$$

Hence, $\overline{Z}_m(v)$ is asymptotically optimal for $p < 4$. Stein (1995a) also shows how to modify this predictor so that it is asymptotically optimal for $p \geq 4$ using a generalization of the procedure outlined in the next section.

Exercises

10 Prove (12).

11 Show that the function U_δ defined in the proof of Theorem 4 is in $\mathcal{L}_\delta(f)$.

12 Provide the details for (16).

13 Show by example that the conclusion of Theorem 4 may not hold if only $f(\boldsymbol{\omega}) \asymp |\boldsymbol{\omega}|^{-p}$ as $|\boldsymbol{\omega}| \to \infty$ is assumed.

5.4 Improving on the sample mean

Let us consider improving upon $\overline{Z}_m$ as a predictor of $I(Z)$ when Z is smooth. In 5.2, we showed that if $f(\boldsymbol{\omega}) \asymp (1 + |\boldsymbol{\omega}|)^{-p}$ with $p > 4$, then for the integral on the right side of (1), frequencies in $A_d(m)$ produce a term of order m^{-4} in the mse and frequencies outside $A_d(m)$ a term of order m^{-p}. We see that we need to find a better approximation to $\int_{[0,1]^d} \exp(i\boldsymbol{\omega}^T\mathbf{x})d\mathbf{x}$ at low frequencies than is given by $\sum_{\mathbf{j} \in \mathcal{G}_m} m^{-d} \exp\left\{im^{-1}\boldsymbol{\omega}^T(\mathbf{j} - \mathbf{h})\right\}$.

Approximating $\int_0^1 \exp(i\nu t)dt$

For $d = 1$, we seek a more accurate approximation to $\int_0^1 \exp(i\nu t)dt = \exp(i\nu/2)\,\mathrm{sinc}(\nu/2)$ than

$$\frac{1}{m}\sum_{j=1}^{m} \exp\left\{im^{-1}\nu(j - \tfrac{1}{2})\right\} = \frac{\exp\left(\frac{i\nu}{2}\right)\sin\left(\frac{\nu}{2}\right)}{m\sin\left(\frac{\nu}{2m}\right)}$$

for $|\nu| \le m\pi$. Although there is more than one way to do this, let us consider functions of the form

$$\phi_m(\nu; \mathbf{a}_q) = \frac{1}{m}\sum_{j=1}^{m} \exp\left\{\frac{i\nu(j-\frac{1}{2})}{m}\right\} + \frac{1}{m}\sum_{j=1}^{q} a_j \left[\exp\left\{\frac{i\nu(j-\frac{1}{2})}{m}\right\} + \exp\left\{\frac{i\nu(m-j+\frac{1}{2})}{m}\right\}\right],$$

where $\mathbf{a}_q = (a_1, \ldots, a_q)$ and $m > 2q$. Straightforward calculation yields

$$\phi_m(\nu; \mathbf{a}_q) = \exp\left(\frac{i\nu}{2}\right)\left[\frac{1}{m}\sin\left(\frac{\nu}{2}\right)\left\{\csc\left(\frac{\nu}{2m}\right) + 2\sum_{j=1}^{m} a_j \sin\left(\frac{j-\frac{1}{2}}{m}\nu\right)\right\} + \frac{2}{m}\cos\left(\frac{\nu}{2}\right)\sum_{j=1}^{m} a_j \cos\left(\frac{j-\frac{1}{2}}{m}\nu\right)\right].$$

Using

$$\csc(t) = t^{-1} + \sum_{\ell=1}^{k} \frac{(-1)^{\ell-1}2(2^{2\ell-1}-1)B_{2\ell}}{(2\ell)!} t^{2\ell-1} + O\left(|t|^{2k+1}\right)$$

for $|t| \le \pi/2$, where B_n is the nth Bernoulli number (Abramowitz and Stegun 1965, 4.3.68), we get that if

$$\sum_{j=1}^{q} a_j(2j-1)^{2r} = 0 \quad \text{for} \quad r = 0, \ldots, s \tag{17}$$

and

$$\sum_{k=1}^{q} a_k(2k-1)^{2r-1} = -\frac{2^{2r-1}-1}{2r} B_{2r} \quad \text{for} \quad r = 1, \ldots, s, \tag{18}$$

then

$$\left|\exp\left(\frac{i\nu}{2}\right)\operatorname{sinc}\left(\frac{\nu}{2}\right) - \phi_m(\nu; \mathbf{a}_q)\right| \ll \frac{|\nu|^{2s+1}}{m^{2s+2}} \tag{19}$$

for $|\nu| \le \pi m$ (Exercise 14). Furthermore, by taking $q = 2s+1$, (17) and (18) give $2s+1$ equations in the $2s+1$ components of $\mathbf{a}_{2s+1}$ and this system of linear equations has a unique solution (Exercise 15), denoted by $\tilde{\mathbf{a}}_{2s+1}$. For example, $\tilde{\mathbf{a}}_3 = (1/12, -1/8, 1/24)$ and $\tilde{\mathbf{a}}_5 = (101/640, -2213/5760, 143/384, -349/1920, 103/5760)$.

We see that by modifying just the weights assigned to observations near the ends of the interval, we are able to get a sharper approximation to $\exp(i\nu/2)\operatorname{sinc}(\nu/2)$ at low frequencies. Another way to think about these modifications is in terms of the Euler–Maclaurin formula, which, for a function h on $[0,1]$, gives approximations to $\int_0^1 h(t)\,dt$ in terms of the values of

h at j/n for $j = 0, \dots, n$ and derivatives of h at 0 and 1. For h possessing $2k-1$ derivatives on $[0,1]$, define the kth order Euler–Maclaurin rule

$$R_k(h) = \frac{1}{2n}\{h(0) + h(1)\} + \frac{1}{n}\sum_{j=1}^{n-1} h\left(\frac{j}{n}\right) - \sum_{\ell=1}^{k} \frac{B_{2\ell}}{(2\ell)!n^{2\ell}}\{h^{(2\ell-1)}(1) - h^{(2\ell-1)}(0)\},$$

where the last sum is set to 0 if $k = 0$ so that $k = 0$ corresponds to the trapezoidal rule. If h has a bounded derivative of order $2k+2$ on $[0,1]$, then

$$\left|\int_0^1 h(t)dt - R_k(h)\right| \le \frac{|B_{2k+2}|}{(2k+2)!n^{2k+2}} \sup_{0\le t\le 1} |h^{(2k+2)}(t)|$$

(Abramowitz and Stegun 1965, p. 886). We can use $R_k(h)$ to integrate a stochastic process if it possesses $2k-1$ mean square derivatives. Alternatively, as in Benhenni and Cambanis (1992), we can approximate the derivatives at the endpoints using finite differences and avoid the need to observe these derivatives. This approach gives rules very similar in spirit to the ones described in the preceding paragraph.

Approximating $\int_{[0,1]^d} \exp(i\boldsymbol{\omega}^T\mathbf{x})d\mathbf{x}$ in more than one dimension

As the number of dimensions increases, the use of $\overline{Z}_m$ to predict $I(Z)$ becomes increasingly problematic. We need to find a better approximation at low frequencies to $\int_{[0,1]^d} \exp(i\boldsymbol{\omega}^T\mathbf{x})d\mathbf{x}$ than is given by $\sum_{\mathbf{j}\in\mathcal{G}_m} m^{-d}\exp\{im^{-1}\boldsymbol{\omega}^T(\mathbf{j}-\mathbf{h})\}$. For $s > 1$, $\prod_{\alpha=1}^d \phi_m(\omega_\alpha; \tilde{\mathbf{a}}_{2s+1})$ provides such an approximation. More specifically, for $m \ge 4s+2$, define the predictor

$$\overline{Z}_{m,s} = m^{-d}\sum_{\mathbf{j}\in\mathcal{G}_m} b_{\mathbf{j}} Z\big(m^{-1}(\mathbf{j}-\mathbf{h})\big),$$

where $b_{\mathbf{j}} = \prod_{q=1}^{d} \beta_{j_q}$, $\beta_j = 1$ if $2s+2 \le j \le m-2s-1$ and $\beta_j = \beta_{m-j} = 1 + a_j$ for $1 \le j \le 2s+1$. Note that $\overline{Z}_{m,0} = \overline{Z}_m$. We then get

$$\begin{aligned}
&\int_{A_d(m)} \left| \prod_{\alpha=1}^{d} \operatorname{sinc}\left(\frac{\omega_\alpha}{2}\right) - \prod_{\alpha=1}^{d} \phi_m(\omega_\alpha; \tilde{\mathbf{a}}_{2s+1}) \right|^2 f(\boldsymbol{\omega})\, d\boldsymbol{\omega} \\
&\quad \ll \int_{A_d(m)} \left| \frac{\boldsymbol{\omega}}{m} \right|^{4s+4} \prod_{\alpha=1}^{d} \operatorname{sinc}^2\left(\frac{\omega_\alpha}{2}\right) (1 + |\boldsymbol{\omega}|)^{-p}\, d\boldsymbol{\omega} \\
&\quad \ll m^{-4s-4} \int_{0<\omega_d<\dots<\omega_1<\pi m} (1+\omega_1)^{4s+2-p} \prod_{\alpha=2}^{d} \frac{1}{1+\omega_\alpha^2}\, d\boldsymbol{\omega} \\
&\quad \ll m^{-4s-4} \left(1 + \langle m \rangle^{4s+3-p}\right) \\
&\quad = o(m^{-p})
\end{aligned}$$

if $4s+4 > p$.

Asymptotic properties of modified predictors

By using $\overline{Z}_{m,s}$ for $4s+4 > p$, we again get the high frequencies of f dominating the mse as we did for $\overline{Z}_m$ when $p < 4$. In particular, by an argument similar to the proof of Theorem 2, we get the following generalization of that result.

Theorem 5. *Suppose for some $p > d$, $f(\boldsymbol{\omega}) \asymp |\boldsymbol{\omega}|^{-p}$ as $|\boldsymbol{\omega}| \to \infty$. Then for $4s+4 > p$,*

$$\operatorname{var}\left\{I(Z) - \overline{Z}_{m,s}\right\} \sim \int_{A_d(m)} {\sum_{\mathbf{j}}}' f(\boldsymbol{\omega} + 2\pi m \mathbf{j}) \prod_{\alpha=1}^{d} \operatorname{sinc}^2\left(\frac{\omega_\alpha}{2}\right) d\boldsymbol{\omega}.$$

Theorems 4 and 5 imply the following.

Corollary 6. *If $f(\boldsymbol{\omega}) \asymp (1+|\boldsymbol{\omega}|)^{-p}$ and $4s+4 > p$, then $\overline{Z}_{m,s}$ is an asymptotically optimal predictor for $I(Z)$.*

We also have an analogue to Theorem 3.

Theorem 7. *If f satisfies (11), then for $4s+4 > p$,*

$$\operatorname{var}\left\{I(Z) - \overline{Z}_{m,s}\right\} \sim m^{-p} (2\pi)^{d-p} {\sum_{\mathbf{j}}}' \overline{f}(\mathbf{j}).$$

The fact that, independent of d, $\overline{Z}_m$ is asymptotically optimal for $p < 4$ and converges at a suboptimal rate for $p > 4$ is noteworthy (see Exercise 9 of 5.2 for the case $p = 4$). We must have $p > d$ for f to be integrable so that the range of p for which $\overline{Z}_m$ is asymptotically optimal narrows as d increases from 1 to 3 and $\overline{Z}_m$ is not asymptotically optimal for any p when $d \ge 4$. Since the suboptimality of $\overline{Z}_m$ for $p \ge 4$ is due to poorly chosen weights given to observations near an edge of $[0,1]^d$ and, as d increases, more of the observations are near an edge of $[0,1]^d$, it is not surprising that the constraints on an isotropic random field Z become increasingly severe as d increases in order for $\overline{Z}_m$ to be asymptotically optimal.

Are centered systematic samples good designs?

If asymptotically optimal predictors are used, Ritter (1995) showed that centered systematic sampling cannot do too much worse than any other sampling design for random fields that are not too far from isotropic. More specifically, suppose $\{D_j\}_{j=1}^{\infty}$ is a sequence of finite subsets of $[0,1]^d$ with D_j containing n_j points and $n_j \to \infty$ as $j \to \infty$. If $f(\boldsymbol{\omega}) \asymp |\boldsymbol{\omega}|^{-p}$ as $|\boldsymbol{\omega}| \to \infty$, then the sequence of BLPs of $I(Z)$ based on the points in D_j cannot have mses that are $o\big(n_j^{-p}\big)$ as $j \to \infty$ (Ritter 1995). Thus, centered systematic sampling achieves the best possible rate of convergence to 0 for the mse. Furthermore, results in Stein (1995b) suggest that if the random field is isotropic, centered systematic samples will sometimes do very nearly as well asymptotically as the best possible designs. If the random field possesses a tensor product autocovariance function (see 2.11), then BLPs based on centered systematic sampling can be badly suboptimal (Ylvisaker 1975; Ritter 1995).

Exercises

14 Verify (19) for $|\nu| \leq \pi m$.

15 Show that for $q = 2s + 1$, there is a unique solution to (17) and (18).

16 Show that if $f(\boldsymbol{\omega}) = o\big(|\boldsymbol{\omega}|^{-d}\big)$ as $|\boldsymbol{\omega}| \to \infty$, then for $4s + 4 > d$, $\operatorname{var}\{I(Z) - \overline{Z}_{m,s}\} = o\big(m^{-d}\big)$. Thus, under this mild condition on f, the mse of the BLP based on the centered systematic sample is $o(m^{-d})$ as $m \to \infty$ and hence is better asymptotically than taking a uniform simple random sample on $[0,1]^d$ of size m^d and averaging the observations.

17 Continuation of 16. For $d = 1$, show by example that if no conditions are placed on f, then $m \operatorname{var}\{I(Z) - \overline{Z}_m\}$ may not tend to 0 as $m \to \infty$.

18 Prove Theorem 5.

19 Prove Theorem 7.

5.5 Numerical results

This section looks at some results for finite m and $d = 1$. Stein (1993c) provides some numerical results for $d = 2$. As a first example, suppose

TABLE 1. Mean squared errors for predicting $I(Z)$ when $K(t) = \exp(-4|t|)(1 + 4|t| + 16t^2/3)$.

m	$\mathrm{mse}(\hat{Z}_m)$	$\mathrm{mse}(\overline{Z}_m)$	$\mathrm{mse}(\overline{Z}_{m,1})$
16	2.78×10^{-8}	3.53×10^{-7}	3.70×10^{-8}
32	3.48×10^{-10}	2.14×10^{-8}	4.16×10^{-10}
48	2.64×10^{-11}	4.21×10^{-9}	3.02×10^{-11}

$K(t) = \exp(-4|t|)(1 + 4|t| + 16t^2/3)$, for which the corresponding spectral density is $f(\omega) = 2^{13}/\{3\pi(16 + \omega^2)^3\}$. The results in Table 1 for predicting $I(Z) = \int_0^1 Z(t)\,dt$ at least qualitatively agree with the asymptotics. Specifically, since $p = 6 > 4$, the simple average $\overline{Z}_m$ is badly suboptimal, particularly for larger m. The modified predictor $\overline{Z}_{m,1}$ performs much better, although even for $m = 48$, it has mse 14% larger than that of the BLP $\hat{Z}_m$. Note that all integrals required to obtain these results can be computed analytically (Exercise 20), so that numerical integration is not needed.

Theorem 5 shows that asymptotically there is no penalty for using $\overline{Z}_{m,s}$ with s larger than necessary. For finite m, using s too large does tend to give larger mses. Table 2 shows what happens when predicting $I(Z)$ and the autocovariance function is $K(t) = e^{-|t|}$. Here, $f(\omega) \asymp (1 + |\omega|)^{-2}$ so that $\overline{Z}_{m,s}$ is asymptotically optimal for all nonnegative integers s. We see that $\overline{Z}_m$ is very nearly optimal for all m considered, $\overline{Z}_{m,1}$ does somewhat worse but is still within 2% of optimal even for $m = 16$, and $\overline{Z}_{m,2}$ does noticeably worse, although it is within 5% of optimal for $m = 48$. Comparing Tables 1 and 2, it is apparent that the penalty for choosing s too small is much more severe than for choosing s too large, as the asymptotic results predict.

The fact that it is possible to find an asymptotically optimal predictor for $I(Z)$ by choosing any integer s such that $4s + 4 > p$ and then using $\overline{Z}_{m,s}$ indicates that prediction of integrals is particularly insensitive to misspecification of the spectral density. The results of Chapter 4 show that all prediction problems are insensitive to misspecification of low frequency behavior under fixed-domain asymptotics. The results here indicate that integrals may be predicted nearly optimally without knowing the high frequency behavior of the spectral density well, either.

TABLE 2. Mean squared errors for predicting $I(Z)$ when $K(t) = e^{-|t|}$.

m	$\hat{Z}_m$	$\overline{Z}_m$	$\overline{Z}_{m,1}$	$\overline{Z}_{m,2}$
16	6.499×10^{-4}	6.510×10^{-4}	6.595×10^{-4}	7.474×10^{-4}
32	1.627×10^{-4}	1.628×10^{-4}	1.638×10^{-4}	1.748×10^{-4}
48	7.232×10^{-5}	7.234×10^{-5}	7.265×10^{-5}	7.591×10^{-5}

Exercises

20 For $K(t) = \exp(-a|t|)\left(1 + a|t| + \frac{1}{3}a^2t^2\right)$, find cov $\left\{Z(s), \int_0^1 Z(t)\,dt\right\}$ for $0 \leq s \leq 1$ and var $\left\{\int_0^1 Z(t)\,dt\right\}$.

21 Reproduce the results in Table 1. Try to extend these results to larger m. You may run into numerical problems for m not much larger than 64. For example, for $m = 64$, S-Plus gives the condition number (the ratio of the largest to smallest eigenvalue) of the covariance matrix of the observations as 4.6×10^9 and refuses to calculate its QR decomposition due to its apparent near singularity. A good project would be to develop methods other than using higher-precision arithmetic to ameliorate these numerical difficulties.

6

Predicting With Estimated Parameters

6.1 Introduction

Chapters 3 and 4 examined the behavior of pseudo-BLPs. Although the results given there provide an understanding of how linear predictors depend on the spectral density of a stationary random field, they do not directly address the more practically pertinent problem of prediction when parameters of a model must be estimated from the same data that are available for prediction. The reason I have avoided prediction with estimated parameters until now is that it is very hard to obtain rigorous results for this problem. The basic difficulty is that once we have to estimate any parameters of the covariance structure, "linear" predictors based on these estimates are no longer actually linear since the coefficients of the predictors depend on the data.

The sort of theory one might hope to develop is that, as the number of observations increases, it is generally possible to obtain:

(A) asymptotically optimal predictors, and
(B) asymptotically correct assessments of mean squared prediction errors

even when certain unknown parameters are estimated. Such general results do exist for predicting future values of a time series observed on the integers with finite-dimensional parameter spaces (Toyooka 1982 and Fuller 1996, Section 8.5). Gidas and Murua (1997) prove that if a continuous time series is observed at $\delta, 2\delta, \ldots, T\delta$, where both δ^{-1} and δT tend to infinity, then (A) and (B) are generally possible for predictions a fixed amount of time after

$T\delta$. The results in all of these works require that the unknown autocovariance function can be consistently estimated as the number of observations increases. Under fixed-domain asymptotics, the results on equivalence of Gaussian measures in 4.2 show that there can quite naturally be parameters that cannot be consistently estimated as the number of observations increases. Indeed, Yakowitz and Szidarovszky (1985, Section 2.4) essentially claim that the impossibility of consistently estimating the autocovariance function based on observations in a fixed region implies that (A) and (B) are unachievable. The results in Chapter 4 show that, at least for Gaussian random fields, this line of reasoning is inadequate. Specifically, Theorems 8 and 10 in 4.3 demonstrate that there is no need to distinguish between equivalent Gaussian measures in order to obtain asymptotically optimal predictions.

These theorems do not by themselves imply (A) and (B). Indeed, direct analogues to Theorems 8 and 10 in 4.3 will not generally be possible for predictions based on estimated models. The problem, as I discuss in 6.8, has to do with the uniformity of these results over all possible predictions. If one restricts the class of predictands appropriately, then I expect that rigorous results in support of (A) and (B) are obtainable. Putter and Young (1998) provide the first step of an approach to proving (A) and (B) for predictions based on estimated parameters, although much remains to be done to obtain any such result when using, as I advocate, the Matérn model for the autocovariance function.

This chapter provides theorems, heuristic derivations, numerical calculations and a simulated example concerning the estimation of autocovariance functions and prediction of random fields based on these estimates. Section 6.2 describes Matheron's notion of microergodicity, which is closely related to equivalence and orthogonality of measures and which plays a crucial role in thinking about whether (A) and (B) should be possible under fixed-domain asymptotics. Section 6.3 demonstrates a crucial flaw in an argument due to Matheron (1971, 1989) that purports to show that (B) is unachievable for predicting integrals of sufficiently smooth random fields.

Section 6.4 describes maximum likelihood and restricted maximum likelihood estimation for the parameters of the covariance function of a Gaussian random field. In many settings, as the number of observations increases, maximum likelihood estimates are asymptotically normal with mean equal to the true value of the parameter vector and covariance matrix given by the inverse of the Fisher information matrix. Section 6.4 briefly describes such standard asymptotic results and explains why they often do not hold under fixed-domain asymptotics.

Section 6.5 advocates the Matérn class as a canonical class of autocovariance functions for spatial interpolation problems. Recall from 2.10 that the general form of the Matérn spectral density of an isotropic random field on $\mathbb{R}^d$ is $f(\boldsymbol{\omega}) = \phi(\alpha^2 + |\boldsymbol{\omega}|^2)^{-\nu-d/2}$. The critical parameter here is ν, which controls the degree of differentiability of the underlying random field. Any class of models that does not include a parameter allowing for a varying degree of differentiability of the random field is, in my opinion, untenable for general usage when interpolating random fields. Thus, in particular, a standard semivariogram model such as the spherical (see 2.10) should not be used unless there is some credible a priori reason to believe the semivariogram must behave linearly near the origin. The same criticism applies to the exponential model, even though it is a special case of the Matérn model with $\nu = \frac{1}{2}$.

Section 6.6 investigates numerically the Fisher information matrix for the parameters of the Matérn model in various settings, including cases in which there are measurement errors. An important finding of 6.6 is that evenly spaced observations can lead to great difficulty in estimating the parameters of the Matérn model.

Theorem 1 in Section 6.7 derives fixed-domain asymptotic properties of maximum likelihood estimates for a class of periodic random fields closely related to the Matérn class. I would expect that similar results hold for estimating the parameters of the (nonperiodic) Matérn class itself, but cannot prove such a claim.

Section 6.8 considers some properties of the commonly used plug-in method for prediction and assessment of mses, in which unknown parameters of the autocovariance function are estimated and then these estimates are treated as if they were the truth. In particular, I give an approximate frequentist formulation of Jeffreys's law relating the additional information a predictand has about unknown parameters beyond that contained in the observations to the effect on the prediction of having to estimate these parameters. This approximation should be compared to the exact Bayesian formulation of Jeffreys's law given in 4.4. The approximation is easily computed and provides the basis of a numerical study on the effect of estimation on subsequent predictions.

Section 6.9 considers an example based on simulated data showing serious problems with some commonly used methods in spatial statistics when the process under investigation is differentiable.

Section 6.10 describes and advocates the Bayesian approach as the best presently available method for accounting for the effect of the uncertainty in the unknown parameters on predictions. However, it turns out that the prior distributions on unknown parameters that are a necessary part of any Bayesian analysis need to be chosen with some care.

6.2 Microergodicity and equivalence and orthogonality of Gaussian measures

Matheron (1971, 1989) discusses fixed-domain asymptotics and its relationship to issues of statistical inference. In these works he considers the notion of microergodicity for random fields observed on a bounded domain. For a class of probability models $\{P_{\boldsymbol{\theta}} : \boldsymbol{\theta} \in \Theta\}$ for a random field on a given bounded domain R and a function h on Θ, he effectively defines $h(\boldsymbol{\theta})$ to be microergodic if, for any $\boldsymbol{\theta} \in \Theta$, $h(\boldsymbol{\theta})$ can be determined correctly with probability 1 based on observing a single realization of the random field on R. It immediately follows that if there exists $\boldsymbol{\theta}_0, \boldsymbol{\theta}_1 \in \Theta$ such that $h(\boldsymbol{\theta}_0) \neq h(\boldsymbol{\theta}_1)$ but $P_{\boldsymbol{\theta}_0} \equiv P_{\boldsymbol{\theta}_1}$, then $h(\boldsymbol{\theta})$ is not microergodic. On the other hand, if for all $\boldsymbol{\theta}, \boldsymbol{\theta}' \in \Theta$, $h(\boldsymbol{\theta}) \neq h(\boldsymbol{\theta}')$ implies $P_{\boldsymbol{\theta}} \perp P_{\boldsymbol{\theta}'}$, it is at least plausible that one can determine the correct value of $h(\boldsymbol{\theta})$ with probability 1. This would follow if Θ were a countable set (Exercise 3), but can be false if the parameter space is uncountable (Exercise 4). Since Matheron does not give a precise mathematical definition of microergodic, for convenience, I define $h(\boldsymbol{\theta})$ to be microergodic if for all $\boldsymbol{\theta}, \boldsymbol{\theta}' \in \Theta$, $h(\boldsymbol{\theta}) \neq h(\boldsymbol{\theta}')$ implies $P_{\boldsymbol{\theta}} \perp P_{\boldsymbol{\theta}'}$.

In practice, determining when $h(\boldsymbol{\theta})$ can be estimated well based on a large number of observations of the random field Z spread throughout R is more important than determining microergodicity, although the two problems are related. Suppose $\mathbf{x}_1, \mathbf{x}_2, \ldots$ is a dense sequence of points in R and $\mathbf{Z}_n = (Z(\mathbf{x}_1), \ldots, Z(\mathbf{x}_n))^T$. Let $\hat{\boldsymbol{\theta}}_n$ be an estimator of $\boldsymbol{\theta}$ based on $\mathbf{Z}_n$. Then $h(\hat{\boldsymbol{\theta}}_n)$ is said to be a consistent estimator of $h(\boldsymbol{\theta})$ if $h(\hat{\boldsymbol{\theta}}_n)$ converges in probability to $h(\boldsymbol{\theta})$ under $P_{\boldsymbol{\theta}}$ for all values of $\boldsymbol{\theta} \in \Theta$. We might generally expect that if $h(\boldsymbol{\theta})$ is microergodic, then there exists a sequence of estimators $\hat{\boldsymbol{\theta}}_n$ such that $h(\hat{\boldsymbol{\theta}}_n)$ is consistent for $h(\boldsymbol{\theta})$. However, such a result cannot be true without some further assumptions (see Exercise 4) and, even when it is true, is often difficult to prove. Wald's classic paper on the consistency of maximum likelihood estimates (Wald 1949; Ferguson 1996, Chapter 17) provides considerable insight into the issues involved in proving consistency of estimators.

For Gaussian measures, we can use the results in 4.2 to determine which functions of a parameter are microergodic. For example, suppose Z is a stationary Gaussian process, $R = [0,1]$ and the class of probability models is $P_{\boldsymbol{\theta}} = G_R(0, K_{\boldsymbol{\theta}})$, where $\boldsymbol{\theta} = (\theta_1, \theta_2)$, $\Theta = (0,\infty) \times (0,\infty)$ and $K_{\boldsymbol{\theta}}(t) = \theta_1 e^{-\theta_2|t|}$ is the class of autocovariance functions. From (24) in 4.2, it follows that $P_{\boldsymbol{\theta}} \equiv P_{\boldsymbol{\theta}'}$ if and only if $\theta_1\theta_2 = \theta_1'\theta_2'$ and they are otherwise orthogonal. Thus, $\theta_1\theta_2$ is microergodic, but neither θ_1 nor θ_2 are. Ying (1991) provides detailed results on the estimation of $\boldsymbol{\theta}$ for this model based on observations in a bounded interval. As an example of a microergodic quantity when there is no finite-dimensional model for the autocovariance function, suppose that Z is a stationary Gaussian process, $R = [0,1]$ and let Θ index the class of all autocovariance functions K on $\mathbb{R}$ for which $K'(0^+)$ exists and is in $(-\infty, 0)$. Theorem 1 in 4.2 shows that for all such autocovariance functions K, $K'(0^+)$ can be determined with probability 1 and hence is microergodic.

If we consider some class of models for a stationary mean 0 Gaussian process on $R = [0,1]$ with $K''(0)$ existing and finite, then whether $K''(0)$ is microergodic depends on the class of models. If the class of models is all autocovariance functions for which $K''(0)$ exists and is finite, then $K''(0)$ is not microergodic. To see this, consider $K_0(t) = e^{-|t|}(1+|t|)$ and $K_1(t) = \frac{1}{8}e^{-2|t|}(1+2|t|)$, for which $K_0''(0) = -1$ and $K_1''(0) = -\frac{1}{2}$. Since, by (24) in 4.2, $G_R(0, K_0) \equiv G_R(0, K_1)$, $K''(0)$ is not microergodic. If the class of models is $K_{\boldsymbol{\theta}}(t) = \theta_1 e^{-\theta_2|t|}(1+\theta_2|t|)$, $\boldsymbol{\theta} = (\theta_1, \theta_2) \in (0,\infty)\times(0,\infty)$, then $-\theta_1\theta_2^2 = K''(0)$ is still not microergodic, but $2\theta_1\theta_2^3 = K'''(0^+)$ is (Exercise 5). On the other hand, if the class of models is $K_\theta(t) = \theta K(t)$ where $\theta \in (0,\infty)$, K is an autocovariance function possessing a spectral density and $K''(0)$ exists, then θ and hence $K_\theta''(0)$ is microergodic, which follows from Exercise 6. However, assuming the autocovariance function is known up to a scalar multiplier is highly restrictive. Finally, if $K_{\boldsymbol{\theta}}(t) = \theta_1 e^{-\theta_2 t^2}$ with $\boldsymbol{\theta} = (\theta_1, \theta_2) \in (0,\infty) \times (0,\infty)$, then $\boldsymbol{\theta}$ is microergodic (Exercise 8) and hence so is $K_{\boldsymbol{\theta}}''(0)$. This last example makes use of the unusual properties of processes with analytic autocovariance functions and should be considered atypical.

Observations with measurement error

This subsection argues that measurement errors should generally have no effect on which parameters of a model for a continuous random field are microergodic. To be more specific, consider the infinite sequence of observations $Y_i = Z(\mathbf{x}_i) + U_i$ for $i = 1, 2, \ldots$ where Z is a mean square continuous Gaussian random field on $\mathbb{R}^d$, the U_is are independent $N(0,\sigma^2)$ random variables that are independent of Z and $\mathbf{x}_1, \mathbf{x}_2, \ldots$ is a dense sequence in some set $R \subset \mathbb{R}^d$ such that every point in R is a limit point of R. Furthermore, suppose the mean and covariance function for Z on R are known up to some finite-dimensional parameter $\boldsymbol{\theta} \in \Theta$ and denote by $P_{\boldsymbol{\theta}}$ the Gaussian measure for Z on R as a function of $\boldsymbol{\theta}$.

If $h(\boldsymbol{\theta})$ is microergodic when Z is observed everywhere on R, then as discussed in the preceding subsection, we would commonly expect that $h(\boldsymbol{\theta})$ can be consistently estimated based on $\mathbf{Z}_n = (Z(\mathbf{x}_1), \ldots, Z(\mathbf{x}_n))^T$ as $n \to \infty$. If so, then we should generally have that $h(\boldsymbol{\theta})$ is consistently estimable based on $\mathbf{Y}_n = (Y_1, \ldots, Y_n)^T$ as $n \to \infty$. To see why this should be the case, note that for any fixed j, $Z(\mathbf{x}_j)$ can be predicted arbitrarily well in terms of $\mathbf{Y}_n$ as $n \to \infty$ (see the proof of Theorem 6 in 4.2). Thus, if $\boldsymbol{\theta}$ can be estimated well based on $\mathbf{Z}_{n_0}$, then by choosing n sufficiently large, $\mathbf{Y}_n$ can be used to predict each component of $\mathbf{Z}_{n_0}$ arbitrarily well, which suggests that $\boldsymbol{\theta}$ can also be estimated well based on $\mathbf{Y}_n$. Furthermore, σ^2 will also be consistently estimable based on $\mathbf{Y}_n$ as $n \to \infty$. Exercise 9 asks you to prove this by using the fact that if $|\mathbf{x}_i - \mathbf{x}_j|$ is small, then $E(Y_i - Y_j)^2$ equals $2\sigma^2$ plus something small.

These arguments suggest that whatever parameters can be consistently estimated when noise-free observations are available can still be consistently estimated from noisy observations. In addition, the measurement error variance is always consistently estimable. However, we should also expect good estimators of the microergodic parameters of Z based on $\mathbf{Y}_n$ to be substantially less precise than good estimators based on $\mathbf{Z}_n$. Stein (1990c, 1993b) considers estimating parameters of periodic stochastic processes observed with measurement error.

Exercises

1 For a stochastic process Z observed on $[0,1]$, define the empirical semivariogram for $\hat{\gamma}(t)$ for $0 \leq t < 1$ by

$$\hat{\gamma}(t) = \frac{1}{2(1-t)} \int_0^{1-t} \{Z(s+t) - Z(s)\}^2 ds.$$

Show that if Z is mean square differentiable, then

$$\hat{\gamma}''(0) = \int_0^1 \{Z'(t)\}^2 dt.$$

2 For a weakly stationary process Z on $\mathbb{R}$ with continuous semivariogram γ, show that $\mathrm{var}\{\int_0^1 Z(t)\,dt\} = 0$ if and only if the spectrum's support is contained in the set $\{\pi j : j \in \mathbb{Z}\backslash\{0\}\}$. Next, show that for Z a stationary Gaussian process with continuous semivariogram, $\mathrm{var}\left\{\int_0^1 Z(t)^2 dt\right\} = 0$ if and only if $\mathrm{var}\{Z(0)\} = 0$. Finally, combining this result with Exercise 1, show that for a stationary mean square differentiable Gaussion process Z, $\mathrm{var}\{\hat{\gamma}''(0)\} = 0$ if and only if γ is identically 0.

3 For $\boldsymbol{\theta} \in \Theta$, let $\{\Omega, \mathcal{F}, P_{\boldsymbol{\theta}}\}$ be a family of probability models on a measurable space $(\Omega, \mathcal{F})$. Consider a function h on Θ whose range is at most countable and suppose that $h(\boldsymbol{\theta}) \neq h(\boldsymbol{\theta}')$ implies $P_{\boldsymbol{\theta}} \perp P_{\boldsymbol{\theta}'}$. Show that there is a measurable function X on Ω such that for all $\boldsymbol{\theta} \in \Theta$, $X = h(\boldsymbol{\theta})$ with probability 1 under $P_{\boldsymbol{\theta}}$.

4 Let Θ be the set of all subsets of the positive integers. Suppose $X_1, X_2, \ldots$ is an infinite sequence of binary random variables. If $\theta \in \Theta$ is not the empty set, then for $j = 1, 2, \ldots$, $X_j = 1$ if $j \in \theta$ and $X_j = 0$ otherwise, so that the sequence of random variables is in fact deterministic. If θ is the empty set, then the X_js are independent and identically distributed with $\Pr(X = 0) = \Pr(X = 1) = \frac{1}{2}$. Define $h(\theta)$ to equal 1 if θ is the empty set and 0 otherwise. Show that θ and hence $h(\theta)$ is microergodic as defined in this section. Show that there does not exist a function $\hat{h}$ of $X_1, X_2, \ldots$ such that $\hat{h} = h(\theta)$ with probability 1 under P_θ for all $\theta \in \Theta$. Conclude that it is not possible to estimate $h(\theta)$ consistently based on $X_1, \ldots, X_n$ as $n \to \infty$.

5 Suppose Z is a mean 0 stationary Gaussian process on $\mathbb{R}$ and $K_{\boldsymbol{\theta}}(t) = \theta_1 e^{-\theta_2|t|}(1 + \theta_2|t|)$, $\boldsymbol{\theta} = (\theta_1, \theta_2) \in (0,\infty) \times (0,\infty)$. For $R = [0,1]$, show $-\theta_1\theta_2^2 = K''(0)$ is not microergodic, but $2\theta_1\theta_2^3 = K'''(0^+)$ is.

6 Suppose Z is a mean 0 stationary Gaussian process on $\mathbb{R}$ with autocovariance function of the form $K_\theta(t) = \theta K(t)$, where K possesses a spectral density with respect to Lebesgue measure and is not identically 0. Show that for $R = [0,1]$ and $\theta \neq \theta'$, $G_R(0, \theta K) \perp G_R(0, \theta' K)$.

7 Suppose Z is a mean 0 stationary Gaussian process on $\mathbb{R}$ with autocovariance function of the form $K_\theta(t) = \theta K(t)$. If the support of the spectrum corresponding to K is discrete, find necessary and sufficient conditions for θ to be microergodic.

8 Suppose Z is a mean 0 stationary Gaussian process on $\mathbb{R}$ with autocovariance function of the form $K_{\boldsymbol{\theta}}(t) = \theta_1 e^{-\theta_2 t^2}$ with $\boldsymbol{\theta} = (\theta_1, \theta_2) \in (0,\infty) \times (0,\infty)$. For $R = [0,1]$, show that $\boldsymbol{\theta}$ and hence $K_{\boldsymbol{\theta}}''(0)$ is microergodic.

9 For $\mathbf{Y}_n$ as defined in the last subsection, show that it is possible to estimate the measurement error variance σ^2 consistently as $n \to \infty$. Hint: consider an average of $\frac{1}{2}(Y_i - Y_j)^2$ over selected pairs (i,j) for which $|\mathbf{x}_i - \mathbf{x}_j|$ is small.

6.3 Is statistical inference for differentiable processes possible?

Matheron (1989, p. 90) states that for an isotropic random field observed on a bounded region whose isotropic semivariogram γ satisfies $\gamma(h) \sim Ch^2$ as $h \downarrow 0$ for some $C > 0$, "statistical inference is impossible." Let us examine what he means by this statement and what is wrong with his reasoning.

Matheron correctly notes that, at least for Gaussian random fields, it is not generally possible to recover C based on observations in a bounded region. The reader who has understood Chapter 4 should think "So what? We do not need to know C in order to obtain asymptotically optimal predictions in this bounded region nor to accurately assess their mses." However, Matheron (1971, Section 2-10-3) gives an example that appears to undermine this argument. I describe only a special case of his example, which is sufficient to show his error. Suppose Z is a stationary Gaussian process on $\mathbb{R}$ and we wish to predict $I(Z) = \int_0^1 Z(t)dt$ based on observing $Z\big((i-0.5)n^{-1}\big)$ for $i = 1, \ldots, n$. Matheron studies the mse of the predictor $\overline{Z}_n = n^{-1}\sum_{i=1}^n Z\big((i-0.5)n^{-1}\big)$, and although his analysis is incorrect, some of his conclusions are still relevant. Theorem 1 in 5.2 shows that if Z has spectral density f satisfying $f(\omega) = o(|\omega|^{-4})$ as $|\omega| \to \infty$, then the mean squared prediction error is asymptotically of the form τn^{-4}, where $\tau = (1/144)\int_{-\infty}^{\infty} \omega^2 \sin^2(\omega/2) f(\omega)d\omega$. (Matheron's results imply that the mse will be of order n^{-5} for sufficiently smooth processes, but this error does not affect the basic thrust of his argument. See Exercise 11 for further details.) Now τ depends nontrivially on the low frequency behavior of f, so it is apparent that τ will only be microergodic under very special assumptions on the class of models. In particular, if we only assume $f(\omega) = o(|\omega|^{-4})$ as $|\omega| \to \infty$, then τ cannot be determined based on observations on $[0,1]$. Hence, based on observing $Z\big((i-0.5)n^{-1}\big)$ for $i = 1, \ldots, n$, it is impossible to get an asymptotically correct assessment of the mean squared prediction error of $\overline{Z}_n$.

The careful reader of Chapter 5 should see the problem in this reasoning: if $f(\omega) = o(\omega^{-4})$ as $\omega \to \infty$, then $\overline{Z}_n$ is not asymptotically optimal, which was shown in 5.4. Jeffreys's law does not guarantee that we will be able to accurately assess the mse of poor predictors! What we should expect to be able to do is to find an asymptotically optimal predictor of $I(Z)$ and to assess its mse accurately.

An example where it is possible

Let us look at a specific case of how one could both predict $I(Z)$ well and accurately assess the mean squared prediction error when a differentiable and stationary Gaussian process Z is observed at $(i-0.5)n^{-1}$ for $i = 1, \ldots, n$. Suppose the spectral density f of Z satisfies $f(\omega) \sim \phi\omega^{-6}$ as $\omega \to \infty$ for some unknown positive constant ϕ. A more challenging and realistic problem would be to assume $f(\omega) \sim \phi\omega^{-\rho}$ as $\omega \to \infty$ with both ϕ and ρ unknown (see Section 6.7 for a related problem), but even the simpler problem when it is known that $\rho = 6$ is an example of what Matheron considers "impossible." Istas and Lang (1997) show how one can go about consistently estimating both ρ and ϕ under certain additional regularity conditions on the autocovariance function. See Constantine and Hall (1994) and Kent and Wood (1997) for related work on estimating ρ when $\rho < 3$.

By Corollary 6 in 5.4, $\overline{Z}_{n,1}$ as defined in 5.4 is asymptotically optimal and Theorem 7 in 5.4 yields

$$\operatorname{var}\{I(Z) - \overline{Z}_{n,1}\} \sim \frac{\phi}{16\pi^5 n^6} \sum_{j=1}^{\infty} j^{-6} = \frac{\pi\phi}{15{,}120 n^6},$$

where the last step uses 23.2.16 in Abramowitz and Stegun (1965). Thus, if we can estimate ϕ consistently as $n \to \infty$, then we can obtain an asymptotically valid estimate of the mse of the asymptotically optimal predictor $\overline{Z}_{n,1}$. I believe it is impossible to estimate ϕ consistently from the sample semivariogram, although I do not know how to prove this claim. However, it is possible to estimate ϕ consistently by taking an appropriately normalized sum of squared third differences of the observations. More specifically, defining the operator Δ_ϵ by $\Delta_\epsilon Z(t) = \epsilon^{-1}\{Z(t+\epsilon) - Z(t)\}$, then $f(\omega) \sim \phi\omega^{-6}$ as $\omega \to \infty$ implies

$$E\left[\{(\Delta_\epsilon)^3 Z(0)\}^2\right] = \int_{-\infty}^{\infty} f(\omega) \left\{\frac{2}{\epsilon} \sin\left(\frac{\epsilon\omega}{2}\right)\right\}^6 d\omega$$

$$\sim \int_{-\infty}^{\infty} \phi\omega^{-6} \left\{ \frac{2}{\epsilon} \sin\left(\frac{\epsilon\omega}{2}\right) \right\}^6 d\omega$$
$$= \frac{11\pi\phi}{10\epsilon} \tag{1}$$

as $\epsilon \downarrow 0$ (Exercise 12). Consider the following estimator of ϕ based on observing Z at $(i-0.5)n^{-1}$ for $i = 1, \ldots, n$,

$$\hat{\phi}_n = \frac{10}{11\pi n^2} \sum_{j=1}^{n-3} \left\{ (\Delta_{1/n})^3 Z\left(\frac{j-0.5}{n}\right) \right\}^2. \tag{2}$$

Equation (1) implies that $E\hat{\phi}_n \to \phi$ as $n \to \infty$ and it is furthermore possible to show that $\operatorname{var} \hat{\phi}_n \to 0$ as $n \to \infty$ (Exercise 13), so that $\hat{\phi}_n$ is a consistent estimator of ϕ.

Exercises

10 For a weakly stationary process Z on $\mathbb{R}$ with spectral density f and autocovariance function K with $\int_{-\infty}^{\infty} \omega^2 f(\omega)\, d\omega < \infty$, show that

$$\int_{-\infty}^{\infty} \omega^2 \sin^2\left(\frac{\omega}{2}\right) f(\omega)\, d\omega = \frac{1}{2}\{K''(1) - K''(0)\}.$$

It immediately follows from Theorem 1 in Chapter 5 that if $f(\omega) = o(|\omega|^{-4})$,

$$\operatorname{var}\{I(Z) - \overline{Z}_n\} \sim \frac{1}{288n^4}\{K''(1) - K''(0)\}.$$

Prove this result using an argument in the time domain under the additional assumption that $K^{(4)}$ exists and is continuous.

11 For a weakly stationary process Z on $\mathbb{R}$, define $X_{in} = \int_{(i-1)/n}^{i/n} \{Z(t) - Z((i-0.5)/n)\}dt$ for $i = 1, \ldots, n$, so that $I(Z) - \overline{Z}_n = \sum_{i=1}^n X_{in}$. Matheron (1971) obtains an incorrect rate of convergence for $\operatorname{var}\{I(Z) - \overline{Z}_n\}$ because he calculates the mse by ignoring the correlations between the X_{in}s for $i = 1, \ldots, n$. Suppose Z has spectral density f satisfying $f(\omega) \sim \phi\omega^{-\rho}$ as $\omega \to \infty$ for some $\phi > 0$ and $\rho > 1$. For what values of ρ is $\operatorname{var}\{I(Z) - \overline{Z}_n\} \sim \sum_{i=1}^n \operatorname{var}(X_{in})$? For what values of ρ is $\operatorname{var}\{I(Z) - \overline{Z}_n\} \asymp \sum_{i=1}^n \operatorname{var}(X_{in})$? Note that Wackernagel (1995, p. 60) also ignores the correlations of the X_{in}s when approximating $\operatorname{var}\{I(Z) - \overline{Z}_n\}$.

12 Fill in the details for (1).

13 For $\hat{\phi}_n$ as defined in (2), prove that $\operatorname{var} \hat{\phi}_n \to 0$ as $n \to \infty$. The argument is somewhat reminiscent of Theorem 1 in Chapter 4, although quite a bit simpler because of the assumption that $f(\omega) \sim \phi\omega^{-6}$ as

$\omega \to \infty$. The key step in the proof is to show

$$\begin{aligned}
&\operatorname{cov}\{(\Delta_{1/n})^3 Z(0), (\Delta_{1/n})^3 Z(t)\} \\
&\quad = \int_{-\infty}^{\infty} \frac{\phi}{\omega^6} \left\{ 2n \sin\left(\frac{\omega}{2n}\right) \right\}^6 e^{i\omega t} d\omega + o(n^2)
\end{aligned}$$

uniformly in t.

6.4 Likelihood Methods

Maximum likelihood estimation plays a central role throughout statistics and is no less appropriate or useful for estimating unknown parameters in models for random fields. This section describes maximum likelihood estimation and a variant known as restricted maximum likelihood estimation for estimating the parameters of Gaussian random fields. Kitanidis (1997) provides an elementary introduction to the use of likelihood methods in spatial statistics.

Suppose Z is a Gaussian random field on $\mathbb{R}^d$ with mean and covariance structure as in 1.5: $Z(\mathbf{x}) = \mathbf{m}(\mathbf{x})^T \boldsymbol{\beta} + \varepsilon(\mathbf{x})$, where $\mathbf{m}$ is a known vector-valued function, $\boldsymbol{\beta}$ is a vector of unknown coefficients and ε has mean 0 with covariance function $\operatorname{cov}\{\varepsilon(\mathbf{x}), \varepsilon(\mathbf{y})\} = K_{\boldsymbol{\theta}}(\mathbf{x}, \mathbf{y})$ for an unknown parameter $\boldsymbol{\theta}$. Observe $\mathbf{Z} = (Z(\mathbf{x}_1), \dots, Z(\mathbf{x}_n))^T$. The likelihood function is just the joint density of the observations viewed as a function of the unknown parameters. A maximum likelihood estimate (MLE) of the unknown parameters is any vector of values for the parameters that maximizes this likelihood function. It is completely equivalent and often somewhat easier to maximize the logarithm of the likelihood function, often called the log likelihood. Let $\mathbf{K}(\boldsymbol{\theta})$ be the covariance matrix of $\mathbf{Z}$ as a function of $\boldsymbol{\theta}$ and assume $\mathbf{K}(\boldsymbol{\theta})$ is nonsingular for all $\boldsymbol{\theta}$. Define $\mathbf{M} = \left(\mathbf{m}(\mathbf{x}_1) \dots \mathbf{m}(\mathbf{x}_n)\right)^T$ and assume it is of full rank. Then (see Appendix A) the log likelihood function is

$$\ell(\boldsymbol{\theta}, \boldsymbol{\beta}) = -\frac{n}{2} \log(2\pi) - \tfrac{1}{2} \log \det\{\mathbf{K}(\boldsymbol{\theta})\} - \tfrac{1}{2} (\mathbf{Z} - \mathbf{M}\boldsymbol{\beta})^T \mathbf{K}(\boldsymbol{\theta})^{-1} (\mathbf{Z} - \mathbf{M}\boldsymbol{\beta}).$$

One way to simplify the maximization of this function is to note that for any given $\boldsymbol{\theta}$, $\ell(\boldsymbol{\theta}, \boldsymbol{\beta})$ is maximized as a function of $\boldsymbol{\beta}$ by

$$\hat{\boldsymbol{\beta}}(\boldsymbol{\theta}) = \mathbf{W}(\boldsymbol{\theta})^{-1} \mathbf{M}^T \mathbf{K}(\boldsymbol{\theta})^{-1} \mathbf{Z}, \tag{3}$$

where $\mathbf{W}(\boldsymbol{\theta}) = \mathbf{M}^T \mathbf{K}(\boldsymbol{\theta})^{-1} \mathbf{M}$. Thus, the MLE of $(\boldsymbol{\theta}, \boldsymbol{\beta})$ can be found by maximizing

$$\begin{aligned}
\ell(\boldsymbol{\theta}, \hat{\boldsymbol{\beta}}(\boldsymbol{\theta})) = {} & -\frac{n}{2} \log(2\pi) - \tfrac{1}{2} \log \det\{\mathbf{K}(\boldsymbol{\theta})\} \\
& - \tfrac{1}{2} \mathbf{Z}^T \{\mathbf{K}(\boldsymbol{\theta})^{-1} - \mathbf{K}(\boldsymbol{\theta})^{-1} \mathbf{M} \mathbf{W}(\boldsymbol{\theta})^{-1} \mathbf{M}^T \mathbf{K}(\boldsymbol{\theta})^{-1}\} \mathbf{Z}
\end{aligned} \tag{4}$$

(Exercise 14). Maximizing the likelihood over some parameters while holding others fixed is called profiling and the function $\ell(\boldsymbol{\theta}, \hat{\boldsymbol{\beta}}(\boldsymbol{\theta}))$ is called the profile log likelihood for $\boldsymbol{\theta}$ (McCullagh and Nelder 1989, p. 254).

Restricted maximum likelihood estimation

The MLE has a minor defect in this setting. If we knew $\boldsymbol{\beta} = \boldsymbol{\beta}_0$, we would presumably estimate $\boldsymbol{\theta}$ by maximizing $\ell(\boldsymbol{\theta}, \boldsymbol{\beta}_0)$ as a function of $\boldsymbol{\theta}$. By construction (Exercise 15),

$$\{\mathbf{Z}-\mathbf{M}\hat{\boldsymbol{\beta}}(\boldsymbol{\theta})\}^T\mathbf{K}(\boldsymbol{\theta})^{-1}\{\mathbf{Z}-\mathbf{M}\hat{\boldsymbol{\beta}}(\boldsymbol{\theta})\} \leq (\mathbf{Z}-\mathbf{M}\boldsymbol{\beta}_0)^T\mathbf{K}(\boldsymbol{\theta})^{-1}(\mathbf{Z}-\mathbf{M}\boldsymbol{\beta}_0) \quad (5)$$

for all $\boldsymbol{\theta}$, so that $\mathbf{M}\hat{\boldsymbol{\beta}}(\boldsymbol{\theta})$ is always "closer" to $\mathbf{Z}$ than is $\mathbf{M}\boldsymbol{\beta}_0$. As a consequence, the MLE of $\boldsymbol{\theta}$ will tend to underestimate the variation in the process, at least relative to what we would get if we knew $\boldsymbol{\beta} = \boldsymbol{\beta}_0$. For example, if $\mathbf{K}(\boldsymbol{\theta}) = \theta\mathbf{V}$, where $\mathbf{V}$ is known and θ is a scalar, then $\hat{\boldsymbol{\beta}}(\theta)$ does not depend on θ and $\hat{\theta} = n^{-1}(\mathbf{Z} - \mathbf{M}\hat{\boldsymbol{\beta}})^T\mathbf{V}^{-1}(\mathbf{Z} - \mathbf{M}\hat{\boldsymbol{\beta}})$, which has expected value $(n-p)n^{-1}\theta$, where p is the rank of $\mathbf{M}$. Common practice would be to estimate θ unbiasedly by $n(n-p)^{-1}\hat{\theta}$. For more complicated models for the covariance function, obvious adjustments for the bias of the MLE are not available.

An alternative approach to estimating $\boldsymbol{\theta}$ is to consider the likelihood function of the contrasts, the linear combinations of the observations whose joint distribution does not depend on $\boldsymbol{\beta}$ (see 1.5). By construction, this likelihood will not depend on $\boldsymbol{\beta}$ and hence we can obtain an estimate of $\boldsymbol{\theta}$ by maximizing this function over just $\boldsymbol{\theta}$. This approach is commonly known as restricted maximum likelihood (REML) estimation and was described by Patterson and Thompson (1971) in the context of variance component estimation, who called the method modified maximum likelihood. Kitanidis (1983) was the first to propose applying REML to the estimation of spatial covariances. The idea is that if little is known about $\boldsymbol{\beta}$ a priori, the contrasts should contain essentially all of the information about $\boldsymbol{\theta}$. Furthermore, since the distribution of the contrasts does not depend on $\boldsymbol{\beta}$, the "overfitting" problem that occurs when using ordinary maximum likelihood should not occur. In particular, if $\mathbf{K}(\theta) = \theta\mathbf{V}$, then the REML estimate of θ is the usual unbiased estimate. A number of simulation studies in the time series setting have demonstrated the general superiority of REML estimation to ML estimation (Wilson 1988; McGilchrist 1989; and Tunicliffe-Wilson 1989).

To calculate the log likelihood of the contrasts, consider the set of contrasts $\mathbf{Y} = \{\mathbf{I} - \mathbf{M}(\mathbf{M}^T\mathbf{M})^{-1}\mathbf{M}^T\}\mathbf{Z}$, where we have assumed $\mathbf{M}$ is of full rank p. The random vector $\mathbf{Y}$ forms a basis for all contrasts of $\mathbf{Z}$. There are then $n-p$ linearly independent contrasts, so that any $n-p$ linearly independent components of $\mathbf{Y}$ also form a basis for all contrasts. Now $\mathbf{Y}$ has a singular normal distribution, so writing down its likelihood is not trivial. One solution is to consider the likelihood of $n-p$ linearly independent contrasts. McCullagh and Nelder (1989, p. 247) give the log likelihood for $\boldsymbol{\theta}$ directly in terms of $\mathbf{Y}$:

$$\ell(\boldsymbol{\theta};\mathbf{Y}) = -\frac{n-p}{2}\log(2\pi) - \tfrac{1}{2}\log\ \det\{\mathbf{K}(\boldsymbol{\theta})\} - \tfrac{1}{2}\log\ \det\{\mathbf{W}(\boldsymbol{\theta})\}$$
$$-\tfrac{1}{2}\mathbf{Y}^T\{\mathbf{K}(\boldsymbol{\theta})^{-1} - \mathbf{K}(\boldsymbol{\theta})^{-1}\mathbf{M}\mathbf{W}(\boldsymbol{\theta})^{-1}\mathbf{M}^T\mathbf{K}(\theta)^{-1}\}\mathbf{Y}. \qquad (6)$$

Exercises 7.8–7.13 of McCullagh and Nelder (1989) outline a derivation of this result. Any maximizer $\hat{\boldsymbol{\theta}}$ of this expression is called a REML estimate of $\boldsymbol{\theta}$. The REML estimate of $\boldsymbol{\beta}$ is then given by $\hat{\boldsymbol{\beta}}(\hat{\boldsymbol{\theta}})$, where $\hat{\boldsymbol{\beta}}(\boldsymbol{\theta})$ is defined as in (3).

An advantage of REML over ML estimation in estimating covariance structures for Gaussian random fields is that REML can be applied to estimating generalized autocovariance functions of IRFs (see 2.9), whereas ML cannot. The problem is that a generalized autocovariance function for an IRF does not define the covariance structure of all of the observations, so that the likelihood for all of the observations is also not defined. However, a generalized autocovariance function together with the Gaussian assumption does define the joint density of the contrasts of the observations, so one can use REML for estimating the parameters of a generalized autocovariance function. Indeed, REML and best linear unbiased prediction of IRFs form a coherent conceptual package, since in the modeling, estimation and prediction one only needs to consider contrasts of the random field.

Gaussian assumption

The likelihood functions given in the previous subsections all assume that the random field is Gaussian. This is a highly restrictive assumption so that it is reasonable to be concerned about the performance of likelihood-based methods based on a Gaussian model when the random field is in fact not Gaussian. In particular, such methods will generally perform poorly if there are even a small number of aberrant observations. However, methods that are functions of the empirical semivariogram such as least squares and generalized least squares (Cressie 1993, Section 2.6) will also be sensitive to aberrant values even though they do not explicitly assume that the random field is Gaussian. Cressie and Hawkins (1980) and Hawkins and Cressie (1984) describe "robust" procedures for estimating semivariograms that are less sensitive to distributional assumptions than procedures based on the empirical semivariogram. However, these procedures do not fully take into account the dependencies in the data and thus may be considerably less precise than likelihood-based estimates when the Gaussian assumption is tenable. A good topic for future research would be the development of models and computational methods for calculating likelihood functions for non-Gaussian random fields. Diggle, Tawn and Moyeed (1998) make an important step in this direction and demonstrate that Markov Chain

Monte Carlo methods provide a suitable computational tool for some non-Gaussian models.

Computational issues

One potentially serious obstacle to employing likelihood methods is computing the likelihood function. In general, if there are n observations, calculating the determinant of $\mathbf{W}(\boldsymbol{\theta})$ and quadratic forms in $\mathbf{W}(\boldsymbol{\theta})^{-1}$ each require $O(n^3)$ calculations. In particular, for irregularly scattered observations in more than one dimension, an $O(n^3)$ calculation is usually necessary to calculate the values of the likelihood function exactly. Thus, if there are more than several hundred observations, exact likelihood calculations are often infeasible. However, if the observations are on a regular lattice, then it is possible to compute the likelihood function exactly with fewer calculations (Zimmerman 1989). In this setting, it is also possible to use spectral methods to approximate the likelihood (Whittle 1954; Guyon 1982; Dahlhaus and Künsch 1987; and Stein 1995c), in which case, the approximate likelihood can be calculated very efficiently by making use of the fast Fourier transform (Press, Flannery, Teukolsky and Vetterling 1992, Chapter 12).

Vecchia (1988) describes a general method for efficiently approximating the likelihood function for spatial data. Let $p(z_1, \dots, z_n)$ be the joint density of $(Z(\mathbf{x}_1), \dots, Z(\mathbf{x}_n))$ evaluated at $(z_1, \dots, z_n)$ and write other joint and conditional densities similarly. Next, write

$$p(z_1, \dots, z_n) = p(z_1) \prod_{j=2}^{n} p(z_j \mid z_1, \dots, z_{j-1})$$

and then approximate $p(z_j \mid z_1, \dots, z_{j-1})$ by the conditional density of $Z(\mathbf{x}_j)$ given just the $\min(m, j-1)$ observations among $Z(\mathbf{x}_1), \dots, Z(\mathbf{x}_{j-1})$ that are nearest to $\mathbf{x}_j$ in Euclidean distance, where m is much smaller than n. The smaller the value of m, the more efficient the computation but the worse the approximation to the true joint density. The ordering of the observations affects the results, but Vecchia (1988) found this effect to be small in the examples he studied and suggests ordering by the values of any one of the coordinate axes of the observation locations.

A somewhat related method for approximating the likelihood is to divide the observation region into some number of subregions, calculate the likelihood for each subregion separately and then multiply these likelihoods together. Similar to Vecchia's procedure, smaller subregions lead to easier computation but worse approximations of the likelihood. Stein (1986) recommended such a procedure for minimum norm quadratic estimators (Rao 1973), which also require computing quadratic forms of $n \times n$ inverse covariance matrices. I would recommend using subregions containing at least 100 observations, in which case, it should be feasible to carry out the necessary computations for many thousands of total observations. Neither this approach nor Vecchia's should involve much loss of information about those parameters governing the local behavior of the random field, which are exactly those that will matter most when interpolating. Furthermore, even if it is possible to calculate the exact likelihood for all of the data, calculating the likelihood separately for subregions might be desirable as a way of looking for possible nonstationarities in the random field. For example, Haas (1990, 1995) deliberately uses only observations near the predictand's location when estimating the semivariogram to allow for possible nonstationarities.

To compute MLEs, it is generally necessary to find the maximum of the likelihood numerically. It is common practice to select starting values for the unknown parameters and then use an iterative procedure such as conjugate gradient (Press, Flannery, Teukolsky and Vetterling 1992) to locate a maximum of the function. If the function has more than one local maximum, there is no guarantee that such algorithms will converge to the global maximum. Indeed, if one uses a model for the autocovariance function such as the spherical (see 2.10), which yields likelihood functions that are not twice differentiable, then it is quite possible to obtain likelihood functions that have more than one local maximum (Ripley 1988). However, when using the Matérn model, I am unaware of any examples of likelihood functions with more than one local maximum. What is possible is for the likelihood not to possess a maximum in the interior of the parameter space. For example, the supremum of the likelihood function may in some cases be obtained as the parameter ν, which controls the differentiability of the random field, tends to ∞.

I do not believe the results in Warnes and Ripley (1987) and Ripley (1988) purporting to show multiple maxima in the likelihood when fitting an exponential autocovariance function. Nevertheless, it is worth pointing out that the various purported multiple maxima in their example correspond to parameter values that will give nearly identical predictions and mses when interpolating, since the slopes at the origin of the corresponding semivariograms are nearly the same for all of the local maxima. The example in Warnes and Ripley (1987) does correctly show that the likelihood function can have long ridges along which it is nearly constant, which could lead to numerical problems when using iterative procedures for finding the maximum. Their presence is not a sign of a problem with likelihood methods but rather an entirely correct indication that the data provide essentially no information for distinguishing between parameter values along the ridge. If, rather than just trying to maximize the likelihood, one plots the log likelihood function, or at least some judiciously chosen profile log likelihoods, then these ridges should be detected.

Some asymptotic theory

This subsection summarizes some standard asymptotic theory for MLEs and describes why such results will often not hold under fixed-domain asymptotics. Suppose we observe a random vector $\mathbf{X}$ whose distribution is from a family of distributions $P_{\boldsymbol{\theta}}$ for $\boldsymbol{\theta} \in \Theta \subset \mathbb{R}^p$ and we propose to estimate $\boldsymbol{\theta}$ via maximum likelihood. One way to try to maximize the likelihood is by finding critical points of the likelihood function. Specifically, define the score function $\mathbf{S}(\boldsymbol{\theta}) = \partial\ell(\boldsymbol{\theta};\mathbf{X})/\partial\boldsymbol{\theta}$, the random vector whose components are the partial derivatives of $\ell(\boldsymbol{\theta};\mathbf{X})$ with respect to the components of $\boldsymbol{\theta}$. Assuming these derivatives exist and are continuous, any MLE in the interior of Θ must be a solution to the score equations, $\mathbf{S}(\boldsymbol{\theta}) = \mathbf{0}$. Let $\boldsymbol{\theta}_0$ be the true value of $\boldsymbol{\theta}$ and define the Fisher information matrix $\mathcal{I}(\boldsymbol{\theta}_0) = \mathrm{cov}_{\boldsymbol{\theta}_0}\{\mathbf{S}(\boldsymbol{\theta}_0), \mathbf{S}(\boldsymbol{\theta}_0)^T\}$. Let $\mathbf{i}(\boldsymbol{\theta})$ be the $p \times p$ matrix whose jkth component is $-\partial^2\ell(\boldsymbol{\theta};\mathbf{X})/\partial\theta_j\partial\theta_k$. Under certain regularity conditions, $E_{\boldsymbol{\theta}_0}\mathbf{i}(\boldsymbol{\theta}_0) = \mathcal{I}(\boldsymbol{\theta}_0)$ (Exercise 18). If $\mathcal{I}(\boldsymbol{\theta}_0)$ is "large" (in the sense that its smallest eigenvalue is large) and $\mathcal{I}(\boldsymbol{\theta}_0)^{-1}\mathbf{i}(\boldsymbol{\theta}_0) \approx \mathbf{I}$ with high probability, then standard asymptotic theory suggests that the MLE is approximately $N(\boldsymbol{\theta}_0, \mathcal{I}(\boldsymbol{\theta}_0)^{-1})$ (Ferguson 1996; Ibragimov and Has'minskii 1981).

To give a more careful statement of this result that includes the settings of concern in this work, suppose that $P_{\boldsymbol{\theta}}$ for $\boldsymbol{\theta} \in \Theta$ is a family of probability measures for a random field Z and $\mathbf{X}_1, \mathbf{X}_2, \ldots$ is a sequence of random vectors of observations from Z. As n increases, we should be thinking that $\mathbf{X}_n$ becomes increasingly informative about $\boldsymbol{\theta}$. In many works on asymptotics of MLEs, it is assumed that the observations making up $\mathbf{X}_n$ contain the observations making up $\mathbf{X}_j$ for all $j < n$, but this will not be the case for the example in 6.7, which considers evenly spaced observations under fixed-domain asymptotics. Use the subscript n to indicate a quantity calculated with $\mathbf{X}_n$ as the observation vector. Then if the smallest eigenvalue of $\mathcal{I}_n(\boldsymbol{\theta}_0)$ tends to infinity as $n \to \infty$, we generally have

$$\mathcal{I}_n(\boldsymbol{\theta}_0)^{1/2}(\hat{\boldsymbol{\theta}}_n - \boldsymbol{\theta}_0) \xrightarrow{\mathcal{L}} N(\mathbf{0}, \mathbf{I}), \tag{7}$$

where $\mathcal{I}_n(\boldsymbol{\theta}_0)^{1/2}$ is some matrix square root of $\mathcal{I}_n(\boldsymbol{\theta}_0)$ (Ibragimov and Has'minskii 1981). More informally, one might say $\hat{\boldsymbol{\theta}}_n - \boldsymbol{\theta}_0 \xrightarrow{\mathcal{L}} N(\mathbf{0}, \mathcal{I}_n(\boldsymbol{\theta}_0)^{-1})$.

Although (7) is part of the folklore of statistical theory, it is often difficult to prove rigorously that MLEs have this behavior. It is usually considerably easier to prove that any consistent sequence of solutions of the score equations has asymptotic behavior given by (7); see, for example, Sweeting (1980). Mardia and Marshall (1984) and Cressie and Lahiri (1993) give some results for random fields under increasing-domain asymptotics. Note that under fixed-domain asymptotics, if there is a nonmicroergodic parameter, then $\boldsymbol{\theta}$ cannot be consistently estimated and we should generally expect (7) to be false. Exercises 18–20 examine some fundamental properties of likelihoods, score equations, Fisher information and their relationship to the asymptotics of MLEs.

When $\boldsymbol{\theta}$ is not microergodic, we might expect something like (7) to hold for the microergodic part of $\boldsymbol{\theta}$. To be more precise, suppose we can write $\boldsymbol{\theta} = (\boldsymbol{\mu}^T, \boldsymbol{\tau}^T)^T$, where $\boldsymbol{\mu}$ is microergodic and no nontrivial function of $\boldsymbol{\tau}$ is microergodic. Results in Crowder (1976) suggest that if $\boldsymbol{\tau}$ is just set to some fixed value rather than estimated and $\boldsymbol{\mu}$ is estimated by a consistent sequence of solutions to the score equations $\partial \ell / \partial \boldsymbol{\mu} = \mathbf{0}$, then the asymptotic behavior of these estimates will be the same as if $\boldsymbol{\tau}$ were known. Considering Crowder's results, a reasonable conjecture is that if all components of $\boldsymbol{\theta}$ are estimated by maximizing the likelihood, then the asymptotic behavior of $\hat{\boldsymbol{\mu}}_n$ will be the same as if $\boldsymbol{\tau}$ were known; that is, $\mathcal{I}_n(\boldsymbol{\mu}_0)^{1/2}(\hat{\boldsymbol{\mu}}_n - \boldsymbol{\mu}_0) \xrightarrow{\mathcal{L}} N(\mathbf{0}, \mathbf{I})$. Ying (1991) proves a very special case of this result for an exponential autocovariance function in one dimension.

Exercises

14 Verify (4).

15 Verify (5).

16 Show that if one defines the likelihood in terms of $\mathbf{g}(\boldsymbol{\theta})$, where $\mathbf{g}$ is an invertible function on Θ, then $\mathbf{g}(\hat{\boldsymbol{\theta}})$ is an MLE for $\mathbf{g}(\boldsymbol{\theta})$ if and only if $\hat{\boldsymbol{\theta}}$ is an MLE for $\boldsymbol{\theta}$. Thus, MLEs are invariant under arbitrary invertible transformations of the parameters.

17 Suppose a random vector $\mathbf{X}$ of length n has a density with respect to Lebesgue measure $p(\cdot \mid \boldsymbol{\theta})$ depending on a parameter $\boldsymbol{\theta}$. If $\mathbf{Y} = \mathbf{h}(\mathbf{X})$, where $\mathbf{h}$ is an invertible function from $\mathbb{R}^n$ to $\mathbb{R}^n$ possessing continuous first partial derivatives, show that $\ell(\boldsymbol{\theta}; \mathbf{Y}) - \ell(\boldsymbol{\theta}; \mathbf{X})$ does not depend on $\boldsymbol{\theta}$ and, hence, it does not matter whether we use $\mathbf{X}$ or $\mathbf{Y}$ in finding an MLE for $\boldsymbol{\theta}$. Thus, MLEs are invariant under smooth invertible transformations of the observations.

The next three exercises review basic properties about likelihood functions and provide a heuristic justification of (7). Assume throughout these exercises that $\{P_{\boldsymbol{\theta}} : \boldsymbol{\theta} \in \Theta\}$ is a class of probability models for the observations with true value $\boldsymbol{\theta}_0$ and that switching the order of differentiation and integration is permissible.

18 Show that $E_{\boldsymbol{\theta}_0}\mathbf{S}(\boldsymbol{\theta}_0) = \mathbf{0}$. Show that $\mathcal{I}(\boldsymbol{\theta}_0) = E_{\boldsymbol{\theta}_0}\mathbf{i}(\boldsymbol{\theta}_0)$.

19 Consider random vectors $\mathbf{X}$ and $\mathbf{Y}$ whose joint distribution is specified by $P_{\boldsymbol{\theta}}$. Show that $\mathcal{I}(\boldsymbol{\theta}_0; \mathbf{X}) + \mathcal{I}(\boldsymbol{\theta}_0; \mathbf{Y}) - \mathcal{I}(\boldsymbol{\theta}_0; (\mathbf{X}, \mathbf{Y}))$ is positive semidefinite, where the argument after the semicolon in $\mathcal{I}(\cdot;\cdot)$ indicates the observations for which the Fisher information matrix is to be calculated. Show that $\mathcal{I}(\boldsymbol{\theta}_0; \mathbf{X}) + \mathcal{I}(\boldsymbol{\theta}_0; \mathbf{Y}) = \mathcal{I}(\boldsymbol{\theta}_0; (\mathbf{X}, \mathbf{Y}))$ if $\mathbf{X}$ and $\mathbf{Y}$ are independent for all $\boldsymbol{\theta}$.

20 Suppose $\mathbf{X}_1, \mathbf{X}_2, \ldots$ is as in the last subsection and that the subscript n generically indicates quantities based on $\mathbf{X}_n$. By taking a first-order Taylor series in the score function about $\boldsymbol{\theta}_0$, give a heuristic argument showing that any consistent sequence of solutions $\hat{\boldsymbol{\theta}}_n$ of the score equations should satisfy $\hat{\boldsymbol{\theta}}_n \approx \boldsymbol{\theta}_0 + \mathbf{i}_n(\boldsymbol{\theta}_0)^{-1}\mathbf{S}_n(\boldsymbol{\theta}_0)$. As $n \to \infty$, suppose $\mathcal{I}_n(\boldsymbol{\theta}_0)^{-1}\mathbf{i}_n(\boldsymbol{\theta}_0) \to \mathbf{I}$ in probability (a weak law of large numbers) and $\mathcal{I}_n(\boldsymbol{\theta}_0)^{-1/2}\mathbf{S}_n(\boldsymbol{\theta}_0) \xrightarrow{\mathcal{L}} N(\mathbf{0}, \mathbf{I})$ (a central limit theorem). Show that (7) plausibly follows.

6.5 Matérn model

For statistical methodologies to be broadly and effectively employed, it is important to have canonical models that work reasonably well in a wide range of circumstances. For the purposes of interpolating spatial data in d dimensions, I recommend the following model: $Z(\mathbf{x}) = \mu + \varepsilon(\mathbf{x})$, where μ is an unknown constant and ε is a mean 0 stationary isotropic Gaussian random field with autocovariance function from the Matérn class; that is, with spectral density $\phi(\alpha^2 + |\boldsymbol{\omega}|^2)^{-\nu-d/2}$ for unknown positive parameters ϕ, ν and α (see Sections 2.7 and 2.10). In making this recommendation, I do not mean to imply that all, or even most, spatial data can be reasonably modeled in this fashion. However, by making prudent extensions to this model where appropriate by including, for example, geometric anisotropies (2.10), measurement errors or by taking a pointwise transformation of the observations (often logarithmic), one could distinctly improve on present practice in spatial statistics. Diggle, Tawn and Moyeed (1998) describe a notable extension by combining Gaussian random fields and generalized linear models (McCullagh and Nelder 1989) into a single class of models for spatial data.

The most important reason for adopting the Matérn model is the inclusion of the parameter ν in the model, which controls the rate of decay of the spectral density at high frequencies, or equivalently, the smoothness of the random field. As the results in Chapters 3 and 4, particularly 3.6, indicate, the rate of decrease of the spectral density at high frequencies plays a critical role in spatial interpolation. Unless there is some theoretical or empirical basis for fixing the degree of smoothness of a random field a priori, I can see no justification for the common practice of selecting semivariogram models such as the spherical, exponential or Gaussian that provide no flexibility in this degree of smoothness. Using empirical semivariograms for model selection can work disastrously for smooth processes as the example in 6.9 demonstrates. Empirical semivariograms are less likely to mislead for random fields that are not differentiable. However, I believe that even in these instances far too much faith is generally placed in empirical semivariograms as a tool for model selection.

Since one would never leave out an overall scale parameter, the presence of ϕ in the model is also essential. In addition, although Theorem 8 in 4.3 implies that one could leave μ out of the model with asymptotically negligible effect, it is hard to argue for arbitrarily taking the mean of Z to be 0 unless there is some substantive reason to believe that it is. The serious issue is whether the parameter α is helpful, since it has negligible impact on the high frequency behavior of the spectral density. The results in Chapters 3 and 4 show that varying α will have little effect on interpolations if the observations are sufficiently dense. Furthermore, in three or fewer dimensions, α cannot be consistently estimated based on observations in a fixed domain, which follows from (20) in 4.2. Indeed, Wahba (1990) essentially argues that α should just be set to 0. This leaves us with the model for the spectral density of $\phi|\boldsymbol{\omega}|^{-2\nu-d}$, which is not integrable in a neighborhood of the origin for $\nu > 0$. Thus, this function is not a spectral density for a stationary random field. It is, however, the spectral density of an IRF (intrinsic random function) of order $\lfloor 2\nu \rfloor$ (see 2.9).

Although leaving α out of the model is a defensible position, there are a number of reasons why I mildly favor its inclusion. First, the mathematical arguments for excluding α are asymptotic and hence should not be considered universally compelling. Particularly for predictands located near or outside the boundaries of the observation region, the value of α can matter substantially. Furthermore, if the correlations of the random field die out at a distance much shorter than the dimensions of the domain of the observations, it may be possible to obtain a decent estimate of α. Handcock, Meier and Nychka (1994) give an example concerning measurements of electrical conductivity in soil that provides clear evidence of the need for positive α both to fit the covariance structure of the data well and to provide sensible interpolations. Second, if the available observations include a substantial measurement error, then I suspect that badly misspecifying the low frequency behavior of the spectral density could lead to serious bias in ML or REML estimates of ν even for moderately large sample sizes. Measurement error makes estimating the high frequency behavior of a random field much more difficult, so that the low frequency behavior can then have a larger influence on parameter estimates. This greater influence may produce substantially biased ML estimates of the high frequency behavior if the low frequency behavior is poorly specified. An example of severe systematic error in ML estimates due to misspecification of a model at low frequencies when the underlying process is deterministic is given in Section 6.3 of Stein (1993b). Using the Matérn model of course does not guarantee that the low frequency behavior of the spectral density is correctly specified. However, allowing α to be estimated from the data does provide substantial additional flexibility to the model while only adding one parameter. Further study of this issue is in order.

My final reason for including α is that I find it somewhat unnatural to link the high frequency behavior of the spectral density and the order of the polynomial of the mean of the random field, which setting $\alpha = 0$ requires. Specifically, for the spectral density $\phi|\boldsymbol{\omega}|^{-2\nu-d}$, the corresponding random field must be an IRF of order at least $\lfloor \nu \rfloor$ and hence its mean is implicitly taken to be a polynomial of order at least $\lfloor \nu \rfloor$ with unknown coefficients (see 2.9). I would prefer to be able to assume the mean is constant no matter how large ν is.

The fact that the order of the polynomial mean must increase with ν if one sets $\alpha = 0$ causes a bit of difficulty with REML estimation of ϕ and ν. Specifically, suppose one models Z as a Gaussian IRF with spectral density $\phi|\boldsymbol{\omega}|^{-2\nu-d}$, where the order r of the IRF is the lowest feasible: $r = \lfloor \nu \rfloor$. Then the number of linearly independent contrasts out of n observations is $n - \binom{d+\lfloor \nu \rfloor}{\lfloor \nu \rfloor}$, assuming this number is nonnegative (Exercise 21). This number jumps downward as ν increases at each integer value of ν, which means that the likelihood of the contrasts for, say, $\nu = 0.5$ is not based on the same information as for any $\nu > 1$. If one is fairly certain a priori that $\nu < 1$, then this problem does not arise.

On the whole, I would advise leaving α in the model. However, if examination of the likelihood function yields no substantial evidence against $\alpha = 0$, one can then set $\alpha = 0$, adopt the appropriate order IRF model and end up with a slightly more parsimonious model for the covariance structure. As long as all predictions are interpolations, I do not see that much harm can come from doing so. Furthermore, certain numerical difficulties that may occur with Matérn autocovariance functions when ν is large can be avoided by using $\alpha = 0$. More specifically, for the Matérn model with ν large, the principal irregular term of the autocovariance function (see 2.7) is dominated by many "regular" terms (even order monomials) in a neighborhood of the origin, which may lead to numerical inaccuracies when calculating likelihood functions or BLUPs based on this model.

Exercise

21 Show that the number of monomials of order at most p in d dimensions is $\binom{d+p}{p}$.

6.6 A numerical study of the Fisher information matrix under the Matérn model

The asymptotic theory of MLEs described in 6.4 suggests that calculating the Fisher information matrix $\boldsymbol{\mathcal{I}}$ and its inverse in various settings is a fruitful way of learning about the behavior of MLEs. This section reports numerical calculations of $\boldsymbol{\mathcal{I}}$ and $\boldsymbol{\mathcal{I}}^{-1}$ for observations from a mean 0 Gaussian process Z on $\mathbb{R}$ with spectral density from the Matérn model, $f_{\boldsymbol{\theta}}(\omega) = \phi(\alpha^2 + \omega^2)^{-\nu-1/2}$. I first consider cases without measurement error and then some with measurement error. In interpreting the results, it is helpful to keep in mind that α mainly affects the low frequency behavior of Z whereas ϕ and ν both have a critical impact on the high frequency behavior of Z.

For a Gaussian random vector with known mean $\mathbf{0}$, the Fisher information matrix takes on a fairly simple form. Specifically, if $\mathbf{Y}$ follows a $N(\mathbf{0}, \boldsymbol{\Sigma}(\boldsymbol{\theta}))$ distribution, then the jkth element of $\boldsymbol{\mathcal{I}}(\boldsymbol{\theta})$ is

$$\mathcal{I}_{jk}(\boldsymbol{\theta}) = \tfrac{1}{2}\operatorname{tr}\left\{\boldsymbol{\Sigma}(\boldsymbol{\theta})^{-1}\boldsymbol{\Sigma}_j(\boldsymbol{\theta})\boldsymbol{\Sigma}(\boldsymbol{\theta})^{-1}\boldsymbol{\Sigma}_k(\boldsymbol{\theta})\right\}, \tag{8}$$

where $\boldsymbol{\Sigma}_j(\boldsymbol{\theta}) = \partial\boldsymbol{\Sigma}(\boldsymbol{\theta})/\partial\theta_j$ (Exercise 22). To carry out this calculation for observations from a Gaussian random field under the Matérn model requires differentiating the modified Bessel function $\mathcal{K}_\nu$ with respect to ν. This can be conveniently done when ν is an integer (Abramowitz and Stegun 1965, 9.6.45). In particular, $(\partial/\partial\nu)\mathcal{K}_\nu(t)|_{\nu=1} = t^{-1}\mathcal{K}_0(t)$ and $(\partial/\partial\nu)\mathcal{K}_\nu(t)|_{\nu=2} = 2t^{-2}\mathcal{K}_0(t) + 2t^{-1}\mathcal{K}_1(t)$.

The sets of observation locations on $\mathbb{R}$ I consider include 40 or 80 observations, varying levels of spacings between observations, and evenly spaced or randomly located observations. Specifically, in the evenly spaced case, there are observations at $\delta, 2\delta, \ldots, n\delta$ for $n = 40$ or 80, where δ ranges between 0.02 and 1. When the sample size is 40, the randomly located observations were generated from a single realization of 40 independent and uniformly distributed random variables on $[0, 40]$. Figure 1 shows these 40 values, which I denote by $t_1, \ldots, t_{40}$. When I refer to 40 random locations with spacing δ, I mean the set of observation locations $\{\delta t_1, \ldots, \delta t_{40}\}$. When I refer to 80 random observations with spacing δ, I mean the set of locations $\{\delta t_1, \ldots, \delta t_{40}\} \cup \{\delta(t_1 + 40), \ldots, \delta(t_{40} + 40)\}$. The reason for repeating and shifting the initial 40 locations rather than generating an independent set of 40 random locations on $[40, 80]$ is to make the cases of 40 and 80 random locations more readily comparable. In particular, by Exercise 19 and the stationarity of Z, repeating the same pattern twice yields a value of $\boldsymbol{\mathcal{I}}$ for 80 observations that is at most double the value for 40 observations. The extent to which this value is not doubled then measures the degree of redundancy in the information in the two halves of the 80 observation sample.

No measurement error and ν unknown

Suppose Z is a Gaussian process on $\mathbb{R}$ with known mean 0 and with spectral density from the Matérn class with $(\phi, \nu, \alpha) = (1, 1, 1)$. This process is just barely not mean square differentiable, since Z is mean square differentiable under the Matérn model for $\nu > 1$. Figure 2 shows the autocovariance

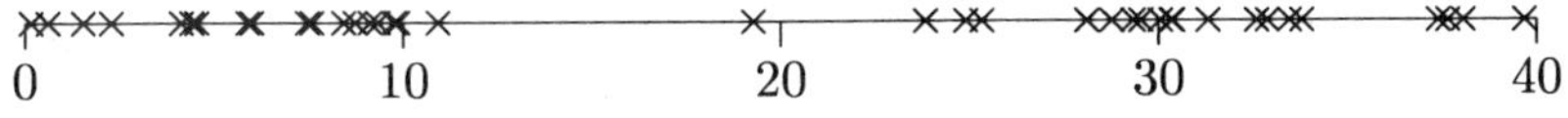

FIGURE 1. Locations of random observations on $[0, 40]$.

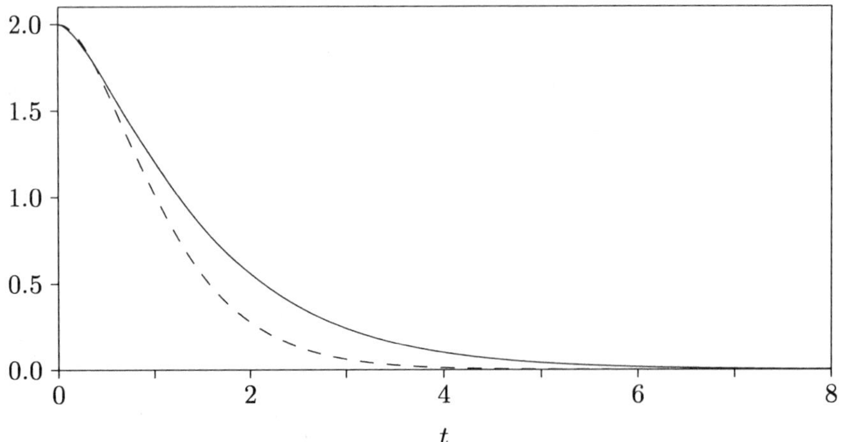

FIGURE 2. Plots of Matérn autocovariance functions used in examples. Solid line corresponds to $\boldsymbol{\theta} = (1, 1, 1)$ and dashed line to $\boldsymbol{\theta} = (24, 2, 2)$.

function for Z. Let $\mathcal{I}_\phi$ indicate the diagonal element of $\boldsymbol{\mathcal{I}}$ corresponding to ϕ, let $\mathcal{I}^\phi$ indicate the diagonal element of $\boldsymbol{\mathcal{I}}^{-1}$ corresponding to ϕ and define $\mathcal{I}_\nu$, $\mathcal{I}^\nu$, $\mathcal{I}_\alpha$ and $\mathcal{I}^\alpha$ similarly. Figure 3 plots $\mathcal{I}_\nu$ and $\mathcal{I}_\alpha$ for the various spacings, sample sizes and patterns. There is no need to plot $\mathcal{I}_\phi$ as it just equals $n/(2\phi^2)$, where n is the sample size. Note that asymptotic theory suggests, for example, that $1/\mathcal{I}_\nu$ is the approximate variance of the MLE of ν if $\phi = 1$ and $\alpha = 1$ are known. For ν, random locations are substantially more informative than evenly spaced locations, especially for larger spacings δ. These results make sense in light of the critical role ν plays in the local behavior of Z. Groups of points that are close together are particularly informative about ν and randomly located observations provide groups of points that are more tightly bunched than evenly spaced points with the same value of δ. For all δ and even or random spacing, doubling the sample size very nearly doubles $\mathcal{I}_\nu$, which means that the information about ν in the observations on $(0, 40\delta]$ is nearly independent of the information in the observations on $(40\delta, 80\delta]$ (see Exercise 19).

For $\mathcal{I}_\alpha$, the picture is rather different. Now, larger values of δ yield greater information, which makes sense for a parameter that mainly affects low frequency behavior. For smaller δ, even and random spacing give nearly the same values for $\mathcal{I}_\alpha$. In addition, for even spacing, when $n = 40$ and $\delta = 0.1$, $\mathcal{I}_\alpha$ is 6.33, whereas by doubling the number of observations ($n = 80$) and halving the spacing ($\delta = 0.05$), $\mathcal{I}_\alpha$ increases only slightly to 6.40. For the nonmicroergodic parameter α, these results are expected, since $\mathcal{I}_\alpha$ should tend to a finite value as the observations in a fixed interval become increasingly dense. For larger δ, even spacing produces somewhat larger values for $\mathcal{I}_\alpha$ than random locations. I do not have a convenient story for this result, although it is not entirely unexpected in light of a theoretical result in Stein (1990b, Section 5) showing that even spacing is asymptotically optimal in a certain sense for estimating α when the spectral density is assumed to be of the form $1/(\alpha^2 + \omega^2)$. Finally, for larger δ, doubling the sample size does approximately double $\mathcal{I}_\alpha$, but for smaller δ, $\mathcal{I}_\alpha$ is less than doubled, which indicates that the information about α in the observations in $(0, 40\delta]$ and in those in $(40\delta, 80\delta]$ is somewhat redundant.

Figures 4 and 5, which show the diagonal elements of $\boldsymbol{\mathcal{I}}^{-1}$ and the approximate correlations of the MLEs implied by (7), deserve similar scrutiny and explanation. One noteworthy result in Figure 4 is that $\mathcal{I}^\alpha$ is not monotonically decreasing as δ increases, despite the fact that $\mathcal{I}_\alpha$ is monotonically increasing, which is related to the fact that the approximate correlations of $\hat{\alpha}$ with $\hat{\phi}$ and $\hat{\nu}$ increase with δ (see Figure 5). Outcomes in Figure 5 deserving notice include the fact that the approximate correlation of $\hat{\phi}$ and $\hat{\nu}$ is essentially independent of sample size and is distinctly lower for all δ when the observations are randomly located.

It is not possible to take asymptotic results such as (7) seriously for at least some of these examples. In particular, for 40 evenly spaced observations and $\delta = 1$, we have $\mathcal{I}^\phi = 5.62$, but it certainly cannot be the case that $\hat{\phi}$ is approximately $N(1, 5.62)$ since $\hat{\phi}$ is always nonnegative. Even in this situation, I believe that $\boldsymbol{\mathcal{I}}^{-1}$ provides at least qualitative insight about the variability of the MLE. An alternative interpretation is to imagine observing N independent realizations of the process Z at the same set of locations. If $\hat{\boldsymbol{\theta}}_N$ is the MLE of $\boldsymbol{\theta}$ based on these N independent and identically distributed random vectors, then $\hat{\boldsymbol{\theta}}_N$ is approximately $N(\boldsymbol{\theta}_0, N^{-1}\boldsymbol{\mathcal{I}}^{-1})$ for N sufficiently large, where $\boldsymbol{\mathcal{I}}$ is the Fisher information matrix for observations from a single realization of the process. For a space-time process observed at a fixed set of spatial locations at sufficiently distant points in time, it may be reasonable to assume that observations from different times are independent realizations of a random field.

No measurement error and ν known

The results of the previous subsection show the random design clearly dominating the evenly spaced design in terms of having smaller values for the diagonal elements of the inverse Fisher information matrix. Before jumping to any conclusions that random designs are always better, it is worthwhile to consider how this result depends on the model selected. In particular, consider the same setting as in the previous subsection but assume that ν is known and only ϕ and α need to be estimated. Figure 6 shows that the evenly spaced designs are now quite competitive with the random designs and even have slightly lower values for $\mathcal{I}^\alpha$ for some δ. Although I do not advocate treating ν as fixed, keep in mind that using the exponential model is the same as using the Matérn model with $\nu = \frac{1}{2}$ assumed known.

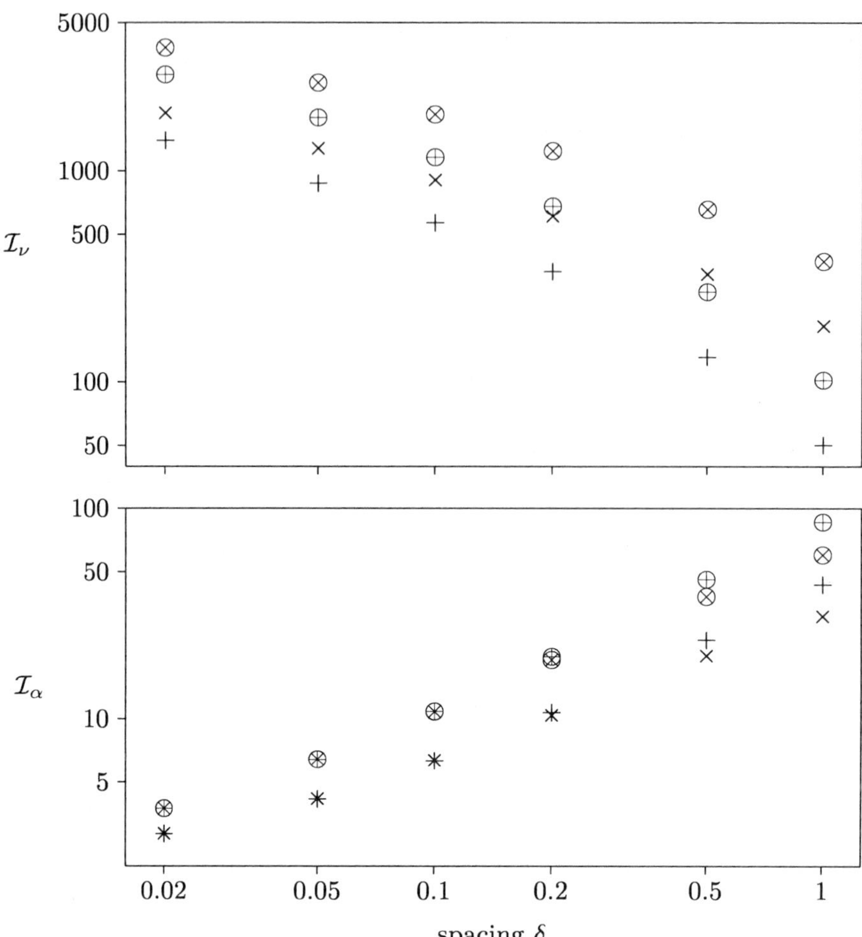

FIGURE 3. Diagonal values of Fisher information matrix for Matérn model with $(\phi, \nu, \alpha) = (1, 1, 1)$.
$+$ indicates 40 evenly spaced observations with spacing δ.
$\oplus$ indicates 80 evenly spaced observations with spacing δ.
$\times$ indicates 40 randomly placed observations on $[0, 40\delta]$.
$\otimes$ indicates the same 40 randomly placed observations on $[0, 40\delta]$ together with each of these observation locations plus 40δ, for a total of 80 observations.

Observations with measurement error

If Z is observed with error then it should be more difficult to estimate the parameters governing the law of Z. To investigate how this loss of information depends on the variance of the measurement error, which I denote by τ, I consider those settings from the previous subsection with 80 observations, evenly spaced and random, and $\delta = 0.1$. In addition to $(\phi, \nu, \alpha) = (1, 1, 1)$, I also consider $(\phi, \nu, \alpha) = (24, 2, 2)$. The autocovariance

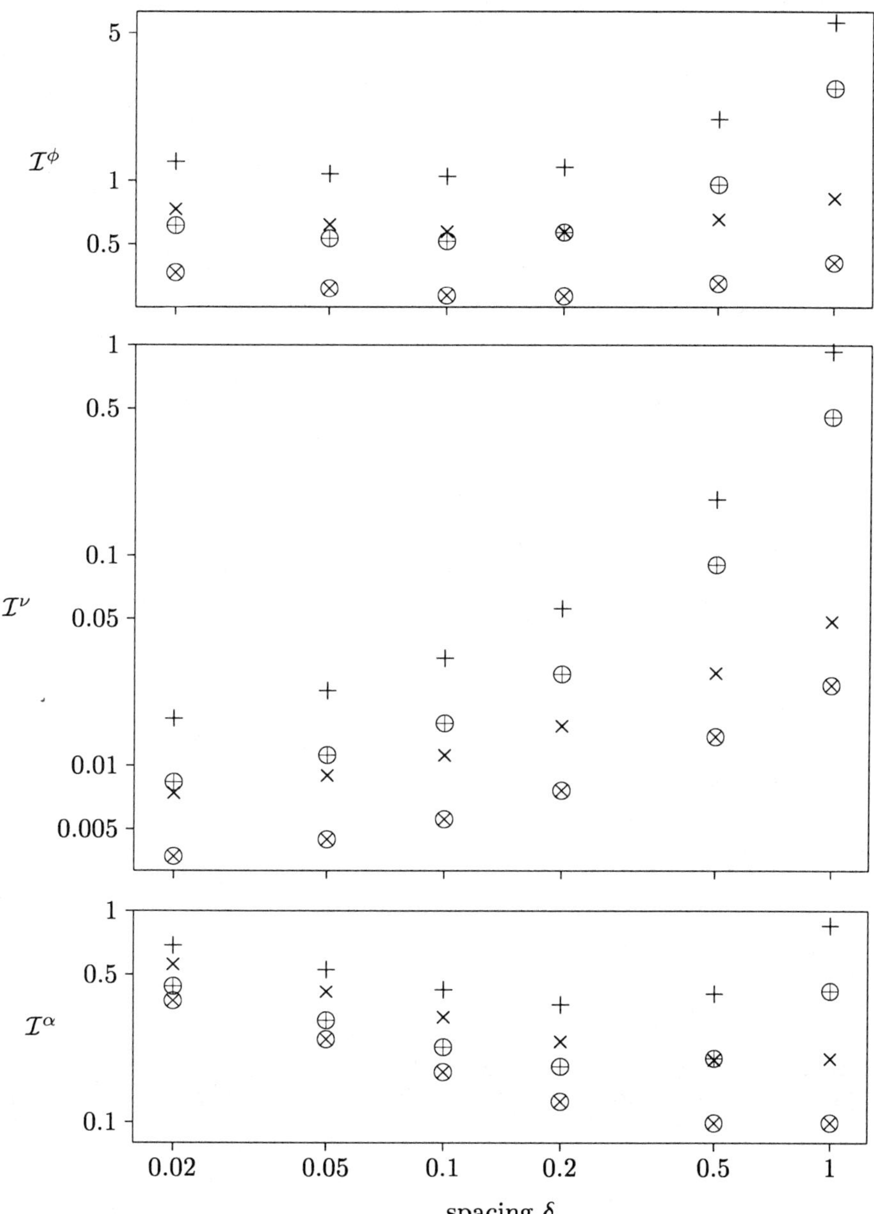

FIGURE 4. Diagonal values of inverse Fisher information matrix for Matérn model with $(\phi, \nu, \alpha) = (1, 1, 1)$. Symbols have same meaning as in Figure 3.

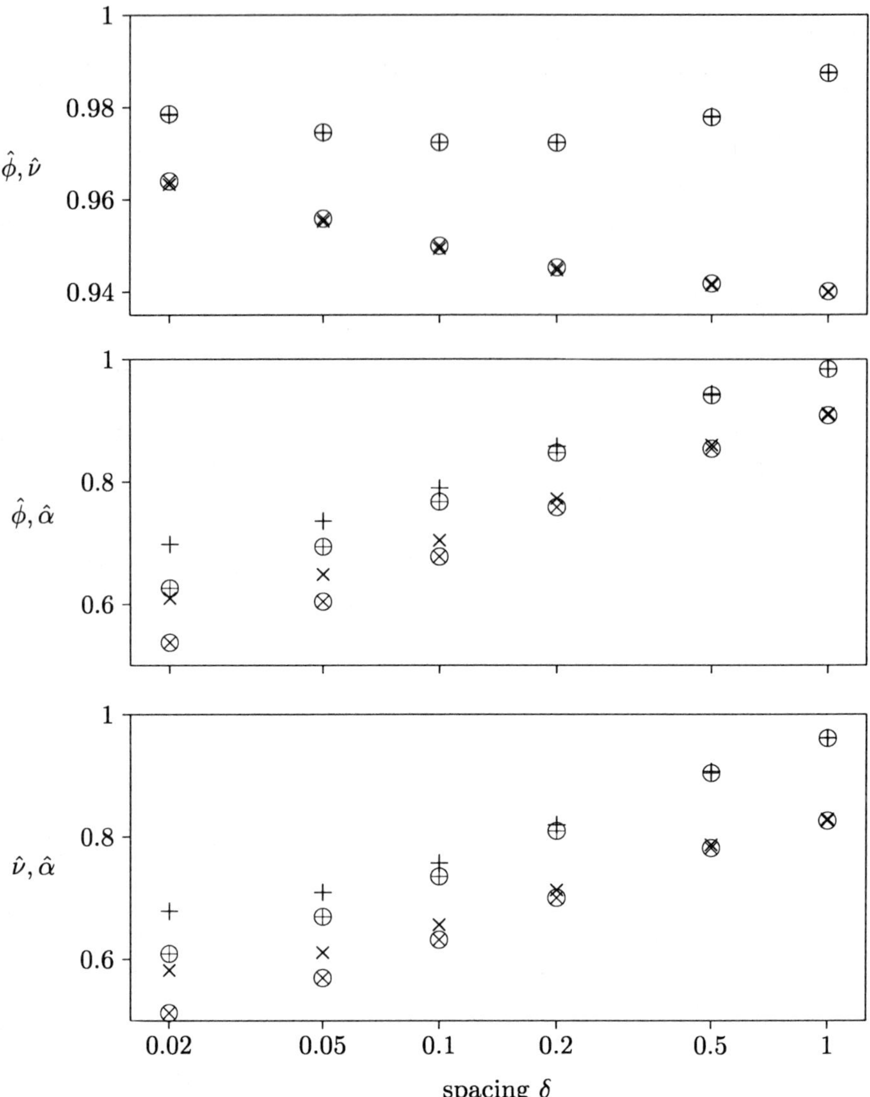

FIGURE 5. Approximate correlations of MLEs of parameters based on inverse Fisher information matrix for Matérn model with parameters $(1, 1, 1)$. Symbols have same meaning as in Figure 3.

functions for these two models are plotted in Figure 2. The values of τ I consider are $10^{-4}, 10^{-3}, 10^{-2}, 10^{-1}$ and 1. Although the value 10^{-4} may seem small, note that it means the standard deviation of the measurement error divided by the standard deviation of the process is 0.7%, which strikes me as quite plausible for many physical quantities.

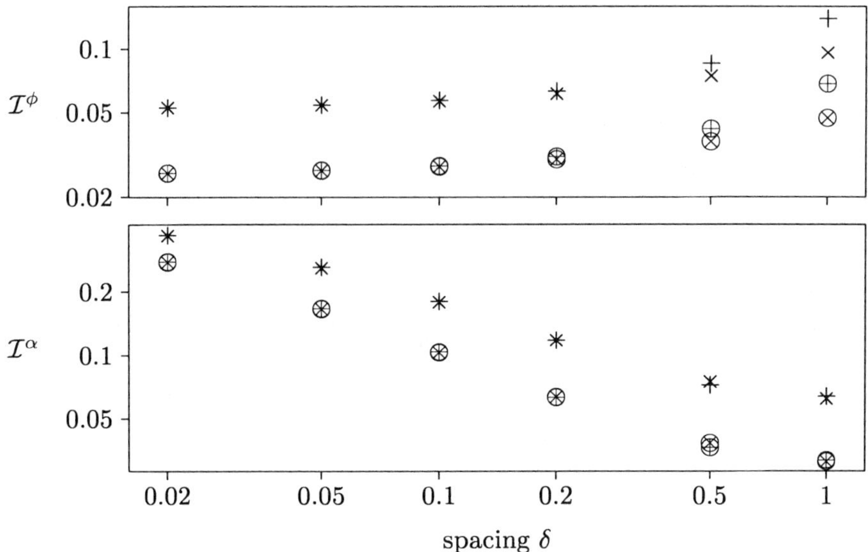

FIGURE 6. Diagonal values of inverse Fisher information matrix for Matérn model with $(\phi, \nu, \alpha) = (1, 1, 1)$ when $\nu = 1$ is known. Symbols have same meaning as in Figure 3.

The reason for including two different values of ν is to see whether increasing τ has more effect on our ability to estimate larger or smaller values of ν. Theoretical results in Stein (1993a) and the intuition that estimating the degree of differentiability of a random process with noisy observations should be harder for smoother processes suggest that τ should have more of an impact on the ability to estimate ν when ν is 2 rather than 1. Results in Tables 1 and 2 support this expectation: for evenly spaced observations, when τ goes from 10^{-4} to 10^{-2}, $\mathcal{I}^\nu$ increases by a factor of 1.55 when $\nu = 1$ but by a factor of 5.13 when $\nu = 2$.

On the other hand, τ is much easier to estimate when $\nu = 2$ than when $\nu = 1$, especially for smaller τ and evenly spaced observations. In particular, for evenly spaced observations and $\tau = 10^{-4}$, $\mathcal{I}^\tau/\tau^2$ is 1,893 for $\nu = 1$ and 1.507 for $\nu = 2$. This large value for $\mathcal{I}^\tau/\tau^2$ for $\nu = 1$ suggests that these data provide essentially no information for distinguishing the true value for τ of 10^{-4} from either $\tau = 0$ or much larger values such as $\tau = 10^{-3}$. Fortunately, in this case we have $f(\omega) \sim \omega^{-3}$ as $\omega \to \infty$, so that $\alpha = 3$ in the notation of Theorem 7 of 3.7, and since $\tau/\delta^{\alpha-1} = 0.01$ is small, this theorem suggests that at least for certain predictions, acting as if $\tau = 0$ will produce nearly optimal predictors.

The other diagonal elements of $\mathcal{I}$ depend on τ as should be expected. Specifically, parameters that are more related to high frequency behavior should be more affected by increasing τ than parameters affecting mostly low frequency behavior. The results in Tables 1 and 2 are in line with this

TABLE 1. Diagonal values of Fisher information matrix and its inverse for $(\phi, \nu, \alpha) = (1, 1, 1)$ for various values of τ based on 80 observations with spacing $\delta = 0.1$. Results for evenly spaced observations indicated by F and random design indicated by R (see Figure 3 for details).

		τ				
		10^{-4}	10^{-3}	10^{-2}	10^{-1}	1
$\phi^2 \mathcal{I}_\phi$	F	39.33	34.51	19.70	8.996	4.002
	R	33.33	24.47	14.96	7.947	3.717
$\nu^2 \mathcal{I}_\nu$	F	1130	927.0	358.5	72.64	11.09
	R	1225	662.6	241.1	61.19	10.17
$\alpha^2 \mathcal{I}_\alpha$	F	10.83	10.83	10.77	10.53	9.517
	R	10.81	10.80	10.73	10.44	9.285
$\tau^2 \mathcal{I}_\tau$	F	0.005481	0.4036	7.625	22.48	31.71
	R	2.862	7.736	17.10	26.17	32.72
$\mathcal{I}^\phi/\phi^2$	F	1.324	1.375	1.750	3.550	12.66
	R	0.4158	0.6510	1.231	3.072	12.25
$\mathcal{I}^\nu/\nu^2$	F	0.07240	0.07693	0.1125	0.3282	1.949
	R	0.01036	0.02227	0.06282	0.2550	1.753
$\mathcal{I}^\alpha/\alpha^2$	F	0.3878	0.3974	0.4657	0.7702	2.729
	R	0.2025	0.2538	0.3689	0.6908	2.076
$\mathcal{I}^\tau/\tau^2$	F	1893	22.03	0.5840	0.09044	0.04618
	R	0.3990	0.1806	0.07989	0.04938	0.03770

heuristic as $\mathcal{I}_\alpha$ decreases only slightly as τ increases but $\mathcal{I}_\phi$ and particularly $\mathcal{I}_\nu$ decrease sharply.

In comparing random and evenly spaced designs, for $\tau = 10^{-4}$ and $\nu = 1$ or 2, the random design does drastically better in terms of the diagonal elements of $\mathcal{I}^{-1}$. The evenly spaced design is much more competitive for larger τ and, for the parameters other than τ, actually has slightly lower values on the diagonal of $\mathcal{I}^{-1}$ for $\nu = 2$ when τ is sufficiently large. However, for $\tau = 1$, it is fair to say that estimating (ϕ, ν, α) is essentially hopeless when $\nu = 1$ or 2. When ν is known, then the other parameters are much easier to estimate. For example, when $\nu = 2$ is known and $\tau = 1$, then for the evenly spaced design, $\mathcal{I}^\phi/\phi^2$ is 1.787 as opposed to 160.1 when ν is unknown and $\mathcal{I}^\alpha/\alpha^2$ is 0.1528 as opposed to 2.462 when ν is unknown.

Conclusions

One overall pattern that emerges from these calculations is that random designs can often yield better parameter estimates than evenly spaced designs of comparable density, sometimes dramatically so. However, if our

TABLE 2. Diagonal values of Fisher information matrix and its inverse for $(\phi, \nu, \alpha) = (24, 2, 2)$ for various values of τ. Observation locations are same as in Table 1.

		τ				
		10^{-4}	10^{-3}	10^{-2}	10^{-1}	1
$\phi^2\mathcal{I}_\phi$	F	30.02	19.05	11.94	7.299	4.038
	R	20.41	14.89	10.33	6.657	3.745
$\nu^2\mathcal{I}_\nu$	F	2791	1223	501.1	193.4	67.27
	R	1579	831.5	394.7	166.4	60.47
$\alpha^2\mathcal{I}_\alpha$	F	58.91	58.54	57.54	54.84	46.63
	R	58.64	58.22	57.17	53.97	44.84
$\tau^2\mathcal{I}_\alpha$	F	2.196	12.12	22.05	28.72	33.14
	R	14.22	20.42	25.74	30.20	33.76
$\mathcal{I}^\phi/\phi^2$	F	2.573	4.003	9.138	30.06	160.1
	R	2.066	4.018	9.796	33.33	179.6
$\mathcal{I}^\nu/\nu^2$	F	0.02673	0.04799	0.1371	0.5790	3.935
	R	0.01973	0.04786	0.1484	0.6457	4.396
$\mathcal{I}^\alpha/\alpha^2$	F	0.1122	0.1486	0.2610	0.6320	2.462
	R	0.09819	0.1494	0.2750	0.6900	2.752
$\mathcal{I}^\tau/\tau^2$	F	1.507	0.1442	0.05976	0.04183	0.03521
	R	0.07993	0.05475	0.04308	0.03657	0.03287

goal is to predict Z at unobserved locations, it does not follow that random designs should be preferred when using the Matérn model. If one wants to predict well throughout some region R, then there is a certain logic to some sort of regular pattern of observations throughout R, although if ν is large, it may be appropriate to include some observations slightly outside R. Presumably, one should try to reach some compromise between designs that lead to good estimates of the unknown parameters as well as accurate predictions of Z based on the available observations and the estimated parameters. See Laslett and McBratney (1990), Pettitt and McBratney (1993) and Handcock (1991) for further discussion of these issues. The example in 6.9 demonstrates that adding even a few closely packed observations to an evenly spaced design can sometimes dramatically improve parameter estimation.

Finally, although all of the examples in this section consider only processes on $\mathbb{R}$, I have run some examples on $\mathbb{R}^2$ with qualitatively similar results. Obviously, there is considerable scope for calculations of $\mathcal{I}$ and $\mathcal{I}^{-1}$ in further settings and for simulations of the actual distributions of MLEs to compare to the asymptotic approximations.

Exercises

22 Verify (8).

23 Repeat the numerical calculations in Tables 1 and 2 for different values of the spacing δ. How do the comparisons between the fixed and random designs change with δ? How do the comparisons between $\nu = 1$ and $\nu = 2$ change with δ?

6.7 Maximum likelihood estimation for a periodic version of the Matérn model

It is generally difficult to determine the asymptotic properties of estimators of the parameters of any model under fixed-domain asymptotics. One situation where it is possible to make progress is for periodic random fields observed on a lattice. This leads us to considering estimation for the class of periodic random fields whose spectral measures place all of their mass on $\mathbb{Z}^d$ and the mass at $\mathbf{j}$ is $\phi(\alpha^2 + |\mathbf{j}|^2)^{-\nu-d/2}$ for $\mathbf{j} \in \mathbb{Z}^d$. The hope is that any asymptotic results we obtain for these periodic models will be similar to those for the Matérn model, although that remains to be proven.

Discrete Fourier transforms

Suppose Z is a mean 0 stationary Gaussian process with spectral measure F and we observe Z at $\delta\mathbf{j}$ for $\mathbf{j} \in \mathcal{G}_m = \{1, \ldots, m\}^d$. The discrete Fourier transform of these observations at a frequency $\boldsymbol{\omega} \in \mathbb{R}^d$ is defined as

$$\hat{Z}(\boldsymbol{\omega}) = \sum_{\mathbf{j} \in \mathcal{G}_m} Z(\delta\mathbf{j}) \exp(-i\boldsymbol{\omega}^T\mathbf{j}).$$

Note that $\hat{Z}(\boldsymbol{\omega})$ has period 2π in each coordinate, so there is no loss in information in restricting attention to frequencies in $(-\pi, \pi]^d$. Consider further restricting to just frequencies $\boldsymbol{\omega}$ in $(-\pi, \pi]^d$ of the form $2\pi m^{-1}\mathbf{p}$, which is equivalent to considering only $\mathbf{p} \in \mathcal{B}_m$, where $\mathcal{B}_m = \{ -\lfloor \frac{1}{2}(m-1)\rfloor, -\lfloor \frac{1}{2}(m-1)\rfloor + 1, \ldots, \lfloor \frac{1}{2}m\rfloor\}^d$. Now, for $\mathbf{j} \in \mathcal{G}_m$,

$$\sum_{\mathbf{p} \in \mathcal{B}_m} \hat{Z}(2\pi m^{-1}\mathbf{p}) \exp(i2\pi m^{-1}\mathbf{j}^T\mathbf{p}) = m^d Z(\delta\mathbf{j}) \tag{9}$$

(Exercise 24), so this further restriction to $\mathbf{p} \in \mathcal{B}_m$ also involves no loss of information. If m is highly composite, $\hat{Z}(2\pi m^{-1}\mathbf{p})$ can be efficiently calculated for all $\mathbf{p} \in \mathcal{B}_m$ using the fast Fourier transform (Press, Flannery, Teukolsky and Vetterling 1992). Indeed, even if m is not highly composite, these calculations can still be done quite efficiently by implementing a d-dimensional version of the fractional fast Fourier transform (Bailey and

Swarztrauber 1991). Alternatively, one can add zeroes to the dataset so that the expanded dataset does have m highly composite, although then one gets $\hat{Z}$ at a different set of frequencies (Bloomfield 1976, p. 73; Priestley 1981, p. 577).

Define the measure F_δ on $A_d(\delta^{-1})$ by $F_\delta(\cdot) = \sum_{\mathbf{k}\in\mathbb{Z}^d} F(\cdot + 2\pi\delta^{-1}\mathbf{k})$. It is then a straightforward calculation (Exercise 25) to show that for $\mathbf{p}, \mathbf{q} \in \mathbb{Z}^d$,

$$
\begin{aligned}
&E\left\{\hat{Z}(2\pi m^{-1}\mathbf{p})\overline{\hat{Z}(2\pi m^{-1}\mathbf{q})}\right\} \\
&= \prod_{u=1}^{d} \exp\{i\pi m^{-1}(q_u - p_u)\} \qquad (10)\\
&\times \int_{A_d(\delta^{-1})} \prod_{u=1}^{d} \frac{\sin^2\left(\frac{1}{2}m\delta\omega_u\right)}{\sin\left(\frac{1}{2}\delta\omega_u - \pi p_u m^{-1}\right)\sin\left(\frac{1}{2}\delta\omega_u - \pi q_u m^{-1}\right)} F_\delta(\mathbf{d}\boldsymbol{\omega}),
\end{aligned}
$$

where the integrand is defined by continuity for those $\boldsymbol{\omega}$ for which the denominator is 0. Here and subsequently in this section, a subscript u indicates the uth component of a vector so that, for example, p_u is the uth component of $\mathbf{p}$.

Periodic case

If Z has period $2\pi m\delta$ in each coordinate, a great simplification occurs in (10). This periodicity implies that F is a discrete measure placing all of its mass on points $(m\delta)^{-1}\mathbf{k}$ for $\mathbf{k} \in \mathbb{Z}^d$, so that F_δ puts all of its mass on points of the form $2\pi(m\delta)^{-1}\mathbf{p}$ for $\mathbf{p} \in \mathcal{B}_m$. Since $\prod_{u=1}^d \sin^2\left(\frac{1}{2}m\delta\omega_u\right)$ has a zero of order $2d$ at all such points, the only way (10) can be nonzero for $\boldsymbol{\omega} = 2\pi(m\delta)^{-1}\mathbf{r}$, $\mathbf{r} \in \mathcal{B}_m$, is if $\sin\left\{\pi m^{-1}(r_u - p_u)\right\} = 0$ and $\sin\left\{\pi m^{-1}(r_u - q_u)\right\} = 0$ for $u = 1, \ldots, d$, which for $\mathbf{p}, \mathbf{q}, \mathbf{r} \in \mathcal{B}_m$ can only occur if $\mathbf{p} = \mathbf{q} = \mathbf{r}$. Thus, for $\mathbf{p}, \mathbf{q} \in \mathcal{B}_m$,

$$
E\left\{\hat{Z}(2\pi m^{-1}\mathbf{p})\overline{\hat{Z}(2\pi m^{-1}\mathbf{q})}\right\} = \begin{cases} m^{2d}F_\delta(2\pi m^{-1}\mathbf{p}), & \mathbf{p} = \mathbf{q} \\ 0, & \mathbf{p} \neq \mathbf{q}. \end{cases}
$$

The fact that for $\mathbf{p} \neq \mathbf{q}$, $E\left\{\hat{Z}(2\pi m^{-1}\mathbf{p})\overline{\hat{Z}(2\pi m^{-1}\mathbf{q})}\right\} = 0$ irrespective of the particular values of $F\big((m\delta)^{-1}\mathbf{k}\big)$ for $\mathbf{k} \in \mathbb{Z}^d$ allows us to obtain a relatively simple expression for the likelihood function. Define $\mathbf{p} > \mathbf{0}$ to mean $\mathbf{p} \neq \mathbf{0}$ and the first nonzero component of $\mathbf{p}$ is positive, define $\mathbf{p} \geq \mathbf{0}$ as $\mathbf{p} > \mathbf{0}$ or $\mathbf{p} = \mathbf{0}$ and say $\mathbf{p} < \mathbf{0}$ if $\mathbf{p} \ngeq \mathbf{0}$. For $\mathbf{p} \in \mathcal{B}_m$, let

$$
X_{\mathbf{p}} = \sum_{\mathbf{j}\in\mathcal{G}_m} Z(\delta\mathbf{j})\cos(2\pi m^{-1}\mathbf{p}^T\mathbf{j})
$$

for $\mathbf{p} \geq \mathbf{0}$ and

$$
X_{\mathbf{p}} = \sum_{\mathbf{j}\in\mathcal{G}_m} Z(\delta\mathbf{j})\sin(2\pi m^{-1}\mathbf{p}^T\mathbf{j})
$$

for $\mathbf{p} < \mathbf{0}$. Using $\hat{Z}(\boldsymbol{\omega}) = \overline{\hat{Z}(-\boldsymbol{\omega})}$, it follows that from $\{X_{\mathbf{p}} : \mathbf{p} \in \mathcal{B}_m\}$ one can recover $\{\hat{Z}(2\pi m^{-1}\mathbf{p}) : \mathbf{p} \in \mathcal{B}_m\}$ and hence the original observations. Thus, by Exercise 17 in 6.4, we can find the MLE for $\boldsymbol{\theta}$ by maximizing the likelihood with $\{X_{\mathbf{p}} : \mathbf{p} \in \mathcal{B}_m\}$ as the vector of observations. Using the cosine transform for $\mathbf{p} \geq \mathbf{0}$ and the sine transform for $\mathbf{p} < \mathbf{0}$ in the definition of $X_{\mathbf{p}}$ is notationally convenient, since the real random variables $X_{\mathbf{p}}$ for $\mathbf{p} \in \mathcal{B}_m$ are then uncorrelated with mean 0 and

$$\operatorname{var}(X_{\mathbf{p}}) = \tfrac{1}{2} m^{2d} F_\delta(2\pi m^{-1}\mathbf{p}) \epsilon_m(\mathbf{p}) \tag{11}$$

for $\mathbf{p} \in \mathcal{B}_m$, where $\epsilon_m(\mathbf{p}) = 1$ unless all components of $2\mathbf{p}$ are 0 or m, in which case, $\epsilon_m(\mathbf{p}) = 2$ (Exercise 26).

Suppose Z is a stationary mean 0 Gaussian random field on $\mathbb{R}^d$ with period 2π in each coordinate and spectral measure with mass $f_{\boldsymbol{\theta}}(\mathbf{j})$ for $\mathbf{j} \in \mathbb{Z}^d$ and no mass elsewhere for $\boldsymbol{\theta} \in \Theta$. If Z is observed at $2\pi m^{-1}\mathbf{j}$ for $\mathbf{j} \in \mathcal{G}_m$, then $\{X_{\mathbf{p}} : \mathbf{p} \in \mathcal{B}_m\}$ is a one-to-one function of the vector of m^d observations $\mathbf{Z}_m$. Furthermore, the $X_{\mathbf{p}}$s are independent mean 0 Gaussian random variables with variances given by (11), so by Exercise 17 in 6.4, the log likelihood for $\boldsymbol{\theta}$ is of the form

$$\begin{aligned} \ell(\boldsymbol{\theta}; \mathbf{Z}_m) = \text{constant} &- \frac{1}{2} \sum_{\mathbf{p} \in \mathcal{B}_m} \log\left\{ m^{2d} \epsilon_m(\mathbf{p}) \sum_{\mathbf{j} \in \mathbb{Z}^d} f_{\boldsymbol{\theta}}(\mathbf{p} + m\mathbf{j}) \right\} \\ &- \frac{1}{2} \sum_{\mathbf{p} \in \mathcal{B}_m} \frac{X_{\mathbf{p}}^2}{m^{2d} \epsilon_m(\mathbf{p}) \sum_{\mathbf{j} \in \mathbb{Z}^d} f_{\boldsymbol{\theta}}(\mathbf{p} + m\mathbf{j})}, \end{aligned} \tag{12}$$

where the constant does not depend on $\boldsymbol{\theta}$ and hence has no impact on the maxima of the function. Call $\hat{\boldsymbol{\theta}}_m$ a maximum likelihood estimator (MLE) of $\boldsymbol{\theta}$ if it maximizes (12) for $\boldsymbol{\theta} \in \Theta$. Suppose that $f_{\boldsymbol{\theta}}(\mathbf{j}) = \phi(\alpha^2 + |\mathbf{j}|^2)^{-\nu - d/2}$, $\boldsymbol{\theta} = (\phi, \nu, \alpha)$ and $\Theta = (0, \infty)^3$. We know that the MLE cannot be consistent for any function of $\boldsymbol{\theta}$ that is not microergodic and we expect that it is consistent for any function of $\boldsymbol{\theta}$ that is microergodic. From (20) in 4.2, we see that ϕ and ν are microergodic in any number of dimensions but that α is microergodic if and only if $d \geq 4$.

Asymptotic results

This section provides asymptotic results for $\mathcal{I}_m = \mathcal{I}_m(\boldsymbol{\theta}_0)$, the Fisher information matrix for $\boldsymbol{\theta}$ based on $\mathbf{Z}_m$. Let $\hat{\boldsymbol{\theta}}_m$ be an MLE for $\boldsymbol{\theta}$ based on $\mathbf{Z}_m$. As indicated in 6.4, we might then expect

$$\mathcal{I}_m^{1/2}(\hat{\boldsymbol{\theta}}_m - \boldsymbol{\theta}_0) \xrightarrow{\mathcal{L}} N(\mathbf{0}, \mathbf{I})$$

if $\boldsymbol{\theta}$ is microergodic. When $d \leq 3$ so that α is not microergodic, as discussed in 6.4, it is possible to give a plausible conjecture about the asymptotic behavior for the MLE of ϕ and ν using results of Crowder (1976).

Let us first give exact expressions for the elements of $\mathcal{I}_m$ in the current setting. Defining

$$\xi_j(\mathbf{p}, m, \alpha, \nu) = \sum_{\mathbf{j}\in\mathbb{Z}^d} \left(\alpha^2 + |\mathbf{p} + m\mathbf{j}|^2\right)^{-\nu-d/2} \log^j \left(\alpha^2 + |\mathbf{p} + m\mathbf{j}|^2\right),$$

and using ℓ as shorthand for $\ell(\boldsymbol{\theta}; \mathbf{Z}_m)$ we have (Exercise 27)

$$\operatorname{var}\left(\frac{\partial\ell}{\partial\phi}\right) = \frac{m^d}{2\phi^2}, \tag{13}$$

$$\operatorname{var}\left(\frac{\partial\ell}{\partial\nu}\right) = \frac{1}{2}\sum_{\mathbf{p}\in\mathcal{B}_m}\left\{\frac{\xi_1(\mathbf{p}, m, \alpha, \nu)}{\xi_0(\mathbf{p}, m, \alpha, \nu)}\right\}^2,$$

$$\operatorname{var}\left(\frac{\partial\ell}{\partial\alpha}\right) = \frac{1}{2}\alpha^2(2\nu + d)^2\sum_{\mathbf{p}\in\mathcal{B}_m}\left\{\frac{\xi_0(\mathbf{p}, m, \alpha, \nu + 1)}{\xi_0(\mathbf{p}, m, \alpha, \nu)}\right\}^2,$$

$$\operatorname{cov}\left(\frac{\partial\ell}{\partial\phi}, \frac{\partial\ell}{\partial\nu}\right) = -\frac{1}{2\phi}\sum_{\mathbf{p}\in\mathcal{B}_m}\frac{\xi_1(\mathbf{p}, m, \alpha, \nu)}{\xi_0(\mathbf{p}, m, \alpha, \nu)},$$

$$\operatorname{cov}\left(\frac{\partial\ell}{\partial\phi}, \frac{\partial\ell}{\partial\alpha}\right) = -\frac{\alpha(2\nu + d)}{2\phi}\sum_{\mathbf{p}\in\mathcal{B}_m}\frac{\xi_0(\mathbf{p}, m, \alpha, \nu + 1)}{\xi_0(\mathbf{p}, m, \alpha, \nu)}$$

and

$$\operatorname{cov}\left(\frac{\partial\ell}{\partial\nu}, \frac{\partial\ell}{\partial\alpha}\right) = \frac{1}{2}\alpha(2\nu + d)\sum_{\mathbf{p}\in\mathcal{B}_m}\frac{\xi_1(\mathbf{p}, m, \alpha, \nu)\xi_0(\mathbf{p}, m, \alpha, \nu + 1)}{\xi_0(\mathbf{p}, m, \alpha, \nu)^2}.$$

Corresponding to common practice in theoretical statistics, (13) and the subsequent equations do not explicitly distinguish between an arbitrary value of a parameter and its true value. A more accurate way to write (13) would be

$$\operatorname{var}_{\boldsymbol{\theta}_0}\left(\frac{\partial\ell}{\partial\phi}\right) = \frac{m^d}{2\phi_0^2},$$

but insisting on this level of explicitness would lead to rather ugly-looking expressions for the remainder of this section.

To state the asymptotic behavior of $\mathcal{I}_m^{-1}$, we first need some notation. Define

$$g_\nu(\mathbf{x}) = \frac{\sum_{\mathbf{j}\in\mathbb{Z}^d}|\mathbf{x} + \mathbf{j}|^{-2\nu-d}\log|\mathbf{x} + \mathbf{j}|}{\sum_{\mathbf{j}\in\mathbb{Z}^d}|\mathbf{x} + \mathbf{j}|^{-2\nu-d}}$$

and

$$h_\nu(\mathbf{x}) = \frac{\sum_{\mathbf{j}\in\mathbb{Z}^d}|\mathbf{x} + \mathbf{j}|^{-2\nu-d-2}}{\sum_{\mathbf{j}\in\mathbb{Z}^d}|\mathbf{x} + \mathbf{j}|^{-2\nu-d}}$$

for $\mathbf{x} \in \mathbb{R}^d$. For the rest of this section, it is convenient to write certain integrals over the unit cube $c_d = \left[-\frac{1}{2}, \frac{1}{2}\right]^d$ as expectations over the random vector $\mathbf{U}$ having uniform distribution on c_d. Thus, for example, $Eg_\nu = E\{g_\nu(\mathbf{U})\} = \int_{c_d} g_\nu(\mathbf{x})\,d\mathbf{x}$ and $\operatorname{cov}(g_\nu, h_\nu) = \operatorname{cov}\{g_\nu(\mathbf{U}), h_\nu(\mathbf{U})\} = \int_{c_d} g_\nu(\mathbf{x})h_\nu(\mathbf{x})\,d\mathbf{x} - Eg_\nu Eh_\nu$. Next, consider two sequences of $k \times \ell$ matrices $\mathbf{A}_m$ and $\mathbf{B}_m$ with elements $a_m(i,j)$ and $b_m(i,j)$, respectively. For a positive sequence $t_1, t_2, \ldots$, I take $\mathbf{A}_m = \mathbf{B}_m\{1 + O(t_m)\}$ to mean $a_m(i,j) = b_m(i,j)\{1 + O(t_m)\}$ for $1 \le i \le k$ and $1 \le j \le \ell$. Similarly, $\mathbf{A}_m \sim \mathbf{B}_m$ means $a_m(i,j) \sim b_m(i,j)$ for $1 \le i \le k$ and $1 \le j \le \ell$. Finally, for symmetric matrices, I use $\cdot$ to indicate elements of the matrix defined by its symmetry.

Theorem 1. *Suppose $d \ge 5$. As $m \to \infty$, for some $\epsilon > 0$,*

$$\mathcal{I}_m^{-1} = \frac{1}{\{\operatorname{var}(g_\nu)\operatorname{var}(h_\nu) - \operatorname{cov}(g_\nu, h_\nu)^2\}m^d}\,\mathbf{Q}_m\left\{1 + O(m^{-\epsilon})\right\}, \tag{14}$$

where $\mathbf{Q}_m$ is symmetric and has jkth element $q_m(j,k)$ with

$$\begin{aligned}
q_m(1,1) &= 2\phi^2\Big[\operatorname{var}(h_\nu)\log^2 m + 2\left\{Eg_\nu Eh_\nu^2 - Eh_\nu E(g_\nu h_\nu)\right\}\log m \\
&\qquad + Eg_\nu^2 Eh_\nu^2 - \left\{E(g_\nu h_\nu)\right\}^2\Big], \\
q_m(1,2) &= \phi\{\operatorname{var}(h_\nu)\log m + Eg_\nu Eh_\nu^2 - Eh_\nu E(g_\nu h_\nu)\}, \\
q_m(2,2) &= \frac{1}{2}\operatorname{var}(h_\nu), \\
q_m(1,3) &= -\frac{\phi m^2}{\alpha(2\nu + d)}\{\operatorname{cov}(g_\nu, h_\nu)\log m + Eg_\nu E(g_\nu h_\nu) - Eg_\nu^2 Eh_\nu\}, \\
q_m(2,3) &= -\frac{m^2\operatorname{cov}(g_\nu, h_\nu)}{\alpha(2\nu + d)}
\end{aligned}$$

and

$$q_m(3,3) = \frac{2m^4\operatorname{var}(g_\nu)}{\alpha^2(2\nu + d)^2}.$$

For $d = 4$,

$$\mathcal{I}_m^{-1} \sim \frac{1}{\operatorname{var}(g_\nu)}\mathbf{R}_m, \tag{15}$$

where $\mathbf{R}_m$ has jkth element $r_m(j,k)$ with

$$\begin{aligned}
r_m(1,1) &= \frac{2\phi^2\log^2 m}{m^4}, \\
r_m(1,2) &= \frac{\phi\log m}{m^4}, \\
r_m(2,2) &= \frac{1}{2m^4}, \\
r_m(1,3) &= -\frac{\phi\operatorname{cov}(g_\nu, h_\nu)\log m}{\pi^2\alpha(2\nu + 4)m^2},
\end{aligned}$$

$$r_m(2,3) = -\frac{\operatorname{cov}(g_\nu, h_\nu)}{\pi^2\alpha(2\nu+4)m^2}$$

and

$$r_m(3,3) = \frac{\operatorname{var}(g_\nu)}{\pi^2\alpha^2(2\nu+4)^2\log m}$$

as $m \to \infty$. Finally, for $d \le 3$, define $\mathcal{I}_m(\phi,\nu)$ as the Fisher information matrix for (ϕ,ν) assuming α is known. Then

$$\mathcal{I}_m(\phi,\nu)^{-1} = \frac{1}{\operatorname{var}(g_\nu)m^d} \tag{16}$$
$$\times \begin{bmatrix} 2(\log^2 m + 2\log m E g_\nu + E g_\nu^2) & \cdot \\ \frac{1}{\phi}\log m + E g_\nu & \frac{1}{2\phi^2} \end{bmatrix} \{1 + O(m^{-\epsilon})\}$$

as $m \to \infty$ for some $\epsilon > 0$.

There are a number of interesting features to these results. First consider $d \ge 4$. We then have that all elements of $\mathcal{I}_m^{-1}$ tend to 0 as $m \to \infty$, which is what we would expect from microergodicity considerations. Thus, I conjecture that the asymptotic normality for $\hat{\boldsymbol{\theta}}_m$ stated in (7) holds for $d \ge 4$, and I assume so from now on. In addition, for convenience, I make statements that presume the limiting covariance matrix of $\mathcal{I}_m^{1/2}(\hat{\boldsymbol{\theta}}_m - \boldsymbol{\theta})$ is the identity matrix, even though this convergence of the moments of the distribution does not necessarily follow from the convergence in distribution of $\mathcal{I}_m^{1/2}(\hat{\boldsymbol{\theta}}_m - \boldsymbol{\theta})$ given in (7). As Cox and Hinkley (1974, p. 282) point out, the convergence in distribution is what is crucial if we want to obtain asymptotically valid confidence intervals for the unknown parameters.

Let us now consider $d \ge 5$ in more detail. In this case, the rates of convergence for each diagonal element of (14) are all different, and hence, so are the asymptotic variances for the components of $\hat{\boldsymbol{\theta}}$. Only $\hat{\nu}$ has asymptotic variance of the "usual" order in parametric inference of m^{-d}, the reciprocal of the number of observations. The estimator $\hat{\phi}$ has asymptotic variance of the slightly larger order $m^{-d}\log^2 m$ and $\hat{\alpha}$ has asymptotic variance of order m^{4-d}. Since g_ν and h_ν do not depend on α, the asymptotic covariance matrix of $(\hat{\phi},\hat{\nu})$ does not depend on α. Furthermore, using corr for correlation, $\operatorname{corr}(\hat{\phi},\hat{\alpha})$ and $\operatorname{corr}(\hat{\nu},\hat{\alpha})$ both tend to 0 as $m \to \infty$, which suggests $(\hat{\phi},\hat{\nu})$ is asymptotically independent of $\hat{\alpha}$. The numerical results in Figure 5 of 6.5 indicate that m may need to be quite large before these correlations are near 0.

Another consequence of (14) to note is that $1 - \operatorname{corr}(\hat{\phi},\hat{\nu}) \sim \gamma \log^{-2} m$, where $\gamma = E h_\nu^2 \left\{\operatorname{var}(g_\nu)\operatorname{var}(h_\nu) - \operatorname{cov}(g_\nu,h_\nu)^2\right\} / \{2\operatorname{var}(h_\nu)^2\}$, which is positive. The fact that the correlation tends to 1 as $m \to \infty$ should not be entirely surprising considering the approximate correlations for $\hat{\phi}$ and $\hat{\nu}$ very near to 1 we found in Figure 5 for the Matérn model. One implication of this correlation tending to 1 is that there is no way to normalize $\hat{\boldsymbol{\theta}}$ componentwise so that it has a nonsingular limiting distribution: for any functions η_1, η_2 and η_3 on the positive integers, $(\eta_1(m)(\hat{\phi}-\phi), \eta_2(m)(\hat{\nu}-\nu), \eta_3(m)(\hat{\alpha}-\alpha))$ cannot converge in law to a nonsingular distribution. However, it is possible to show (Exercise 28) that if (7) holds then

$$(2m^d)^{1/2} \begin{bmatrix} \dfrac{\hat{\phi}-\phi}{2\phi} - (\log m + Eg_\nu)(\hat{\nu}-\nu) \\ \hat{\nu}-\nu \\ \dfrac{\alpha(2\nu+d)}{m^2}(\hat{\alpha}-\alpha) \end{bmatrix} \xrightarrow{\mathcal{L}} N(\mathbf{0}, \boldsymbol{\Sigma}(\nu)), \tag{17}$$

where

$$\boldsymbol{\Sigma}(\nu) = \frac{1}{\operatorname{var}(g_\nu)\operatorname{var}(h_\nu) - \operatorname{cov}(g_\nu, h_\nu)^2} \tag{18}$$

$$\times \begin{bmatrix} \operatorname{var}(g_\nu)\operatorname{var}(h_\nu) - \operatorname{cov}(g_\nu, h_\nu)^2 & \cdot & \cdot \\ \operatorname{cov}(g_\nu, h_\nu)Eh_\nu & \operatorname{var}(h_\nu) & \cdot \\ \operatorname{cov}(g_\nu, h_\nu)Eg_\nu + \operatorname{var}(g_\nu)Eh_\nu & -2\operatorname{cov}(g_\nu, h_\nu) & 4\operatorname{var}(g_\nu) \end{bmatrix}.$$

For $d = 4$, (7) and (15) imply $\operatorname{corr}(\hat{\phi}, \hat{\nu}) \to 1$, although the approximation in (15) is not sharp enough to allow an asymptotic approximation of $1 - \operatorname{corr}(\hat{\phi}, \hat{\nu})$ as in the $d \geq 5$ case. Exercise 37 outlines how to obtain an asymptotic approximation to $1 - \operatorname{corr}(\hat{\phi}, \hat{\nu})$ for $d = 4$.

When $d \leq 3$, α is not microergodic. As discussed in 6.4, we should not expect $\hat{\boldsymbol{\theta}}_m$ to be asymptotically normal in this case, since the proof of such a result requires that the MLE be consistent. However, as I noted in 6.4, we might expect asymptotic normality to hold for the microergodic parameters ϕ and ν. In particular, a reasonable conjecture is that if (ϕ, ν, α) are all estimated by maximizing the likelihood, then the asymptotic behavior of $(\hat{\phi}, \hat{\nu})$ will be the same as if α were known; that is,

$$(2m^d)^{1/2} \begin{bmatrix} \dfrac{\hat{\phi}-\phi}{2\phi} - (\log m + Eg_\nu)(\hat{\nu}-\nu) \\ \operatorname{var}(g_\nu)^{1/2}(\hat{\nu}-\nu) \end{bmatrix} \xrightarrow{\mathcal{L}} N(\mathbf{0}, \mathbf{I}).$$

If the mean of Z were an unknown constant and we used REML to estimate $\boldsymbol{\theta}$, then Theorem 1 would also apply to the Fisher information matrix for the contrasts. Specifically, the likelihood of the contrasts just leaves the term $\mathbf{p} = \mathbf{0}$ out of the sums over $\mathcal{B}_m$ in (12) and it follows that the asymptotic results in Theorem 1 are unchanged (although the result in Exercise 37 for $d = 4$ changes slightly).

If the observations are made with independent Gaussian measurement errors with mean 0 and constant variance σ^2, common sense and the results in 6.6 suggest that the estimation of (ϕ, ν, α) should be more difficult than when there is no measurement error. However, the discussion in 4.2 suggests that whatever parameters can be consistently estimated when there is no measurement error can still be consistently estimated when there is. In the present setting, the $X_{\mathbf{p}}$s, $\mathbf{p} \in \mathcal{B}_m$ are still independent Gaussians when there are measurement errors, so it is in principle possible to carry out an asymptotic analysis similar to the one in Theorem 1. Rather than doing that, I just indicate how we might expect the rates of convergence to change by considering results in Stein (1993). In that work, I essentially studied the $d = 1$ case here except that the parameter α was omitted. Specifically, I assumed that $f_{\boldsymbol{\theta}}(j) = \phi|j|^{-2\nu-1}$ for $j \neq 0$ and $f_{\boldsymbol{\theta}}(0) = 0$. If there are no measurement errors, a minor modification of Theorem 1 here shows that the diagonal elements of the inverse Fisher information matrix for (ϕ, ν) are of the orders of magnitude $m^{-1} \log^2 m$ and m^{-1} as in (16) for $d = 1$. However, if there are measurement errors of variance σ^2 (independent of m), then the diagonal elements of the inverse Fisher information matrix for (ϕ, ν, σ^2) are of the orders $m^{-1/(2\nu)} \log^2 m$, $m^{-1/(2\nu)}$ and m^{-1}, respectively. Thus, the convergence rates for $\hat{\phi}$ and $\hat{\nu}$ are lower when there is measurement error and the effect is more severe for larger values of ν.

Proof of Theorem 1. As an example of how to approximate the elements of $\mathcal{I}_m$ as m increases, consider $\operatorname{cov}(\partial\ell/\partial\phi, \partial\ell/\partial\nu)$. First note that $\xi_1(\mathbf{0}, m, \alpha, \nu)/\xi_0(\mathbf{0}, m, \alpha, \nu) \ll \log m$. Next, for $\mathbf{p} \neq \mathbf{0}$,

$$
\begin{aligned}
\xi_1(\mathbf{p}, m, \alpha, \nu) &= m^{-2\nu-d} \sum_{\mathbf{j} \in \mathbb{Z}^d} \left(\frac{\alpha^2}{m^2} + \left|m^{-1}\mathbf{p} + \mathbf{j}\right|^2 \right)^{-\nu-d/2} \\
&\qquad \times \left\{ 2 \log m + \log \left(\frac{\alpha^2}{m^2} + |m^{-1}\mathbf{p} + \mathbf{j}|^2 \right) \right\} \\
&= 2m^{-2\nu-d} \sum_{\mathbf{j} \in \mathbb{Z}^d} |m^{-1}\mathbf{p} + \mathbf{j}|^{-2\nu-d} \left(\log m + \log |m^{-1}\mathbf{p} + \mathbf{j}| \right) \\
&\qquad + O\left(|\mathbf{p}|^{-2} m^{-2\nu-d} \log m \right)
\end{aligned} \tag{19}
$$

(Exercise 29) and similarly

$$
\xi_0(\mathbf{p}, m, \alpha, \nu) = m^{-2\nu-d} \sum_{\mathbf{j} \in \mathbb{Z}^d} |m^{-1}\mathbf{p} + \mathbf{j}|^{-2\nu-d} + O(|\mathbf{p}|^{-2} m^{-2\nu-d}).
$$

Then (Exercise 30)

$$
\sum_{\mathbf{p} \in \mathcal{B}_m} \frac{\xi_1(\mathbf{p}, m, \alpha, \nu)}{\xi_0(\mathbf{p}, m, \alpha, \nu)} = 2m^d \log m + 2 \sum_{\mathbf{p} \in \mathcal{B}_m}{}' g_\nu(m^{-1}|\mathbf{p}|) + R_m, \tag{20}
$$

where the prime on the sum indicates $\mathbf{p} = \mathbf{0}$ is excluded and

$$
R_m \ll \log m \sum_{\mathbf{p} \in \mathcal{B}_m}{}' |\mathbf{p}|^{-2} \ll (1 + \langle m \rangle^{d-2}) \log m.
$$

The definition of the Riemann integral suggests

$$
m^{-d} \sum_{\mathbf{p} \in \mathcal{B}_m}{}' g_\nu(m^{-1}|\mathbf{p}|) \to E g_\nu
$$

as $m \to \infty$, which is true, but we require a slightly sharper result. Specifically, it is possible to show

$$m^{-d} \sum_{\mathbf{p}\in\mathcal{B}_m}{}' g_\nu(m^{-1}|\mathbf{p}|) = Eg_\nu + O(m^{-\epsilon}) \tag{21}$$

for some $\epsilon > 0$ (Exercise 31). Thus, for some $\epsilon > 0$,

$$m^{-d} \operatorname{cov}\left(\frac{\partial\ell}{\partial\phi}, \frac{\partial\ell}{\partial\nu}\right) = -\frac{1}{\phi}\log m - \frac{1}{\phi}Eg_\nu + O(m^{-\epsilon}).$$

Similarly,

$$m^{-d} \operatorname{var}\left(\frac{\partial\ell}{\partial\nu}\right) = 2\log^2 m + 4\log m Eg_\nu + 2Eg_\nu^2 + O(m^{-\epsilon})$$

for some $\epsilon > 0$.

Next, consider $\operatorname{var}(\partial\ell/\partial\alpha)$. First, since

$$\frac{\xi_0(\mathbf{p}, m, \alpha, \nu+1)}{\xi_0(\mathbf{p}, m, \alpha, \nu)} \asymp (1+|\mathbf{p}|)^{-2},$$

it follows that $\operatorname{var}(\partial\ell/\partial\alpha)$ is bounded in m for $d \leq 3$ and tends to ∞ for $d \geq 4$. More specifically,

$$\operatorname{var}\left(\frac{\partial\ell}{\partial\alpha}\right) \to \frac{1}{2}\alpha^2(2\nu+d)^2 \sum_{\mathbf{p}\in\mathbb{Z}^d} \frac{1}{(\alpha^2+|\mathbf{p}|^2)^2} \tag{22}$$

for $d \leq 3$ (Exercise 32),

$$m^{-d+4} \operatorname{var}\left(\frac{\partial\ell}{\partial\alpha}\right) = \frac{1}{2}\alpha^2(2\nu+d)^2 Eh_\nu^2 + O(m^{-\epsilon}) \tag{23}$$

for some $\epsilon > 0$ when $d \geq 5$ (Exercise 33). The case $d = 4$ requires particular care. It is possible to show that

$$\sum_{\mathbf{p}\in\mathcal{B}_m} \left\{\frac{\xi_0(\mathbf{p}, m, \alpha, \nu+1)}{\xi_0(\mathbf{p}, m, \alpha, \nu)}\right\}^2 \sim \sum_{\mathbf{p}\in\mathcal{B}_m}{}' h_\nu(m^{-1}|\mathbf{p}|)^2, \tag{24}$$

from which it follows

$$(\log m)^{-1} \operatorname{var}\left(\frac{\partial\ell}{\partial\alpha}\right) \to \alpha^2(2\nu+4)^2\pi^2 \tag{25}$$

(Exercise 34). Exercise 37 outlines how to obtain a sharper result for $\operatorname{var}(\partial\ell/\partial\alpha)$ when $d = 4$. Now consider $\operatorname{cov}(\partial\ell/\partial\phi, \partial\ell/\partial\alpha)$. For $d = 1$,

$$\operatorname{cov}\left(\frac{\partial\ell}{\partial\phi}, \frac{\partial\ell}{\partial\alpha}\right) \to -\frac{\alpha(2\nu+1)}{2\phi} \sum_{p=-\infty}^{\infty} \frac{1}{\alpha^2+p^2},$$

for $d \geq 3$,

$$m^{2-d} \operatorname{cov}\left(\frac{\partial\ell}{\partial\phi}, \frac{\partial\ell}{\partial\alpha}\right) = -\frac{\alpha(2\nu+d)}{2\phi} Eh_\nu + O(m^{-\epsilon})$$

for some $\epsilon > 0$ and for $d = 2$,

$$(\log m)^{-1} \operatorname{cov}\left(\frac{\partial \ell}{\partial \phi}, \frac{\partial \ell}{\partial \alpha}\right) \to -\frac{\alpha(2\nu+2)\pi}{\phi}.$$

Finally, consider $\operatorname{cov}(\partial\ell/\partial\nu, \partial\ell/\partial\alpha)$. For $d = 1$,

$$\operatorname{cov}\left(\frac{\partial \ell}{\partial \nu}, \frac{\partial \ell}{\partial \alpha}\right) \to \frac{1}{2}\alpha(2\nu+1) \sum_{p=-\infty}^{\infty} \frac{\log(\alpha^2+p^2)}{\alpha^2+p^2},$$

for $d \geq 3$,

$$m^{2-d} \operatorname{cov}\left(\frac{\partial \ell}{\partial \nu}, \frac{\partial \ell}{\partial \alpha}\right) = \alpha(2\nu+d)\left\{\log m Eh_\nu + E(g_\nu h_\nu)\right\} + O(m^{-\epsilon})$$

for some $\epsilon > 0$ and for $d = 2$,

$$\operatorname{cov}\left(\frac{\partial \ell}{\partial \nu}, \frac{\partial \ell}{\partial \alpha}\right) \sim \alpha(\nu+1) \sum_{\mathbf{p}\in\mathcal{B}_m}{}' \frac{\log|\mathbf{p}|}{|\mathbf{p}|^2} \sim \pi\alpha(\nu+1)\log^2 m.$$

Putting these results together for $d \geq 5$ gives

$$\mathcal{I}_m = m^d \mathbf{S}_m\{1 + O(m^{-\epsilon})\} \tag{26}$$

for some $\epsilon > 0$, where $\mathbf{S}_m$ has jkth element $s_m(j,k)$ given by

$$\begin{aligned}
s_m(1,1) &= \frac{m^d}{2\phi^2}, \\
s_m(1,2) &= -\frac{m^d}{\phi}\log m - \frac{m^d}{\phi} Eg_\nu, \\
s_m(1,3) &= -\frac{\alpha(2\nu+d)m^{d-2}}{2\phi} Eh_\nu, \\
s_m(2,2) &= 2m^d(\log^2 m + 2\log m Eg_\nu + Eg_\nu^2), \\
s_m(2,3) &= m^{d-2}\alpha(2\nu+d)\{\log m Eh_\nu + E(g_\nu h_\nu)\}
\end{aligned}$$

and

$$s_m(3,3) = \frac{1}{2} m^{d-4}\alpha^2(2\nu+d)^2 Eh_\nu^2.$$

It follows that for some $\epsilon > 0$ (Exercise 35),

$$\begin{aligned}
\det(\mathcal{I}_m) = \frac{m^{3d-4}\alpha^2(2\nu+d)^2}{2\phi^2} &\left[Eg_\nu^2 Eh_\nu^2 - (Eg_\nu)^2 Eh_\nu^2 - Eg_\nu^2 (Eh_\nu)^2\right. \\
&\left. + 2Eg_\nu E(g_\nu h_\nu) Eh_\nu - \{E(g_\nu h_\nu)\}^2\right] + O(m^{3d-4-\epsilon}).
\end{aligned} \tag{27}$$

We see that the terms of order $m^{3d-4}\log^2 m$ and $m^{3d-4}\log m$ all exactly cancel, which explains why it was essential to get, for example, $\operatorname{var}(\partial\ell/\partial\nu)$ beyond the leading term $2m^d \log^2 m$ in order to approximate $\mathcal{I}_m^{-1}$. Now (Exercise 36)

$$Eg_\nu^2 Eh_\nu^2 - (Eg_\nu)^2 Eh_\nu^2 - Eg_\nu^2 (Eh_\nu)^2 + 2Eg_\nu E(g_\nu h_\nu) Eh_\nu - \{E(g_\nu h_\nu)\}^2$$

$$= \operatorname{var}(g_\nu)\operatorname{var}(h_\nu) - \operatorname{cov}(g_\nu, h_\nu)^2, \tag{28}$$

which is nonnegative by the Cauchy–Schwarz inequality and can be shown to be positive (Exercise 36). Equation (14) of Theorem 1 follows.

Next consider $d = 4$. The only element of $\mathcal{I}_m$ as given in (26) that is now incorrect is $\operatorname{var}(\partial\ell/\partial\alpha)$, which we have shown is $\alpha^2(2\nu+4)^2\pi^2\log m + o(\log m)$. It follows that

$$\det(\mathcal{I}_m) \sim \left(\frac{\pi\alpha}{\phi}\right)^2 (2\nu+4)^2 \operatorname{var}(g_\nu) m^8 \log m\,,$$

which yields (15). Finally, for $d \le 3$, the proof of (16) in Theorem 1 is immediate from the results already given. □

Exercises

24 Verify (9).

25 Verify (10).

26 Verify (11).

27 Derive the covariance matrix of the score function for the setting in this section.

28 Verify that (18) holds if (7) is true.

29 Verify (19).

30 Verify (20).

31 Verify (21).

32 Verify (22).

33 Verify (23).

34 Verify (24) and show that (25) follows.

35 Verify that (27) follows from (26).

36 Verify (28) and show that the result is positive for all $\nu > 0$.

37 (Only for the truly brave or truly foolish.) Show that for $d = 4$,

$$\operatorname{var}\left(\frac{\partial\ell}{\partial\alpha}\right) = 2\pi^2\log m + C_{\nu,\alpha} + O(m^{-\epsilon}) \tag{29}$$

for some $\epsilon > 0$, where

$$C_{\nu,\alpha} = \int_{c_4\setminus b_4(1/2)} |\mathbf{x}|^{-4}\mathbf{dx} - \pi^2\{1 + 2\log(2\alpha)\} + A_\alpha + B_\nu,$$

$$A_\alpha = \lim_{m\to\infty}\left[\int_{mc_4}(\alpha^2 + |\mathbf{x}|^2)^{-2}\mathbf{dx} - \sum_{\mathbf{p}\in B_m}(\alpha^2 + |\mathbf{p}|^2)^{-2}\right]$$

and finally,

$$B_\nu = \int_{c_4} \left\{ \frac{2\sigma_{2\nu+6}(\mathbf{x})}{|\mathbf{x}|^{2\nu+6}\sigma_{2\nu+4}(\mathbf{x})^2} + \frac{\sigma_{2\nu+6}(\mathbf{x})^2}{\sigma_{2\nu+4}(\mathbf{x})^2} - \frac{2}{|\mathbf{x}|^{2\nu+8}\sigma_{2\nu+4}(\mathbf{x})} - \frac{1}{|\mathbf{x}|^4} \right\} d\mathbf{x},$$

where $\sigma_\mu(\mathbf{x}) = \sum_{\mathbf{j}}' |\mathbf{x}+\mathbf{j}|^{-\mu}$. To obtain (29), prove the following results:

$$\sum_{\mathbf{p}\in B_m} \left\{ \frac{\xi_0(\mathbf{p}, m, \alpha, \nu+1)}{\xi_0(\mathbf{p}, m, \alpha, \nu)} \right\}^2 - \int_{mc_4} \left\{ \frac{\xi_0(\mathbf{x}, m, \alpha, \nu+1)}{\xi_0(\mathbf{x}, m, \alpha, \nu)} \right\}^2 d\mathbf{x} = A_\alpha + O(m^{-\epsilon}),$$

$$\int_{mc_4} \left\{ \frac{\xi_0(\mathbf{x}, m, \alpha, \nu+1)}{\xi_0(\mathbf{x}, m, \alpha, \nu)} \right\}^2 d\mathbf{x} - \int_{c_4} \left(\frac{\alpha^2}{m^2} + |\mathbf{x}|^2 \right)^{-2} d\mathbf{x} = B_\nu + O(m^{-\epsilon}),$$

$$\int_{c_4} \left(\frac{\alpha^2}{m^2} + |\mathbf{x}|^2 \right)^{-2} d\mathbf{x} = \int_{b_4(1/2)} \left(\frac{\alpha^2}{m^2} + |\mathbf{x}|^2 \right)^{-2} d\mathbf{x} + \int_{c_4 \backslash b_4(1/2)} |\mathbf{x}|^{-4} d\mathbf{x} + O(m^{-\epsilon})$$

and

$$\int_{b_4(1/2)} \left(\frac{\alpha^2}{m^2} + |\mathbf{x}|^2 \right)^{-2} d\mathbf{x} = \pi^2\{2\log m - 2\log(2\alpha) - 1\} + O(m^{-\epsilon})$$

for some $\epsilon > 0$. Use (29) to show that $1 - \operatorname{corr}(\hat{\phi}, \hat{\nu}) \sim \gamma / \log^2 m$ and find an expression for γ.

6.8 Predicting with estimated parameters

When unknown parameters of the covariance structure are estimated from the available data, perhaps the most commonly used method for predicting random fields and assessing the mses of these predictions is the plug-in method: estimate the second-order structure in some manner and then proceed as if this estimated second-order structure were the truth (Christensen 1991, Section 6.5; Zimmerman and Cressie 1992). To be more specific, suppose $Z(\mathbf{x}) = \mathbf{m}(\mathbf{x})^T\boldsymbol{\beta} + \varepsilon(\mathbf{x})$ for $\mathbf{x} \in \mathbb{R}^d$, where $\mathbf{m}$ is a known vector-valued function, $\boldsymbol{\beta}$ is a vector of unknown parameters and ε is a mean 0 Gaussian random field with autocovariance function in some parametric family $K_{\boldsymbol{\theta}}$, $\boldsymbol{\theta} \in \Theta$. We observe $\mathbf{Z} = (Z(\mathbf{x}_1), \ldots, Z(\mathbf{x}_n))^T$ and wish to predict $Z(\mathbf{x}_0)$. If $\boldsymbol{\theta}$ were known, we could then predict $Z(\mathbf{x}_0)$ using the BLUP.

Define the vector-valued function $\boldsymbol{\lambda}(\boldsymbol{\theta}) = (\lambda_1(\boldsymbol{\theta}), \ldots, \lambda_n(\boldsymbol{\theta}))^T$ by letting $\boldsymbol{\lambda}(\boldsymbol{\theta})^T\mathbf{Z}$ be the BLUP of $Z(\mathbf{x}_0)$ under the model $K_{\boldsymbol{\theta}}$. Assume the BLUP exists, which is equivalent to assuming $\mathbf{m}(\mathbf{x}_0)$ is in the column space of $(\mathbf{m}(\mathbf{x}_1) \ldots \mathbf{m}(\mathbf{x}_n))$ (see 1.5). If $\hat{\boldsymbol{\theta}}$ is some estimator of $\boldsymbol{\theta}$ based on $\mathbf{Z}$ then $\boldsymbol{\lambda}(\hat{\boldsymbol{\theta}})^T\mathbf{Z}$ is the corresponding plug-in predictor. This plug-in predictor is sometimes called the EBLUP, where the E can be thought of as meaning "estimated" or "empirical" (Zimmerman and Cressie 1992). Zimmerman and Zimmerman (1991) and Cressie and Zimmerman (1992) describe results of some simulation studies showing that plug-in methods can often work well even with fairly small datasets.

Let us now consider quantities analogous to $E_0e_0^2$, $E_0e_1^2$ and $E_1e_1^2$ in Chapters 3 and 4. It is convenient to define $\lambda_0(\boldsymbol{\theta}) = -1$ for all $\boldsymbol{\theta}$ so that the prediction error of the BLUP as a function of $\boldsymbol{\theta}$ is

$$e(\boldsymbol{\theta}) = e(\boldsymbol{\theta}; Z(\mathbf{x}_0)) = \sum_{i=0}^{n} \lambda_i(\boldsymbol{\theta}) Z(\mathbf{x}_i).$$

Define the function

$$M(\boldsymbol{\theta}) = \sum_{i,j=0}^{n} \lambda_i(\boldsymbol{\theta})\lambda_j(\boldsymbol{\theta}) K_{\boldsymbol{\theta}}(\mathbf{x}_i - \mathbf{x}_j), \tag{30}$$

so that if $\boldsymbol{\theta}_0$ is the true value of $\boldsymbol{\theta}$ then $M(\boldsymbol{\theta}_0)$ is the mse of the BLUP.

The error of the plug-in predictor is $e(\hat{\boldsymbol{\theta}})$ and the plug-in estimate of its mse is

$$M(\hat{\boldsymbol{\theta}}) = \sum_{i,j=0}^{n} \lambda_i(\hat{\boldsymbol{\theta}})\lambda_j(\hat{\boldsymbol{\theta}}) K_{\hat{\boldsymbol{\theta}}}(\mathbf{x}_i - \mathbf{x}_j). \tag{31}$$

Note that $M(\hat{\boldsymbol{\theta}})$ does not have a direct interpretation as an expectation over the probability law for the random field Z. Nevertheless, (31) is the natural analogue to what was called $E_1e_1^2$ in Chapters 3 and 4, since it is the presumed mse of our predictor if we ignore the fact that $\hat{\boldsymbol{\theta}}$ is not the same as $\boldsymbol{\theta}_0$.

The natural analogue to $E_0e_1^2$ is

$$E_0 e(\hat{\boldsymbol{\theta}})^2 = E_0 \left\{ \sum_{i=0}^{n} \lambda_i(\hat{\boldsymbol{\theta}}) Z(\mathbf{x}_i) \right\}^2, \tag{32}$$

where E_0 indicates expectation under the true model. Suppose the estimator $\hat{\boldsymbol{\theta}}$ depends on $\mathbf{Z}$ only through its contrasts, which I denote by $\mathbf{Y}$. Zimmerman and Cressie (1992) point out that existing procedures, including ML and REML, do yield estimates that are functions of the contrasts, and I assume that this is the case in the remainder of this section. It follows that $e(\hat{\boldsymbol{\theta}}) - e(\boldsymbol{\theta}_0)$ is also a function of the contrasts and is hence independent of $e(\boldsymbol{\theta}_0)$ when Z is Gaussian. Thus,

$$E_0 e(\hat{\boldsymbol{\theta}})^2 = M(\boldsymbol{\theta}_0) + E_0\{e(\hat{\boldsymbol{\theta}}) - e(\boldsymbol{\theta}_0)\}^2, \tag{33}$$

so that $E_0 e(\hat{\boldsymbol{\theta}})^2 \geq M(\boldsymbol{\theta}_0)$. It may seem obvious that the BLUP should have a smaller mse than a plug-in predictor, which replaces the true $\boldsymbol{\theta}_0$ with an estimator. However, (33) does require that $e(\boldsymbol{\theta}_0)$ and $e(\hat{\boldsymbol{\theta}}) - e(\boldsymbol{\theta}_0)$ be uncorrelated. Although assuming Z to be Gaussian is stronger than necessary for this uncorrelatedness to hold (Christensen 1991, Section 6.5), it is not difficult to construct examples of non-Gaussian processes for which $E_0 e(\hat{\boldsymbol{\theta}})^2 < M(\boldsymbol{\theta}_0)$ (see Exercise 38).

Before proceeding to more difficult problems, it is worth noting that in the simple setting where the autocovariance function K is known up to a scalar multiple, there is a satisfactory finite sample frequentist solution to the prediction problem. More specifically, suppose $K_{\boldsymbol{\theta}} = \theta K$, the rank of $(\mathbf{m}(\mathbf{x}_1) \ldots \mathbf{m}(\mathbf{x}_n))$ is p and $\hat{\theta}$ is the REML estimate of θ, given in 6.4. Since the BLUP of $Z(\mathbf{x}_0)$ does not depend on θ, the EBLUP and the BLUP are the same. It is then a simple extension of standard results on prediction intervals in regression problems with independent Gaussian errors (Seber 1977, Section 5.3) to show that $e(\hat{\theta})/M(\hat{\theta})^{1/2}$ follows a t distribution with $n - p$ degrees of freedom, which can be used to give exact frequentist prediction intervals.

More generally, there is no entirely satisfactory frequentist solution to making inferences based on EBLUPs. Harville and Jeske (1992) and Zimmerman and Cressie (1992) consider a more sophisticated method for estimating $E_0 e(\hat{\boldsymbol{\theta}})^2$ than the plug-in estimator of mse $M(\hat{\boldsymbol{\theta}})$ in (31). Their method involves three separate approximations. First, they derive an exact relationship between $E_0 e_0^2$ and $E_0 e(\hat{\boldsymbol{\theta}})^2$ that holds under highly restrictive conditions and then assume this relationship is approximately true more generally. Next, they further approximate this result as a function of $\boldsymbol{\theta}_0$ using Taylor series. Finally, they replace $\boldsymbol{\theta}_0$ in this expression by $\hat{\boldsymbol{\theta}}$. Unfortunately, Zimmerman and Cressie (1992) report simulation results showing that when neighboring observations are strongly correlated, this approach can sometimes produce worse answers than $M(\hat{\boldsymbol{\theta}})$.

To carry out a simulation study such as the one in Zimmerman and Cressie (1992), it is only necessary to simulate the observations and not the predictands. To be more specific, consider approximating $E_0 e(\hat{\boldsymbol{\theta}})^2$ via simulation under some given Gaussian model for a random field Z, some set of observations and a particular predictand. Use the subscript 0 to indicate the true model. Calculate $E_0 e_0^2$ once and for all. Simulate n realizations of the observations under the true model. For the jth realization, let $\hat{\boldsymbol{\theta}}(j)$ be the estimator for $\boldsymbol{\theta}$, $e_0(j)$ the error of the BLUP and $e(\hat{\boldsymbol{\theta}}(j), j)$ the error of the EBLUP. Then

$$E_0 e_0^2 + \frac{1}{n}\sum_{j=1}^{n}\{e_0(j) - e(\hat{\boldsymbol{\theta}}(j), j)\}^2$$

has expected value $E_0 e(\hat{\boldsymbol{\theta}})^2$ and, furthermore, converges with probability 1 to $E_0 e(\hat{\boldsymbol{\theta}})^2$ as $n \to \infty$ if, for example, the n realizations are independent.

The value of $E_0 e_0^2 + \left\{e(\hat{\boldsymbol{\theta}}) - e(\boldsymbol{\theta}_0)\right\}^2$ for a single realization of the observations has a direct interpretation as a conditional mse. Specifically, since Z is Gaussian, the conditional law of $e(\hat{\boldsymbol{\theta}})$ given the contrasts $\mathbf{Y} = \mathbf{y}$ is $N\left(e(\hat{\boldsymbol{\theta}}(\mathbf{y})) - e_0, E_0 e_0^2\right)$, where, for clarity, I have made explicit the dependence of $\hat{\boldsymbol{\theta}}$ on $\mathbf{y}$. It follows that

$$E_0\left\{e(\hat{\boldsymbol{\theta}}(\mathbf{y}))^2 \mid \mathbf{Y} = \mathbf{y}\right\} = E_0 e_0^2 + \left\{e(\hat{\boldsymbol{\theta}}(\mathbf{y})) - e_0\right\}^2. \tag{34}$$

In some circumstances, quite a bit can be learned by calculating $\left\{e(\hat{\boldsymbol{\theta}}) - e_0\right\}^2$ for a small number of simulations, perhaps one. See 6.9 for an example.

Simulation may also provide the basis for a more accurate assessment of $E_0 e(\hat{\boldsymbol{\theta}})^2$ than is given by the plug-in method. Specifically, one could approximate mses by using repeated simulations of the observations from the estimated model, which Davison and Hinkley (1997, p. 15) call parametric simulation and Efron and Tibshirani (1993, p. 53) call parametric bootstrap. Conceptually, the idea is simple. Obtain the REML estimate $\hat{\boldsymbol{\theta}}$ of the unknown parameters of the autocovariance function from the data and then do repeated simulations of the observations assuming that $K_{\hat{\theta}}$ is the actual autocovariance function. The value of $\boldsymbol{\beta}$ used in the simulations is irrelevant to the error of EBLUPs so we may as well set $\boldsymbol{\beta} = \mathbf{0}$. Let $\hat{\boldsymbol{\theta}}^*(j)$ be the REML estimate of $\boldsymbol{\theta}$ for the jth simulation and compute $e(\hat{\boldsymbol{\theta}}^*(j), j) - e(\hat{\boldsymbol{\theta}}, j)$ for each simulation. Then estimate the distribution of $e(\hat{\boldsymbol{\theta}})$ by convolving the empirical distribution of $e(\hat{\boldsymbol{\theta}}^*(j), j) - e(\hat{\boldsymbol{\theta}}, j)$ for $j = 1, \ldots, n$ with a $N(0, M(\hat{\boldsymbol{\theta}}))$ distribution. More specifically, estimate $\Pr\{e(\hat{\boldsymbol{\theta}}) \le t \mid \boldsymbol{\theta}_0\}$ by

$$\Pr_n(t; \hat{\boldsymbol{\theta}}) = \frac{1}{n}\sum_{j=1}^{n}\Phi\left(\frac{t - \left\{e(\hat{\boldsymbol{\theta}}^*(j), j) - e(\hat{\boldsymbol{\theta}}, j)\right\}}{M(\hat{\boldsymbol{\theta}})^{1/2}}\right),$$

where Φ is the cumulative distribution function of a $N(0, 1)$ random variable. Putter and Young (1998) also recommend parametric simulation in this setting and prove essentially that it will work well whenever plug-in methods work well. The much more interesting question of when it works better than plug-in methods is unresolved for the problem of spatial interpolation.

Jeffreys's law revisited

In 4.4, we studied the close connection between the effect of not knowing certain parameters of a random field model on a prediction and the additional information the predictand provides about the unknown parameters. In particular, (59) in 4.4 provides an exact Bayesian quantification of this notion. This section studies some approximate frequentist analogues to results in 4.4 for Gaussian random fields. In Plausible Approximation 1, this subsection gives an approximate frequentist version of (59) in 4.4 for plug-in predictions when the unknown parameters are estimated by maximum likelihood. Plausible Approximation 2 approximates the Kullback divergence of plug-in predictive distributions from the predictive distribution with the parameters known in terms of the mse of the BLP and the actual and presumed mses of the plug-in predictor. This result is very similar to (58) in 4.4 for pseudo-BLPs.

Throughout this subsection, let $\mathbf{Y}$ be the vector of observations, Z the predictand and suppose the distribution of $(\mathbf{Y}, Z)$ is Gaussian and is known up to a parameter $\boldsymbol{\theta}$ with true value $\boldsymbol{\theta}_0$. In this subsection, $\boldsymbol{\theta}$ refers to all unknown parameters in the distribution of $(\mathbf{Y}, Z)$ and not just to unknown parameters of the covariance structure. Let $\hat{\boldsymbol{\theta}}$ be an estimator of $\boldsymbol{\theta}$ based on $\mathbf{Y}$. Define $p(Z \mid \mathbf{Y}; \boldsymbol{\theta})$ to be the conditional density of Z given $\mathbf{Y}$ as a function of $\boldsymbol{\theta}$. If $\boldsymbol{\theta}_0$ were known, we would use $p(Z \mid \mathbf{Y}; \boldsymbol{\theta}_0)$ to make predictions about Z. The plug-in estimator for this conditional distribution is $p(Z \mid \mathbf{Y}; \hat{\boldsymbol{\theta}})$. A plausible measure of the effect of using the plug-in conditional density rather than the actual conditional density with $\boldsymbol{\theta}_0$ known is

$$D(\boldsymbol{\theta}_0, \hat{\boldsymbol{\theta}}; Z \mid \mathbf{Y}) = E_0 \left\{ \log \frac{p(Z \mid \mathbf{Y}; \boldsymbol{\theta}_0)}{p(Z \mid \mathbf{Y}; \hat{\boldsymbol{\theta}})} \right\}, \tag{35}$$

the Kullback divergence of the plug-in conditional density from the conditional density evaluated at $\boldsymbol{\theta}_0$. Note that the right side of (35) is an expectation over both $\mathbf{Y}$ and Z. The main results of this section are two plausible approximations to $D(\boldsymbol{\theta}_0, \hat{\boldsymbol{\theta}}; Z \mid \mathbf{Y})$.

For random vectors $\mathbf{W}$ and $\mathbf{X}$ with joint distribution indexed by a parameter $\boldsymbol{\theta}$, let $\mathcal{I}(\boldsymbol{\theta}; \mathbf{X})$ be the Fisher information matrix for $\boldsymbol{\theta}$ when $\mathbf{X}$ is observed. Furthermore, define

$$\mathcal{I}(\boldsymbol{\theta}; \mathbf{W} \mid \mathbf{X}) = \operatorname{cov}_{\boldsymbol{\theta}} \left[\frac{\partial}{\partial \boldsymbol{\theta}} \log p(\mathbf{W} \mid \mathbf{X}; \boldsymbol{\theta}), \left\{ \frac{\partial}{\partial \boldsymbol{\theta}} \log p(\mathbf{W} \mid \mathbf{X}; \boldsymbol{\theta}) \right\}^T \right].$$

We have

$$\mathcal{I}(\boldsymbol{\theta}; (\mathbf{W}, \mathbf{X})) = \mathcal{I}(\boldsymbol{\theta}; \mathbf{X}) + \mathcal{I}(\boldsymbol{\theta}; \mathbf{W} \mid \mathbf{X}) \tag{36}$$

(Exercise 39), so that $\mathcal{I}(\boldsymbol{\theta}; \mathbf{W} \mid \mathbf{X})$ is the expected increase in Fisher information for $\boldsymbol{\theta}$ when $\mathbf{W}$ is observed in addition to $\mathbf{X}$. Thus, for example, if $\mathbf{W}$ and $\mathbf{X}$ are independent, $\mathcal{I}(\boldsymbol{\theta}; \mathbf{W} \mid \mathbf{X}) = \mathcal{I}(\boldsymbol{\theta}; \mathbf{W})$ (Exercise 19 in 6.4) and if $\mathbf{W}$ is a function of $\mathbf{X}$, $\mathcal{I}(\boldsymbol{\theta};\mathbf{W}\mid\mathbf{X})$ is a matrix of zeroes. Define $\mathbf{i}(\boldsymbol{\theta};\mathbf{W}\mid\mathbf{X})$ to be the matrix with $-\partial^2\log p(\mathbf{W}\mid\mathbf{X};\boldsymbol{\theta})/\partial\theta_j\partial\theta_k$ as its jkth element so that $E_0\mathbf{i}(\boldsymbol{\theta}_0;\mathbf{W}\mid\mathbf{X})=\mathcal{I}(\boldsymbol{\theta}_0;\mathbf{W}\mid\mathbf{X})$ (see Exercise 18 in 6.4). Finally, let $e(Z;\boldsymbol{\theta})$ be the error of the BLP of Z evaluated as if $\boldsymbol{\theta}$ were the true parameter value.

Plausible Approximation 1. *If $\hat{\boldsymbol{\theta}}$ is the MLE of $\boldsymbol{\theta}$ based on $\mathbf{Y}$ and*

$$\hat{\boldsymbol{\theta}}\approx\boldsymbol{\theta}_0 \quad \textit{with high probability,} \tag{37}$$

then

$$D(\boldsymbol{\theta}_0,\hat{\boldsymbol{\theta}};Z\mid\mathbf{Y})\approx\operatorname{tr}\left\{\mathcal{I}(\boldsymbol{\theta}_0;\mathbf{Y})^{-1}\mathcal{I}(\boldsymbol{\theta}_0;Z\mid\mathbf{Y})\right\}. \tag{38}$$

Plausible Approximation 2. *If $\hat{\boldsymbol{\theta}}$ is some estimator for $\boldsymbol{\theta}$ based on $\mathbf{Y}$ such that*

$$\frac{M(\hat{\boldsymbol{\theta}})}{M(\boldsymbol{\theta}_0)}\approx 1 \quad \textit{with high probability} \tag{39}$$

and

$$\frac{E_0\left\{e(Z;\hat{\boldsymbol{\theta}})-e(Z;\boldsymbol{\theta}_0)\right\}^2}{M(\boldsymbol{\theta}_0)} \quad \textit{is small,} \tag{40}$$

then

$$D(\boldsymbol{\theta}_0,\hat{\boldsymbol{\theta}};Z\mid\mathbf{Y})\approx\frac{1}{4M(\boldsymbol{\theta}_0)^2}\,E_0\left\{M(\hat{\boldsymbol{\theta}})-M(\boldsymbol{\theta}_0)\right\}^2 + \frac{1}{2M(\boldsymbol{\theta}_0)}\,E_0\left\{e(Z;\hat{\boldsymbol{\theta}})-e(Z;\boldsymbol{\theta}_0)\right\}^2. \tag{41}$$

The right side of (38) is a measure of the relative increase in information contained in Z that was not contained in $\mathbf{Y}$. Thus, (38) provides an approximate frequentist analogue to (59) in 4.4 on the connection between prediction and estimation. If the covariance matrix of $(\mathbf{Y},Z)$ is known and its mean vector is linear in $\boldsymbol{\theta}$, then (38) is an equality (Exercise 40). The condition (37) that $\hat{\boldsymbol{\theta}}\approx\boldsymbol{\theta}_0$ with high probability is troubling in the present context, since we would like to be able to apply (38) in settings where there are nonmicroergodic parameters. Considering the very small impact that parameters having negligible effect on the high frequecy behavior of the model can have on interpolation, as is demonstrated in 3.5–3.7 and Stein (1999), I believe that there are circumstances under which (38) can be rigorously justified even when $\boldsymbol{\theta}$ cannot be consistently estimated. More specifically, for interpolation problems, I expect that the uncertainty in the estimation of nonmicroergodic parameters can be ignored in obtaining the leading term in an approximation to $D(\boldsymbol{\theta}_0,\hat{\boldsymbol{\theta}};Z\mid\mathbf{Y})$. White (1973) and Cox and Hinkley (1974, p. 357–358) consider Plausible Approximation 1 when $\mathbf{Y}$ and Z are independent.

Plausible Approximation 2 is a direct analogue to (58) in 4.4 on the behavior of pseudo-BLPs. Note that (39) and (40) do not require $\hat{\boldsymbol{\theta}}\approx\boldsymbol{\theta}_0$.

In particular, under fixed-domain asymptotics, the parameters that cannot be estimated consistently should have asymptotically negligible impact on the mses of interpolants, so it is quite plausible to have (37) false but (39) and (40) true in this setting.

The two approximations in this section are stated in terms of BLPs, plug-in BLPs and, in the case of Plausible Approximation 1, MLEs. It is possible to state analogous results for BLUPs and EBLUPs by letting $\boldsymbol{\theta}$ just refer to the parameters of the covariance structure and replacing the MLE by the REML estimator in Plausible Approximation 1 (Exercise 45).

Heuristic Derivation of Plausible Approximation 1. Define $\mathbf{S}(\boldsymbol{\theta}; Z \mid \mathbf{Y}) = \boldsymbol{\partial} \log p(Z \mid \mathbf{Y}; \boldsymbol{\theta})/\boldsymbol{\partial\theta}$. Under (37) and ignoring terms that are plausibly of lower order,

$$
\begin{aligned}
D(\boldsymbol{\theta}_0, \hat{\boldsymbol{\theta}}; Z \mid \mathbf{Y}) &\approx -E_0 \Big[\log \Big\{1 + (\hat{\boldsymbol{\theta}} - \boldsymbol{\theta}_0)^T \mathbf{S}(\boldsymbol{\theta}_0; Z \mid \mathbf{Y}) \\
&\qquad\qquad - \tfrac{1}{2} (\hat{\boldsymbol{\theta}} - \boldsymbol{\theta}_0)^T \mathbf{i}(\boldsymbol{\theta}_0; Z \mid \mathbf{Y}) (\hat{\boldsymbol{\theta}} - \boldsymbol{\theta}_0)\Big\}\Big] \\
&\approx -E_0 \Big\{(\hat{\boldsymbol{\theta}} - \boldsymbol{\theta}_0)^T \mathbf{S}(\boldsymbol{\theta}_0; Z \mid \mathbf{Y})\Big\} \\
&\quad + \tfrac{1}{2} E_0 \Big\{(\hat{\boldsymbol{\theta}} - \boldsymbol{\theta}_0)^T \mathbf{i}(\boldsymbol{\theta}_0; Z \mid \mathbf{Y}) (\hat{\boldsymbol{\theta}} - \boldsymbol{\theta}_0)\Big\} \\
&\quad + \tfrac{1}{2} E_0 \Big\{(\hat{\boldsymbol{\theta}} - \boldsymbol{\theta}_0)^T \mathbf{S}(\boldsymbol{\theta}_0; Z \mid \mathbf{Y})\Big\}^2 . \qquad (42)
\end{aligned}
$$

Now $\hat{\boldsymbol{\theta}}$ is a function of $\mathbf{Y}$, so

$$
\begin{aligned}
&E_0 \Big\{(\hat{\boldsymbol{\theta}} - \boldsymbol{\theta}_0)^T \mathbf{S}(\boldsymbol{\theta}_0; Z \mid \mathbf{Y})\Big\} \\
&\quad = E_0 \Big[(\hat{\boldsymbol{\theta}} - \boldsymbol{\theta}_0)^T E \left\{\mathbf{S}(\boldsymbol{\theta}_0; Z \mid \mathbf{Y}) \mid \mathbf{Y}\right\}\Big] = 0 \qquad (43)
\end{aligned}
$$

(Exercise 41) and

$$
E_0 \Big\{(\hat{\boldsymbol{\theta}} - \boldsymbol{\theta}_0)^T \mathbf{S}(\boldsymbol{\theta}_0; Z \mid \mathbf{Y})\Big\}^2 = E_0 \Big\{(\hat{\boldsymbol{\theta}} - \boldsymbol{\theta}_0)^T \mathbf{i}(\boldsymbol{\theta}_0; Z \mid \mathbf{Y}) (\hat{\boldsymbol{\theta}} - \boldsymbol{\theta}_0)\Big\} \qquad (44)
$$

(Exercise 42). Next, $(\mathbf{Y}, Z)$ Gaussian implies that

$$
E_0 \left\{\mathbf{i}(\boldsymbol{\theta}_0; Z \mid \mathbf{Y}) \mid \mathbf{Y}\right\} = \boldsymbol{\mathcal{I}}(\boldsymbol{\theta}_0; Z \mid \mathbf{Y}) \qquad (45)
$$

(Exercise 42), so that

$$
\begin{aligned}
&E_0 \Big\{(\hat{\boldsymbol{\theta}} - \boldsymbol{\theta}_0)^T \mathbf{i}(\boldsymbol{\theta}_0; Z \mid \mathbf{Y}) (\hat{\boldsymbol{\theta}} - \boldsymbol{\theta}_0)\Big\} \\
&\quad = \operatorname{tr} E_0 \Big\{(\hat{\boldsymbol{\theta}} - \boldsymbol{\theta}_0)(\hat{\boldsymbol{\theta}} - \boldsymbol{\theta}_0)^T \mathbf{i}(\boldsymbol{\theta}_0; Z \mid \mathbf{Y})\Big\} \\
&\quad = \operatorname{tr} E_0 \Big[(\hat{\boldsymbol{\theta}} - \boldsymbol{\theta}_0)(\hat{\boldsymbol{\theta}} - \boldsymbol{\theta}_0)^T E_0 \left\{\mathbf{i}(\boldsymbol{\theta}_0; Z \mid \mathbf{Y}) \mid \mathbf{Y}\right\}\Big] \\
&\quad \approx \operatorname{tr} \left\{\boldsymbol{\mathcal{I}}(\boldsymbol{\theta}_0; \mathbf{Y})^{-1} \boldsymbol{\mathcal{I}}(\boldsymbol{\theta}_0; Z \mid \mathbf{Y})\right\}, \qquad (46)
\end{aligned}
$$

where I have assumed (37) implies $\hat{\boldsymbol{\theta}} - \boldsymbol{\theta}_0 \approx N(\mathbf{0}, \mathcal{I}(\boldsymbol{\theta}_0; \mathbf{Y})^{-1})$. Then (38) follows from (42)–(44) and (46). □

HEURISTIC DERIVATION OF PLAUSIBLE APPROXIMATION 2. Since $(\mathbf{Y}, Z)$ is Gaussian,

$$D(\boldsymbol{\theta}_0, \hat{\boldsymbol{\theta}}; Z \mid \mathbf{Y}) = \frac{1}{2} E_0 \left\{ \log \frac{M(\hat{\boldsymbol{\theta}})}{M(\boldsymbol{\theta}_0)} - 1 + \frac{e(Z; \hat{\boldsymbol{\theta}})^2}{M(\hat{\boldsymbol{\theta}})} \right\} \tag{47}$$

(Exercise 43). Using (39),

$$E_0 \left\{ \log \frac{M(\hat{\boldsymbol{\theta}})}{M(\boldsymbol{\theta}_0)} \right\} \approx E_0 \left\{ \frac{M(\hat{\boldsymbol{\theta}}) - M(\boldsymbol{\theta}_0)}{M(\boldsymbol{\theta}_0)} \right\} - \frac{1}{2} E_0 \left\{ \frac{M(\hat{\boldsymbol{\theta}}) - M(\boldsymbol{\theta}_0)}{M(\boldsymbol{\theta}_0)} \right\}^2 . \tag{48}$$

Next, since $e(Z; \boldsymbol{\theta}_0)$ is independent of $\mathbf{Y}$ and hence independent of $e(Z; \hat{\boldsymbol{\theta}}) - e(Z; \boldsymbol{\theta}_0)$ and $M(\hat{\boldsymbol{\theta}})$,

$$E_0 \left\{ \frac{e(Z; \hat{\boldsymbol{\theta}})^2}{M(\hat{\boldsymbol{\theta}})} \right\} = E_0 \left\{ \frac{M(\boldsymbol{\theta}_0)}{M(\hat{\boldsymbol{\theta}})} \right\} + E_0 \frac{\{e(Z; \hat{\boldsymbol{\theta}}) - e(Z; \boldsymbol{\theta}_0)\}^2}{M(\hat{\boldsymbol{\theta}})} \tag{49}$$

(Exercise 44). Now

$$E_0 \left\{ \frac{M(\boldsymbol{\theta}_0)}{M(\hat{\boldsymbol{\theta}})} \right\} \approx 1 - E_0 \left\{ \frac{M(\hat{\boldsymbol{\theta}}) - M(\boldsymbol{\theta}_0)}{M(\boldsymbol{\theta}_0)} \right\} + E_0 \left\{ \frac{M(\hat{\boldsymbol{\theta}}) - M(\boldsymbol{\theta}_0)}{M(\boldsymbol{\theta}_0)} \right\}^2 , \tag{50}$$

and by (40),

$$E_0 \frac{\{e(Z; \hat{\boldsymbol{\theta}}) - e(Z; \boldsymbol{\theta}_0)\}^2}{M(\hat{\boldsymbol{\theta}})} \approx \frac{1}{M(\boldsymbol{\theta}_0)} E_0 \left\{ e(Z; \hat{\boldsymbol{\theta}}) - e(Z; \boldsymbol{\theta}_0) \right\}^2 . \tag{51}$$

Combining (47)–(51) yields (41). □

Numerical results

Plausible Approximation 1 suggests that it would be instructive to compute $\Delta\mathcal{I} = \operatorname{tr}\left\{\mathcal{I}(\boldsymbol{\theta}_0; \mathbf{Y})^{-1} \mathcal{I}(\boldsymbol{\theta}_0; Z \mid \mathbf{Y})\right\}$ in various settings to learn about the effect of estimation on subsequent predictions. As I discussed in the previous subsection, I believe that $\Delta\mathcal{I}$ is at least qualitatively informative even in situations where not all components of $\boldsymbol{\theta}$ are microergodic so that the argument for Plausible Approximation 1 does not apply. This subsection numerically examines the behavior of $\Delta\mathcal{I}$ for mean 0 stationary Gaussian processes on $\mathbb{R}$ under two particular Matérn models.

We consider interpolation and extrapolation problems in which the observations are, for the most part, evenly spaced with distance δ between neighboring observations and the predictand is a distance δ' from the nearest observation, where δ' is either δ or 0.5δ. Figure 7 shows some results when $\delta' = \delta$. More specifically, suppose there are observations at δj for $j = 1, \ldots, 40$ and $42, \ldots, 81$ and we wish to predict at 41δ (interpolation) or 0 (extrapolation). For various values of δ and the Matérn models $(\phi, \nu, \alpha) = (1, 1, 1)$ or $(24, 2, 2)$, Figure 7 gives the values for $\Delta\mathcal{I}$. Several general trends emerge. First, $\Delta\mathcal{I}$ depends much more strongly on δ when interpolating than when extrapolating, with the result that $\Delta\mathcal{I}$ is smaller when interpolating for smaller δ but is generally larger for larger δ. Furthermore, the difference between the cases ν known and unknown is larger when interpolating. In particular, for smaller δ, when interpolating, $\Delta\mathcal{I}$ is quite near to 0.0125, which is what we would get if only ϕ were unknown. Thus, to the extent that Plausible Approximation 1 is relevant in this setting, the additional effect of uncertainty about α on interpolation is quite small when δ is small.

Figure 8 considers observations at δj for $j = 1, \ldots, 80$, in which case, there is no way to have an interpolation problem in which the distance from the predictand to the nearest observation equals the distance between neighboring observations. Instead, I consider interpolating at 40.5δ and, to have a comparable extrapolation problem in which the predictand is 0.5δ from the nearest observation, predicting at 0.5δ. Thus, we now have $\delta' = 0.5\delta$. Let us first consider ν unknown. In this case, when interpolating, $\Delta\mathcal{I}$ is in all instances considerably greater than when $\delta' = \delta$. However, when extrapolating, $\Delta\mathcal{I}$ sometimes increases and sometimes decreases from the results with $\delta' = \delta$. Indeed, we now have that when ν is unknown, $\Delta\mathcal{I}$ is always larger when interpolating than extrapolating in the cases examined here. When ν is known, the values of $\Delta\mathcal{I}$ are generally quite similar to those for $\delta' = \delta$ whether interpolating or extrapolating.

Considering the numerical results in 3.5 and the theorems in 3.6 on the effect of misspecifying the spectral density on interpolation and extrapolation, the results here showing that $\Delta\mathcal{I}$ is often larger when interpolating than extrapolating need some explanation. To review, 3.6 studied the effect of misspecifying the spectral density on predicting at 0 based on observations either at δj for all negative integers j (extrapolation) or at δj for all nonzero integers j (interpolation). As $\delta \downarrow 0$, results in 3.6 show that the effect of misspecifying the spectral density at either high or low frequencies on the actual mse of a pseudo-BLP is smaller when interpolating than when extrapolating (Theorems 3 and 5 in 3.6). Furthermore, the effect of misspecifying the spectral density at low frequencies on the assessment of mse of pseudo-BLPs is also smaller when interpolating (Theorem 6 in 3.6). However, because of the difficulty of comparing spectral densities with different high frequency behavior when evaluating mses, evaluating the effects of such misspecifications on the assessment of mses when interpolating and extrapolating is problematic, although Theorem 4 in 3.6 attempts to address this problem. Thus, when ν is unknown, so that the estimated high frequency behavior of the spectral density will be different from the actual high frequency behavior, we should not necessarily expect $\Delta\mathcal{I}$ to be smaller

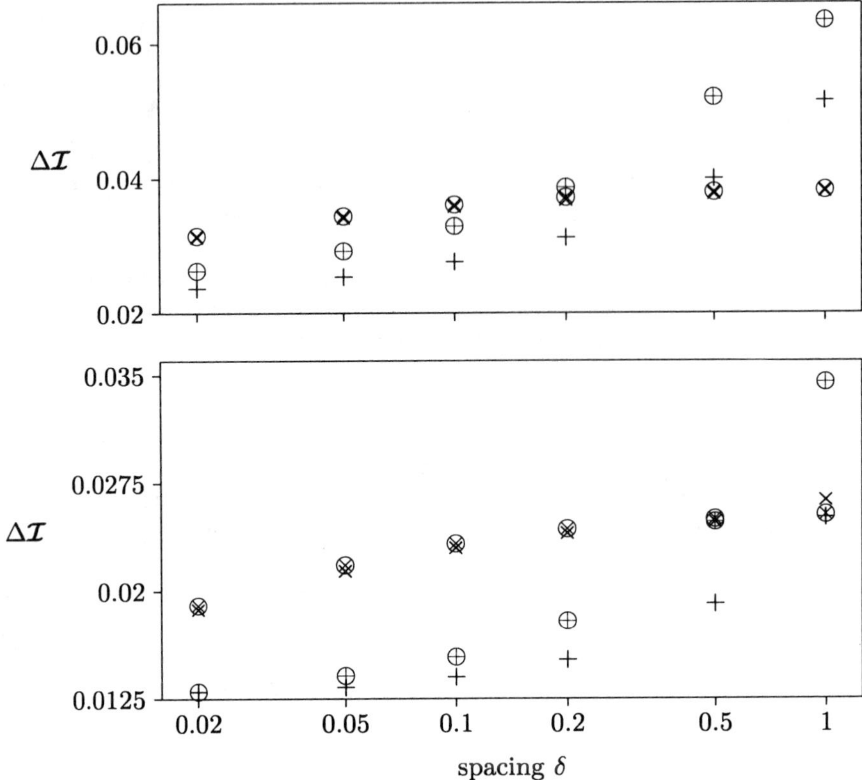

FIGURE 7. Values of $\Delta\mathcal{I}$ for Matérn model with $(\phi, \nu, \alpha) = (1, 1, 1)$ or $(\phi, \nu, \alpha) = (24, 2, 2)$. In the top figure, all parameters are considered unknown and in the bottom figure ν is considered known and ϕ and α unknown. The observations are at δj for $j = 1, \ldots, 40$ and $j = 42, \ldots, 81$. The predictands are at 41δ (interpolation) and 0 (extrapolation).

$+$ indicates $(\phi, \nu, \alpha) = (1, 1, 1)$ and the predictand at 41δ.

$\oplus$ indicates $(\phi, \nu, \alpha) = (24, 2, 2)$ and the predictand at 41δ.

$\times$ indicates $(\phi, \nu, \alpha) = (1, 1, 1)$ and the predictand at 0.

$\otimes$ indicates $(\phi, \nu, \alpha) = (24, 2, 2)$ and the predictand at 0.

If only ϕ is unknown, then $\Delta\mathcal{I} = 0.0125$ in all cases.

when interpolating than extrapolating even for δ small. Nevertheless, we should note that the setting studied in 3.6 is most comparable to the $\delta' = \delta$ case in Figure 7, for which the numerical results here show that for sufficiently small δ, $\Delta\mathcal{I}$ is smaller when interpolating than when extrapolating, whether or not ν is known.

Plausible Approximations 1 and 2 give us a way of gaining an understanding as to why $\Delta\mathcal{I}$ is always larger for interpolating than for extrapolation when $\delta' = 0.5 = \delta$ with ν unknown. Specifically, to the extent that they are both relevant, a large value of $\Delta\mathcal{I}$ indicates either a large inefficiency for the EBLUP or a large error in the plug-in assessment of mse of the

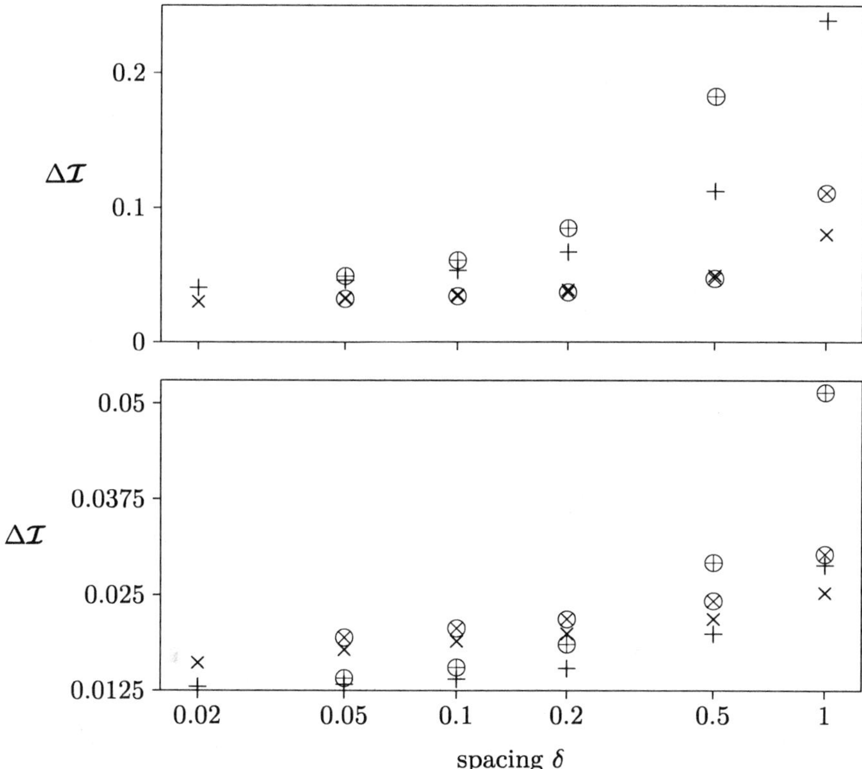

FIGURE 8. Values of $\Delta\mathcal{I}$ for Matérn model with $(\phi, \nu, \alpha) = (1, 1, 1)$ or $(\phi, \nu, \alpha) = (24, 2, 2)$. In the top figure, all parameters are considered unknown and in the bottom figure ν is considered known and ϕ and α unknown. The observations are at δj for $j = 1, \ldots, 80$. The predictands are at 40.5δ (interpolation) and 0.5δ (extrapolation).

$+$ indicates $(\phi, \nu, \alpha) = (1, 1, 1)$ and the predictand at 40.5δ.

$\oplus$ indicates $(\phi, \nu, \alpha) = (24, 2, 2)$ and the predictand at 40.5δ.

$\times$ indicates $(\phi, \nu, \alpha) = (1, 1, 1)$ and the predictand at 0.5δ.

$\otimes$ indicates $(\phi, \nu, \alpha) = (24, 2, 2)$ and the predictand at 0.5δ.

No results are given for $(\phi, \nu, \alpha) = (24, 2, 2)$ and $\delta = 0.02$ due to numerical difficulties. For $(\phi, \nu, \alpha) = (24, 2, 2)$, $\delta = 1$ and the predictand at 40.5, $\Delta\mathcal{I} = 0.6278$, which is omitted from the top figure.

EBLUP or both. Table 3 shows the effect of misspecifying ν on the efficiency of pseudo-BLPs when the true spectral density is the Matérn model with $(\phi, \nu, \alpha) = (24, 2, 2)$ and the presumed spectral density is the Matérn model with $(\phi, \nu, \alpha) = (24, y, y)$ for values of y near 2. The observations and predictands are as in Figures 7 and 8 with $\delta = 0.2$; results for other values of δ are qualitatively similar. In all cases, the effect of misspecifying the model is much larger when extrapolating than when interpolating, particularly so when $\delta' = 0.5\delta$. Furthermore, the difference between interpolation and

TABLE 3. Relative increases in mse due to using Matérn model with parameters $\nu = \alpha = y$ instead of the correct values $\nu = \alpha = 2$. Observations and predictands are as in Figures 7 and 8 with $\delta = 0.2$. Distance from predictand to nearest observation is denoted by δ', so that $\delta' = 0.2$ for the setting in Figure 7 and $\delta' = 0.1$ for the setting in Figure 8.

	Interpolation		Extrapolation	
y	$\delta' = 0.2$	$\delta' = 0.1$	$\delta' = 0.2$	$\delta' = 0.1$
1.6	2.57×10^{-2}	8.24×10^{-3}	5.59×10^{-2}	6.04×10^{-2}
1.7	1.32×10^{-2}	4.10×10^{-3}	3.00×10^{-2}	3.18×10^{-2}
1.8	5.43×10^{-3}	1.62×10^{-3}	1.27×10^{-2}	1.33×10^{-2}
1.9	1.26×10^{-3}	3.65×10^{-4}	3.06×10^{-3}	3.14×10^{-3}
2	0	0	0	0
2.1	1.10×10^{-3}	3.00×10^{-4}	2.84×10^{-3}	2.84×10^{-3}
2.2	4.11×10^{-3}	1.10×10^{-3}	1.10×10^{-2}	1.09×10^{-2}
2.3	8.72×10^{-3}	2.27×10^{-3}	2.40×10^{-2}	2.34×10^{-2}
2.4	1.46×10^{-2}	3.72×10^{-3}	4.14×10^{-2}	4.00×10^{-2}

extrapolation turns out to be much larger when the misspecified Matérn model has parameter values $(24, y, 2)$, so that only ν, and not both ν and α, is incorrect. These results clearly show that the larger values for $\Delta\mathcal{I}$ in Figure 8 when interpolating and ν is unknown cannot be attributed to inefficiencies in the interpolant due to having to estimate ν. In conjunction with Plausible Approximation 2, this finding leads me to attribute the relatively large values for $\Delta\mathcal{I}$ in Figures 7 and 8 when interpolating to inaccuracies in the plug-in assessment of mse of the EBLUP, although a full-scale simulation study would provide a more definitive way of resolving this matter.

Some issues regarding asymptotic optimality

Section 4.3 gave results showing the uniform asymptotic optimality of pseudo-BLPs and pseudo-BLUPs under a fixed but misspecified autocovariance function when the corresponding spectral density has the correct high frequency behavior. It is not possible to obtain a directly comparable result for EBLUPs using the Matérn model for the autocovariance function or any other model that includes a parameter controlling the rate of decay of the spectral density at high frequencies. The fundamental obstacle is the mismatch in Hilbert spaces for the true and estimated models. Specifically, if $\hat{\nu} < \nu$, then there will be elements in $\mathcal{H}_R(F)$ (the Hilbert space generated by $Z(\mathbf{x})$ for $\mathbf{x} \in R$ under the inner product defined by F) that are not in $\mathcal{H}_R(\hat{F})$, so that it will not even be possible to define an EBLUP for all elements in $\mathcal{H}_R(F)$. It may be possible to obtain uniformly asymptotically optimal point predictions, that is, replacing the supremum over $\mathcal{H}_{-n}$ in Theorems 8, 10 and 12 in 4.3 by a supremum over all $Z(\mathbf{x})$ for $\mathbf{x} \in R$. However, I do not think uniform asymptotically correct assessment of mses is possible even if one restricts to point predictions. The problem is that for any fixed set of observations, as the predictand location tends towards one of the observations, even a tiny error in $\hat{\nu}$ will lead to unboundedly large relative errors in the assessment of mses. If there are measurement errors, then I believe it is possible to obtain uniformly asymptotically correct assessment of mses, since the problems in assessing mse for a predictand very near to an observation should no longer occur. The approach taken in Putter and Young (1998) may be helpful in solving these problems, but the results in that work are not nearly strong enough to provide any answers at present.

Exercises

38 Consider the setting of Exercise 6 in 2.4 and assume, for simplicity, that the mean of Z is known to be 0. Argue that under any reasonable choice for the estimated autocovariance function of the process, the plug-in predictor has a smaller mse than the BLP.

39 Verify (36).

40 Show that (38) is an equality if the covariance matrix of $(\mathbf{Y}, Z)$ is known and its mean vector is linear in $\boldsymbol{\theta}$.

41 Suppose $\mathbf{X}$ and $\mathbf{Y}$ are random vectors whose joint distribution is in the parametric family $P_{\boldsymbol{\theta}}$ for some $\boldsymbol{\theta} \in \Theta$. Let $\mathbf{S}(\boldsymbol{\theta}; (\mathbf{X}, \mathbf{Y}))$ be the score function for $\boldsymbol{\theta}$ based on the observation $(\mathbf{X}, \mathbf{Y})$. Under suitable regularity conditions, prove $E_{\boldsymbol{\theta}_0}\{\mathbf{S}(\boldsymbol{\theta}_0; (\mathbf{X}, \mathbf{Y}))|\mathbf{X}\} = \mathbf{0}$ with probability 1. This result can be thought of as expressing a martingale property of the score function. Show that (43) follows.

42 Verify (44) and (45).

43 Verify (47).

44 Verify (49).

45 State and provide heuristic derivations of Plausible Approximations 1 and 2 appropriate for BLUPs, EBLUPs and REML estimates assuming $\boldsymbol{\theta}$ contains just the unknown parameters for the covariance structure.

6.9 An instructive example of plug-in prediction

As I have suggested previously, it is for differentiable random fields that I find present practice in spatial statistics to be seriously flawed. This section considers an example based on simulated data that more explicitly demonstrates some of the problems that can occur when predicting smooth random fields.

Suppose Z is a stationary Gaussian process on $\mathbb{R}$ with mean 0 and autocovariance function $K(t) = e^{-0.4|t|}(1 + 0.4|t|)$ so that Z is exactly once mean square differentiable (see 2.7). We observe this process at the 20 locations $-9.5, -8.5, \ldots, 8.5, 9.5$ and wish to predict it at $-10, -9, \ldots, 10$ and at ± 10.5 when the mean of Z and its autocovariance function are unknown. Figure 9 plots the simulated values of the observations and indicates the locations of the predictands; the actual simulated values are given in Table 4. We see that our predictions include both interpolations and extrapolations.

The empirical semivariogram (see 2.9) is a staple of the kriging literature as a tool for selecting models for semivariograms and estimating the parameters in these models. Figure 10 plots the empirical semivariogram up to distance 10 for the simulated realization in Figure 9. Distances greater than 10 are not plotted because of the severe lack of reliability of empirical semivariograms at distances more than half the dimensions of the observation region, which corresponds to common geostatistical practice (Journel and Huijbregts 1978, p. 194). Figure 10 also plots the actual semivariogram. It is critical to note the rather large and apparently systematic differences between the actual and empirical semivariograms at the shorter distances. Far from being unusual, this phenomenon should be expected in light of the strong correlations that exist in the empirical semivariogram at different distances. For example, using $\hat{\gamma}$ to indicate the empirical semivariogram, $\operatorname{corr}\{\hat{\gamma}(1), \hat{\gamma}(2)\} = 0.981$, $\operatorname{corr}\{\hat{\gamma}(1), \hat{\gamma}(3)\} = 0.938$ and $\operatorname{corr}\{\hat{\gamma}(2), \hat{\gamma}(3)\} = 0.880$ (Exercise 46). Thus, the empirical semivariogram can appear quite regular and still be substantially in error. Of course, the fact that the empirical semivariogram has correlated values is well known (Cressie 1985, 1993), but I believe that the consequences of these potentially large correlations are not generally sufficiently appreciated. If one had a regression problem with observations that were equal to the underlying regression function plus independent errors that was as smooth as the empirical semivariogram in Figure 10, then it would be sound to conclude that the regression function could be well estimated. It is a difficult psychological adjustment to look at Figure 10 and recognize that the strong correlations present can easily yield a smooth empirical semivariogram so different from the actual semivariogram.

Considering the apparent quadratic behavior at the origin of the empirical semivariogram in Figure 10, it would now be within the realm of accepted present practice in spatial statistics to fit a Gaussian semivariogram, $\gamma(t; \phi, \alpha) = \phi(1 - e^{-\alpha t^2})$ to the data. Although Goovaerts (1997) recommends never using the Gaussian semivariogram without a measurement error term, note that there is no measurement error term here and the true semivariogram is quadratic near the origin. Since, as noted in 3.5, software in spatial statistics commonly includes the Gaussian as the only semivariogram model that is quadratic near the origin, it would be hard

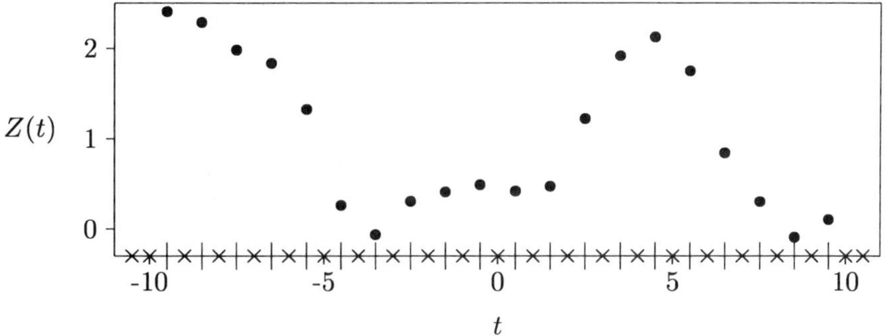

FIGURE 9. Simulated realization of Gaussian process with mean 0 and autocovariance function $K(t) = e^{-0.4|t|}(1 + 0.4|t|)$. The $\times$s on the horizontal axis are the locations of predictands and the $|$s are the locations of the observations.

TABLE 4. Simulated values of process pictured in Figure 9. The last three rows are the additional observations used towards the end of this section.

t	$Z(t)$	t	$Z(t)$
−9.5	2.3956811	0.5	0.4109609
−8.5	2.2767195	1.5	0.4647669
−7.5	1.9736058	2.5	1.2113779
−6.5	1.8261141	3.5	1.9055446
−5.5	1.3136954	4.5	2.1154852
−4.5	0.2550507	5.5	1.7372076
−3.5	−0.0741740	6.5	0.8333657
−2.5	0.2983559	7.5	0.2932142
−1.5	0.4023333	8.5	−0.1024508
−0.5	0.4814850	9.5	0.0926624
−0.25	0.4267716		
0.0	0.4271087		
0.25	0.4461579		

to fault the practitioner who adopted the Gaussian model in this case. Of course, the Gaussian model is severely in error in the sense that it implies Z has analytic realizations when in fact the process has only one derivative. Nevertheless, it is instructive to compare plug-in predictors based on the REML estimate and an "eyeball" estimate that fits the empirical semivariogram very well at the shorter distances.

If we suppose Z is Gaussian with mean μ and semivariogram of the form $\phi(1 - e^{-\alpha t^2})$ with (μ, ϕ, α) unknown, the REML estimate of $\boldsymbol{\theta} = (\phi, \alpha)$ is $\hat{\boldsymbol{\theta}} = (0.667, 0.247)$. Figure 11 replots the empirical semivariogram together with $\gamma(t; \hat{\boldsymbol{\theta}})$. The REML estimate yields a fitted semivariogram with slightly larger curvature near the origin than the empirical semivariogram. Figure 11 also plots $\gamma(t; \tilde{\boldsymbol{\theta}})$ for my eyeball estimate $\tilde{\boldsymbol{\theta}} = (1, 0.12)$; this eye-

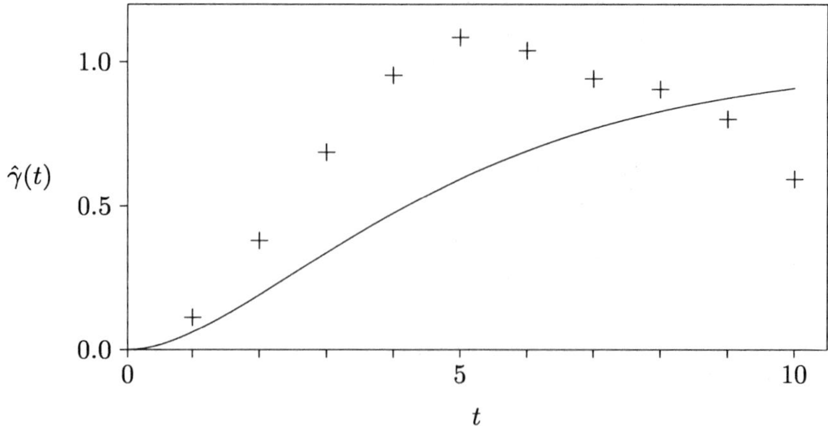

FIGURE 10. Empirical and actual semivariograms for data shown in Figure 9. Smooth curve is the actual semivariogram and +s are the empirical semivariogram.

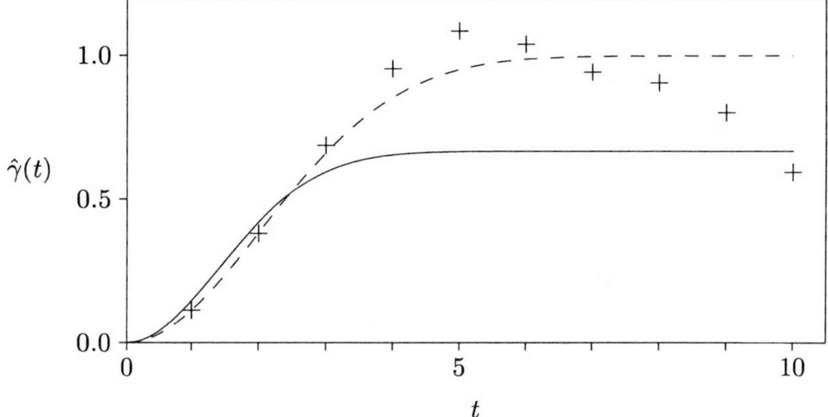

FIGURE 11. Empirical and estimated semivariograms for data shown in Figure 9. Using the Gaussian model for the semivariogram, solid line indicates REML estimate and dashed line indicates eyeball estimate.

ball estimate matches the empirical semivariogram distinctly better than the REML estimate at the shorter distances. Furthermore, comparing Figures 10 and 11 shows that the eyeball estimate is visually closer to the true semivariogram than the REML estimate. Is this evidence that the REML estimate is inferior to the eyeball fit in this example?

Behavior of plug-in predictions

This subsection considers predicting Z at the locations indicated in Figure 9 assuming the mean of Z is an unknown constant. Figure 12 plots the BLUPs under the true model as well as the plug-in predictors or EBLUPs using the Gaussian model and $\hat{\boldsymbol{\theta}}$ or $\tilde{\boldsymbol{\theta}}$ to estimate $\boldsymbol{\theta}$. Near the middle of the observation region, both EBLUPs are almost identical to the BLUPs. As one gets nearer to the edges of the observation region and, particularly, outside the observation region, the EBLUPs, especially using the eyeball estimate $\tilde{\boldsymbol{\theta}}$, can be substantially different from the BLUPs. Define $e_0(t)$ as the error of the BLUP of $Z(t)$ and $e(t;\hat{\boldsymbol{\theta}})$ and $e(t;\tilde{\boldsymbol{\theta}})$ as the errors of the EBLUPs of $Z(t)$. Take E_0 to mean expectation under the true model for Z and let $\mathbf{y}$ be the observed value of the contrasts $\mathbf{Y}$ of the observations. Define $M(t;\boldsymbol{\theta})$ as in (30) with t taking the place of $\mathbf{x}_0$, so that, for example, $M(t;\hat{\boldsymbol{\theta}})$ is the plug-in estimate of the mse when predicting $Z(t)$ using the REML estimate $\hat{\boldsymbol{\theta}}$. Finally, take

$$C(t;\hat{\boldsymbol{\theta}},\mathbf{y}) = E_0\left\{e(t;\hat{\boldsymbol{\theta}})^2 \mid \mathbf{Y}=\mathbf{y}\right\} \tag{52}$$

and $C(t;\tilde{\boldsymbol{\theta}},\mathbf{y}) = E_0\left\{e(t;\tilde{\boldsymbol{\theta}})^2 \mid \mathbf{Y}=\mathbf{y}\right\}$ to be the conditional mses of the EBLUPs given the contrasts. Table 5 gives the mses of the BLUPs, the conditional mses of the EBLUPs and the plug-in estimates of the mse. By symmetry, $E_0 e_0(t)^2$, $M(t;\hat{\boldsymbol{\theta}})$ and $M(t;\tilde{\boldsymbol{\theta}})$ are even functions of t, hence the pairing of predictands in Table 5. In contrast, $C(t;\hat{\boldsymbol{\theta}},\mathbf{y})$ and $C(t;\tilde{\boldsymbol{\theta}},\mathbf{y})$ depend on the particular values of the observations around t and are not symmetric in t. The values for the conditional mses support our conclusions from Figure 12. Specifically, for predictands near the middle of the observation range, the conditional mses of the EBLUPs are not much larger than the unconditional mses of the BLUPs. On the other hand, near or outside the boundaries of the observation region, $C(t;\hat{\boldsymbol{\theta}},\mathbf{y})/M(t;\boldsymbol{\theta}_0)$ and especially $C(t;\tilde{\boldsymbol{\theta}},\mathbf{y})/M(t;\boldsymbol{\theta}_0)$ can be large. Thus, although the REML estimate of $\hat{\boldsymbol{\theta}}$ produces poor plug-in predictions when extrapolating, the eyeball estimate $\tilde{\boldsymbol{\theta}}$ produces much worse plug-in extrapolations.

The most striking result in Table 5 is the severely overoptimistic values for the plug-in estimates of mse. Not only are they much smaller than the conditional mses of the corresponding plug-in predictors, they also are generally much smaller than the mses of the BLUPs. Furthermore, although the plug-in mses based on either estimate share this overoptimism, the problem is much worse for the eyeball estimate; in some cases, $M(t;\tilde{\boldsymbol{\theta}})/C(t;\tilde{\boldsymbol{\theta}},\mathbf{y})$ is less than 10^{-6}. The next subsection considers why even $\hat{\boldsymbol{\theta}}$ produces such unrealistic plug-in mses.

Cross-validation

One method that is sometimes suggested for diagnosing misfits of semivariograms is cross-validation (Cressie 1993, pp. 101–104). Specifically, suppose we have observations $Z(\mathbf{x}_1),\ldots,Z(\mathbf{x}_n)$ and plan to predict at other locations using an EBLUP where the mean of Z is taken to be an unknown

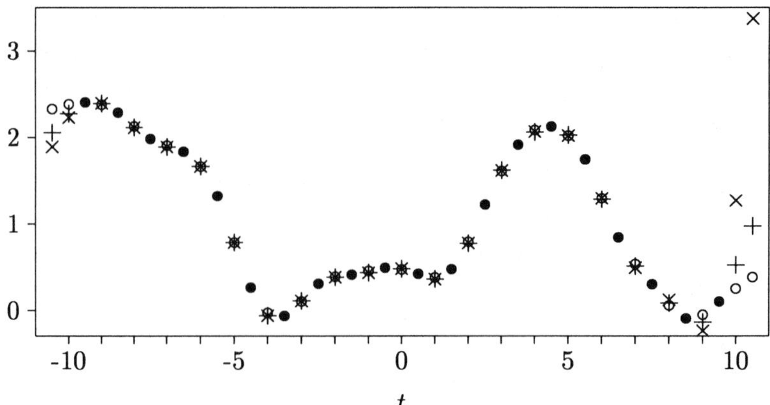

FIGURE 12. BLUPs and EBLUPs for simulation in Figure 9. The •s are observed values, ◦s are BLUPs, +s are EBLUPs using the REML estimate $\hat{\boldsymbol{\theta}}$ and ×s are EBLUPs using the eyeball estimate $\tilde{\boldsymbol{\theta}}$.

constant. Let $\hat{\gamma}$ be an estimate of the semivariogram and define $\hat{Z}_{-j}(\mathbf{x}_j)$ as the EBLUP of $Z(\mathbf{x}_j)$ using all of the observations other than $Z(\mathbf{x}_j)$ and $\hat{\gamma}$ for the semivariogram. Furthermore, let $\hat{\sigma}_{-j}(\mathbf{x}_j)^2$ be the plug-in estimate of the mse of $\hat{Z}_{-j}(\mathbf{x}_j)$ as a predictor of $Z(\mathbf{x}_j)$. Because we know the actual value of $Z(\mathbf{x}_j)$, we can compare $Z(\mathbf{x}_j) - \hat{Z}_{-j}(\mathbf{x}_j)$ with $\hat{\sigma}_{-j}(\mathbf{x}_j)$ for $j = 1, \ldots, n$. For example, Cressie (1993, p. 102) notes that if $\hat{\gamma}$ is a good estimate of the semivariogram, we should expect

$$\frac{1}{n}\sum_{j=1}^{n} \frac{\left\{Z(\mathbf{x}_j) - \hat{Z}_{-j}(\mathbf{x}_j)\right\}^2}{\hat{\sigma}_{-j}(\mathbf{x}_j)^2} \tag{53}$$

to be near 1, so that a value of (53) far from 1 is a sign of a poorly fit semivariogram. However, it does not follow that a value near 1 means that the fit is adequate. Actually, Cressie (1993) recommends computing the square root of (53), but since I have been reporting mean squared errors rather than root mean squared errors throughout this work, I use (53) here for consistency.

For the example in this section, (53) equals 210.4 for the eyeball estimate of the semivariogram and equals 1.223 for the REML estimate. Thus, cross-validation correctly identifies the eyeball estimate as a poor fit but does not detect a problem with the Gaussian model for the REML estimate, which is in stark contrast with the results in Table 5. Of course, the results in Table 5 would not be available in practice since they depend on knowing the true model.

Let us consider why Table 5 and cross-validation give such different conclusions for the REML estimate. First, as evidence that the results of this simulation are not a fluke, I ran four further simulations for the same setting and got values for (53) of 1.172, 0.953, 1.589 and 1.475 when fitting

TABLE 5. Performance of two EBLUPs when using the Gaussian model for the semivariogram. The two parameter estimates are the REML $\hat{\boldsymbol{\theta}}$ and an eyeball estimate $\tilde{\boldsymbol{\theta}}$ (see Figure 11). $M(t;\hat{\boldsymbol{\theta}})$ is, for example, the plug-in value for the mse of the EBLUP based on $\hat{\boldsymbol{\theta}}$ (see (31)) and $C(t;\hat{\boldsymbol{\theta}},\mathbf{y})$ is the conditional mse of the EBLUP based on $\hat{\boldsymbol{\theta}}$ (see (52)). All figures other than in the first column are 1,000 times their actual values, so that, for example, $M(10.5;\boldsymbol{\theta}_0)$ equals 0.0766.

t	$M(t;\boldsymbol{\theta}_0)$	$C(t;\hat{\boldsymbol{\theta}},\mathbf{y})$	$M(t;\hat{\boldsymbol{\theta}})$	$C(t;\tilde{\boldsymbol{\theta}},\mathbf{y})$	$M(t;\tilde{\boldsymbol{\theta}})$
−10.5	76.6	147	91.2	259	5.86
10.5		436		9067	
−10	18.2	28.1	12.2	38.9	0.414
10		97.2		1071	
−9	3.31	3.79	0.447	4.00	4.53×10^{-3}
9		9.25		35.3	
−8	2.78	2.81	0.106	2.82	3.75×10^{-4}
8		4.10		7.98	
−7	2.74	2.80	0.0489	2.80	7.21×10^{-5}
7		3.00		3.90	
−6	2.74	2.85	0.0315	2.86	2.27×10^{-5}
6		2.75		2.80	
−5	2.74	2.81	0.0244	2.80	9.91×10^{-6}
5		2.98		2.81	
−4	2.74	3.38	0.0210	3.36	5.48×10^{-6}
4		2.98		2.85	
−3	2.74	2.96	0.0192	2.94	3.61×10^{-6}
3		3.03		2.91	
−2	2.74	2.75	0.0182	2.75	2.74×10^{-6}
2		2.82		2.78	
−1	2.74	2.88	0.0177	2.91	2.35×10^{-6}
1		2.83		2.87	
0	2.74	2.99	0.0176	3.05	2.23×10^{-6}

the parameters of the Gaussian semivariogram using REML. REML is trying its best to fit the poorly chosen model to the observations, which are spaced 1 unit apart. In doing so, it yields plug-in mses that are not too far off if the predictand is no closer than 1 unit from the nearest observation, which is the case when cross-validating. A further indication that this fitted model is not so bad when the predictand is no closer than 1 unit from any observation is that when using all 20 observations to predict at 10.5 or -10.5, which are also 1 unit away from the nearest observation, the plug-in mses for the REML estimate are relatively accurate (see Table 5). However, when trying to predict at locations that are only 0.5 units away from the nearest observation, the badly misspecified Gaussian semivariogram implies a highly unrealistic correlation structure between the observations and predictand, which is a problem that cross-validation based on evenly spaced observations cannot uncover. One should not conclude from these simulations that cross-validation is not a useful technique in spatial statistics, merely that it is not a foolproof method for detecting problems with a model, particularly if the observations are evenly spaced.

Application of Matérn model

Let us reconsider this simulated dataset using the Matérn model for the spectral density, $f_{\boldsymbol{\theta}}(\omega) = \phi(\alpha^2 + |\boldsymbol{\omega}|^2)^{-\nu-d/2}$, where $\boldsymbol{\theta} = (\phi, \nu, \alpha)$ and $\Theta = (0, \infty)^3$. This model includes the truth, $\boldsymbol{\theta} = (0.128/\pi, 1.5, 0.4)$, so we might expect to do much better using likelihood methods than we did when we used the Gaussian semivariogram model. The REML estimate of $\boldsymbol{\theta}$ is $\hat{\boldsymbol{\theta}} = (5.389, 3.787, 1.262)$. Note that $\hat{\phi} = 5.389$ is much larger than the true value $\phi = 0.128/\pi = 0.0407$, which, from Theorem 1 in 6.7, is what we should expect when $\hat{\nu} > \nu$. If we use the parameterization suggested by Handcock and Wallis (1994) and described in 2.10, the parameter estimates do not look quite so bad. Specifically, $\boldsymbol{\eta} = (\sigma, \nu, \rho) = (1, 1.5, 1.624)$ and $\hat{\boldsymbol{\eta}} = (0.871, 3.787, 3.084)$.

Table 6 gives conditional and plug-in mses for EBLUPs based on using the Matérn model and the REML estimates. Contrasting the results in Tables 5 and 6, the plug-in mses are somewhat more accurate than for the REML estimates and the Gaussian model, but they are still off by about one order of magnitude for the interpolations. However, even this apparent improvement is somewhat fortunate. Let $p\ell(\nu)$ be the profile log likelihood of the contrasts; that is, as a function of ν, the supremum over ϕ and α of the log likelihood of the contrasts. Recall (see 2.10) that the Gaussian model is obtained by letting $\nu \to \infty$ in the Matérn model, so that $p\ell(\infty)$ equals the log likelihood of the contrasts under the Gaussian model evaluated at the REML estimates for that model (Exercise 47). The plot of $p\ell(\nu)$ in Figure 13 shows that there is little basis in the data for choosing between $\nu = 3.787$ and any larger value. Indeed, $p\ell(3.787) - p\ell(\infty) = 0.504$, which corresponds to a likelihood ratio of $e^{0.504} = 1.655$ and indicates that the Gaussian model provides a quite good fit to the available data, even though it is a terrible model for the process.

Table 6 also gives prediction information when ν is set to 15 and the likelihood of the contrasts is maximized with respect to ϕ and α. Taking $\nu = 15$ assumes the process is 14 but not quite 15 times mean square differentiable. When extrapolating or interpolating near the edges of the

TABLE 6. Performance of two EBLUPs when use Matérn model for the semivariogram. The two parameter estimates are $\hat{\boldsymbol{\theta}}$, the REML, and $\hat{\boldsymbol{\theta}}(15)$, obtained by arbitrarily setting the parameter ν to 15 and then maximizing the likelihood of the contrasts over the other parameters. As in Table 5, all figures other than in the first column are 1,000 times the actual figures.

t	$M(t;\boldsymbol{\theta}_0)$	$C(t;\hat{\boldsymbol{\theta}},\mathbf{y})$	$M(t;\hat{\boldsymbol{\theta}})$	$C(t;\hat{\boldsymbol{\theta}}(15),\mathbf{y})$	$M(t;\hat{\boldsymbol{\theta}}(15))$
−10.5	76.6	102	94.7	135	92.4
10.5		145		316	
−10	18.2	22.1	16.0	26.6	13.2
10		33.6		71.5	
−9	3.31	3.56	1.20	3.73	0.584
9		4.63		7.52	
−8	2.78	2.79	0.607	2.81	0.173
8		3.01		3.68	
−7	2.74	2.82	0.505	2.81	0.0986
7		2.77		2.89	
−6	2.74	2.89	0.481	2.87	0.0754
6		2.78		2.77	
−5	2.74	2.77	0.475	2.80	0.0662
5		2.94		3.03	
−4	2.74	3.18	0.474	3.35	0.0621
4		2.88		3.00	
−3	2.74	2.87	0.473	2.94	0.0601
3		2.91		3.04	
−2	2.74	2.75	0.473	2.75	0.0593
2		2.79		2.82	
−1	2.74	2.85	0.473	2.88	0.0589
1		2.82		2.83	
0	2.74	2.94	0.473	3.00	0.0587

observation region, the plug-in mses are quite similar whether one uses the Gaussian model and REML or the Matérn model with $\nu = 15$ and REML. However, when interpolating near the middle of the observation region, the Gaussian model gives plug-in mses less than $\frac{1}{3}$ as large as the Matérn model with $\nu = 15$.

The large uncertainty about ν combined with its critical impact on assessment of mses of interpolants implies that it is essentially impossible to

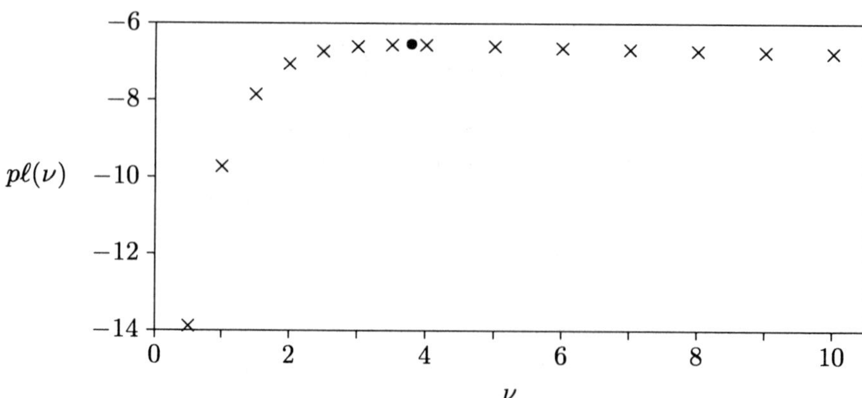

FIGURE 13. Profile log likelihood of the contrasts for ν using the 20 simulated observations. The • indicates $(\hat{\nu}, p\ell(\hat{\nu}))$.

obtain defensible mses from these data without strong prior information about ν. The fact that the plot of $p\ell(\nu)$ alerts us to this situation is a great strength of using the Matérn model together with likelihood methods. Note, though, that it is essential to study the likelihood function and not just find its maximum.

Considering the numerical results in 6.6, the fact that the likelihood function provides so little information about ν is not surprising. As those results show, evenly spaced observations make estimation of ν particularly difficult. I simulated three additional observations at $-0.25, 0$ and 0.25 (see Table 4) and for these 23 observations, recomputed the profile log likelihood of the contrasts (see Figure 14). We now obtain the much better estimate for ν of 1.796 and, in addition, have strong evidence against large values for ν. In particular, $\exp\{p\ell(1.796) - p\ell(\infty)\} = 1.27 \times 10^5$. Table 7 shows that the plug-in predictors now perform well for both interpolations and extrapolations and the plug-in estimates of mse are all reasonable. Table 7 also shows properties of the plug-in predictors if the Gaussian model is fit to these data using REML. Figure 15 plots the estimated autocovariance functions under both the Matérn and Gaussian models. Note that the estimated Gaussian model has far greater curvature near the origin after the additional three points are included in the analysis. I leave it to the reader to explain why the plug-in estimates of mse under the Gaussian model are now far too conservative for the predictions at $\pm 4, \ldots, \pm 10$ but are still badly overoptimistic at ± 1 (Exercise 48).

Conclusions

The reader could argue that I have only shown the detailed results of one simulation, which could be misleading. This is possible, although the simulation shown was the first one I ran. To the reader who doubts the

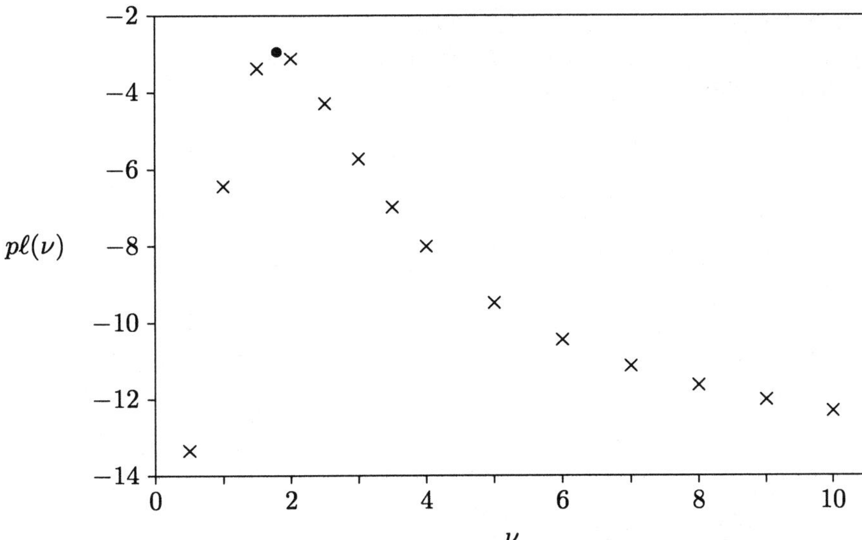

FIGURE 14. Profile log likelihood of the contrasts for ν using the original 20 simulated observations plus the three additional observations. The • indicates $(\hat{\nu}, p\ell(\hat{\nu}))$.

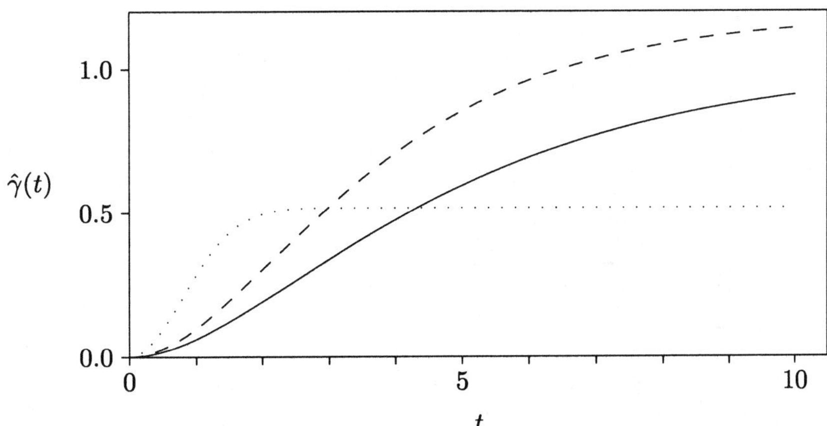

FIGURE 15. True and estimated semivariograms using 3 additional observations. The solid line indicates the truth, the dashed line the REML estimate under the Matérn model and the dotted line the REML estimate under the Gaussian model.

representativeness of this single simulation, I strongly recommend running some of your own.

The main lesson from this example is that standard practice in spatial statistics is seriously flawed for differentiable processes. Specifically, plotting the empirical semivariogram and then selecting a model that, to the eye, appears to fit its general shape can lead to severe model misspeci-

TABLE 7. Performance of two EBLUPs with additional three observations when using Matérn or Gaussian model for the semivariogram and estimating parameters by REML. All figures other than in the first column are 1,000 times the actual values.

		Matérn model		Gaussian model	
t	$M(t;\boldsymbol{\theta}_0)$	$C(t;\hat{\boldsymbol{\theta}},\mathbf{y})$	$M(t;\hat{\boldsymbol{\theta}})$	$C(t;\hat{\boldsymbol{\theta}},\mathbf{y})$	$M(t;\hat{\boldsymbol{\theta}})$
−10.5	76.6	79.1	116	680	411
10.5		79.9		201	
−10	18.2	18.6	25.7	140	142
10		19.0		48.5	
−9	3.31	3.34	3.94	9.71	28.9
9		3.37		5.58	
−8	2.78	2.78	3.09	3.70	22.1
8		2.78		3.23	
−7	2.74	2.75	3.01	2.77	20.9
7		2.74		2.85	
−6	2.74	2.76	3.00	2.75	20.7
6		2.74		2.85	
−5	2.74	2.74	3.00	2.86	20.4
5		2.75		3.18	
−4	2.74	2.80	3.00	3.59	19.3
4		2.74		6.35	
−3	2.74	2.77	2.99	4.01	15.2
3		2.74		18.3	
−2	2.70	2.72	2.91	6.38	6.05
2		2.70		21.3	
−1	2.21	2.31	2.11	8.87	0.280
1		2.27		21.3	
0	0	0	0	0	0

fication. The problem is that the empirical semivariogram is a poor tool for distinguishing exactly how smooth a differentiable process is. Furthermore, for differentiable processes, one needs to be careful about judging the quality of a parametric estimate of the semivariogram by how well it fits the empirical semivariogram. Finally, evenly spaced observations can cause substantial difficulties in predicting if the parameter ν in the Matérn model needs to be estimated. The inclusion of even a few additional observations that have smaller spacing than the rest of the observations can dramatically improve the estimation of the semivariogram.

Exercises

46 If the random vector $\mathbf{Z}$ of length n has distribution $N(\mathbf{0}, \boldsymbol{\Sigma})$, show that for fixed, symmetric $n \times n$ matrices $\mathbf{A}$ and $\mathbf{B}$, $\text{cov}(\mathbf{Z}^T\mathbf{AZ}, \mathbf{Z}^T\mathbf{BZ}) = 2\text{tr}(\mathbf{A}\boldsymbol{\Sigma}\mathbf{B}\boldsymbol{\Sigma})$ (see Appendix A for relevant results on multivariate normal distributions). Using this result, write a computer program to calculate the correlations for the empirical semivariogram $\hat{\gamma}$ at distances $1, 2, \ldots, 10$ for n observations evenly spaced 1 unit apart. Use your program to calculate these correlations when $n = 20, 40$ and 60 and (i) $K(t) = e^{-0.4|t|}(1 + 0.4|t|)$, (ii) $K(t) = e^{-0.2|t|}(1 + 0.2|t|)$ and (iii) $K(t) = e^{-0.75t^2}$. Comment on the results. See Genton (1998) for further examples of the correlations in empirical semivariograms.

47 Suppose we model Z as a Gaussian random field with isotropic autocovariance function from the Matérn model and mean function known except for a vector of linear regression coefficients. We observe Z at some finite set of locations and let $p\ell(\nu)$ be the profile log likelihood of the contrasts as a function of ν. Show that

$$\lim_{\nu \to \infty} p\ell(\nu) = \tilde{\ell}(\hat{\xi}_1, \hat{\xi}_2),$$

where $\tilde{\ell}(\xi_1, \xi_2)$ is the log likelihood of the contrasts under the model $K(r) = \xi_1 e^{-\xi_2 r^2}$ for the isotropic autocovariance function and $(\hat{\xi}_1, \hat{\xi}_2)$ is the REML estimate for (ξ_1, ξ_2).

48 Tell a story explaining the last two columns of Table 7.

6.10 Bayesian approach

The Bayesian approach to prediction provides a general methodology for taking into account the uncertainty about parameters on subsequent predictions. In particular, as described in 4.4, if $\mathbf{Y}$ is the vector of observations, Z the predictand and $\boldsymbol{\theta}$ the vector of unknown parameters, the Bayesian solution to making inferences about Z is to use the predictive density

$$p(Z \mid \mathbf{Y}) = \int_{\Theta} p(Z \mid \boldsymbol{\theta}, \mathbf{Y}) p(\boldsymbol{\theta} \mid \mathbf{Y})\, \mathbf{d}\boldsymbol{\theta}, \tag{54}$$

where $p(\boldsymbol{\theta} \mid \mathbf{Y})$ is the posterior density for $\boldsymbol{\theta}$ given by

$$p(\boldsymbol{\theta} \mid \mathbf{Y}) = \frac{p(\mathbf{Y} \mid \boldsymbol{\theta}) p(\boldsymbol{\theta})}{\int_{\Theta} p(\mathbf{Y} \mid \boldsymbol{\theta}') p(\boldsymbol{\theta}')\, \mathbf{d}\boldsymbol{\theta}'} \tag{55}$$

with $p(\boldsymbol{\theta})$ the prior density for $\boldsymbol{\theta}$. Although some scientists and statisticians are uncomfortable with basing inferences on what is necessarily a somewhat arbitrarily chosen prior distribution for the unknown parameters, it strikes me as a rather small additional leap of faith beyond that required in assuming, for example, that the spatial process under consideration is a realization of an isotropic Gaussian random field with isotropic autocovariance function of some particular form. See Berger (1985), for example, for further discussion concerning the theory behind the Bayesian approach to statistics and Gelman, Carlin, Stern and Rubin (1995) for a recent treatment of the practical application of Bayesian methods.

This section briefly addresses a few issues that arise in selecting prior distributions when using the Matérn model. Suppose Z is an isotropic Gaussian random field on $\mathbb{R}^d$ with spectral density from the Matérn class. Since, in principle, prior densities should represent the investigator's uncertainties about the unknown parameters prior to having collected the data, it is helpful if the parameterization for the model is chosen so that the individual parameters have natural and easily understood interpretations. To this end, the parameterization of Handcock and Wallis (1994) (see 2.10) is a sensible choice. To review, the isotropic spectral density is of the form

$$g_{\eta}(u) = \frac{\sigma c(\nu, \rho)}{\left(\frac{4\nu}{\rho^2} + u^2\right)^{\nu+d/2}}, \tag{56}$$

where $\boldsymbol{\eta} = (\sigma, \nu, \rho)$ and

$$c(\nu, \rho) = \frac{\Gamma\left(\nu + \frac{d}{2}\right)(4\nu)^{\nu}}{\pi^{d/2}\Gamma(\nu)\rho^{2\nu}}.$$

The parameter ν measures the differentiability of the random field, $\sigma = \text{var}\{Z(\mathbf{x})\}$ and ρ measures how quickly the correlations of the random field decay with distance.

Let us now consider placing a prior density on $\boldsymbol{\eta} = (\sigma, \nu, \rho)$ assuming, for simplicity, that the mean of Z is known to be 0. Because of the conceptual and practical difficulties of converting one's knowledge about unknown parameters into a probability distribution, one possible solution is to select the "flat" prior density $p(\boldsymbol{\eta}) = 1$ on $(0, \infty)^3$. Since this function integrates to ∞, it is not a probability density on the parameter space and is called an improper prior (Berger 1985, p. 82). Improper priors often yield proper (integrable) posterior distributions, but this is not the case in the present setting. Figures 13 and 14 showing the profile log likelihoods for ν for the two simulated datasets considered in 6.9 illustrate the problem: the profile likelihoods do not tend to 0 as $\nu \to \infty$. Indeed, for any fixed positive values of σ and ρ and any vector of observations $\mathbf{Z} = (Z(\mathbf{x}_1), \ldots, Z(\mathbf{x}_n))$ with observed value $\mathbf{z}$, the likelihood tends to a positive limit as $\nu \to \infty$ (see Exercise 47 in 6.9 for a related result). It follows that $\int_{\Theta} p(\mathbf{z} \mid \boldsymbol{\eta}) d\boldsymbol{\eta} = \infty$, so that (55) does not give a meaningful result. We could try to use just the numerator of (55) as the (nonintegrable) posterior for $\boldsymbol{\eta}$, but the resulting predictive distribution for $Z(\mathbf{x}_0)$ is then also not integrable. The same problem would occur for any improper prior of the form $p(\sigma, \nu, \rho) = p(\sigma, \rho)p(\nu)$ with $p(\nu)$ not integrable at ∞. A similar but more subtle problem can occur when using a prior whose marginal density for ρ is not integrable at ∞ (see Exercise 50).

For the random field $Z(\mathbf{x}) = \mathbf{m}(\mathbf{x})^T\boldsymbol{\beta}+\varepsilon(\mathbf{x})$, where ε is a mean 0 isotropic Gaussian random field with spectral density g_η as given by (56) and $\boldsymbol{\beta}$ is a vector of length q of unknown regression coefficients, Handcock and Wallis (1994) suggest the prior density

$$p(\boldsymbol{\beta}, \boldsymbol{\eta}) = \frac{1}{\sigma(1+\rho)^2(1+\nu)^2}, \tag{57}$$

which is identical to assuming that

$$p(\boldsymbol{\beta}, \sigma, \nu/(1+\nu), \rho/(1+\rho)) = \sigma^{-1} \tag{58}$$

on $\mathbb{R}^q \times (0, \infty) \times (0, 1)^2$ (Exercise 51). Thus, for every fixed $\boldsymbol{\beta}$ and σ, $p(\boldsymbol{\beta}, \boldsymbol{\eta})$ is an integrable function of ρ and ν. Exercise 52 asks you to show that if there are at least $q + 1$ observations, the posterior density for $\boldsymbol{\beta}$ and $\boldsymbol{\eta}$ obtained using (57) as the prior is proper.

In using the prior (57), one should bear in mind that, unlike ν, ρ is not dimensionless and has units of distance. Therefore, the meaning of the marginal prior density $p(\rho) = (1 + \rho)^{-2}$ depends on the units used to measure distance. If we want our results to be the same whether we measure distances in meters or kilometers, we should normalize distances in some manner. One possible normalization is to set the distance between the two most distant observations to 1. In conjunction with (57), this normalization provides an "automatic" prior that could be employed by users who are either ill-equipped for or uninterested in developing a prior that reflects their knowledge of the random field under study.

Application to simulated data

This subsection compares posterior predictive densities to plug-in predictive densities for the 20 initial and 3 additional simulated observations considered in 6.9. For the posterior predictive distribution I use the prior described in the preceding subsection. Mark Handcock calculated the posterior predictive distributions using programs reported in Handcock and Wallis (1994). The plug-in predictive distributions $p(\cdot \mid \hat{\nu}, \hat{\alpha})$ are based on taking $e(\hat{\boldsymbol{\theta}})/M(\hat{\boldsymbol{\theta}})^{1/2}$ to follow a standard t distribution with 22 degrees of freedom. As described in 6.8, using this t distribution does provide for an appropriate accounting of the effect of the uncertainty in σ on the predictions. However, this plug-in predictive distribution does not take direct account of the effect of uncertainty in the parameters ν and α on the predictions.

Figure 16 gives these two predictive distributions for the predictions at -10.5 (an extrapolation) and -1 (an interpolation); comparisons at other locations are qualitatively similar. As expected, the posterior predictive densities show somewhat greater spread than the plug-in densities. Based on a single simulation, it is not possible to conclude that one procedure is better than the other. Although the posterior predictive density is more appropriate from the Bayesian perspective, I would guess that it also generally has better frequentist properties than the plug-in predictive density. The parametric simulation procedure described in 6.8 is likely to be a better competitor to the posterior predictive density than is the plug-in predictive density. A resolution of these issues requires either a large advance in higher-order asymptotic theory for spatial prediction problems, or more realistically, a large and well-designed simulation study to compare the properties of different methods for predicting with estimated covariance structures.

I have deliberately chosen not to provide posterior predictive densities based on just the 20 evenly spaced initial simulated observations in 6.9. My reason for this omission is that I find it inappropriate to even undertake such a calculation when, as demonstrated in Figure 13, the uncertainty in ν is so great. The problem, as the results in Tables 5 and 6 demonstrate, is that the assessment of mses of prediction varies dramatically over a range of ν values for which the data provide almost no basis for distinguishing. As a consequence, the posterior predictive density will depend strongly on the prior placed on the parameters of the Matérn model. Since prior distributions on these parameters will necessarily be chosen rather arbitrarily, there will consequently be considerable arbitrariness in the posterior predictive densities. If asked to produce a predictive density in this situation I would point to the results in Tables 5 and 6 and refuse to do so. If that strategy failed I would show how the posterior predictive density varies dramatically with the choice of prior and suggest that whoever is asking for the predictive density choose which prior to use.

Exercises

49 Consider the random field $Z(\mathbf{x}) = \mathbf{m}(\mathbf{x})^T\boldsymbol{\beta} + \varepsilon(\mathbf{x})$, where ε is a mean 0 Gaussian random field with covariance function from some model $K_{\boldsymbol{\theta}}$. For a vector of observations $\mathbf{Z} = (Z(\mathbf{x}_1), \ldots, Z(\mathbf{x}_n))^T$ and the improper prior density $p(\boldsymbol{\beta}, \boldsymbol{\eta}) = 1$, show that (6) gives the logarithm of the marginal posterior density for $\boldsymbol{\theta}$ (Harville 1974).

50 This problem gives a simple example of how one can end up with an improper posterior density by using an improper marginal prior density on the parameter ρ in (56). Suppose Z is a stationary Gaussian process

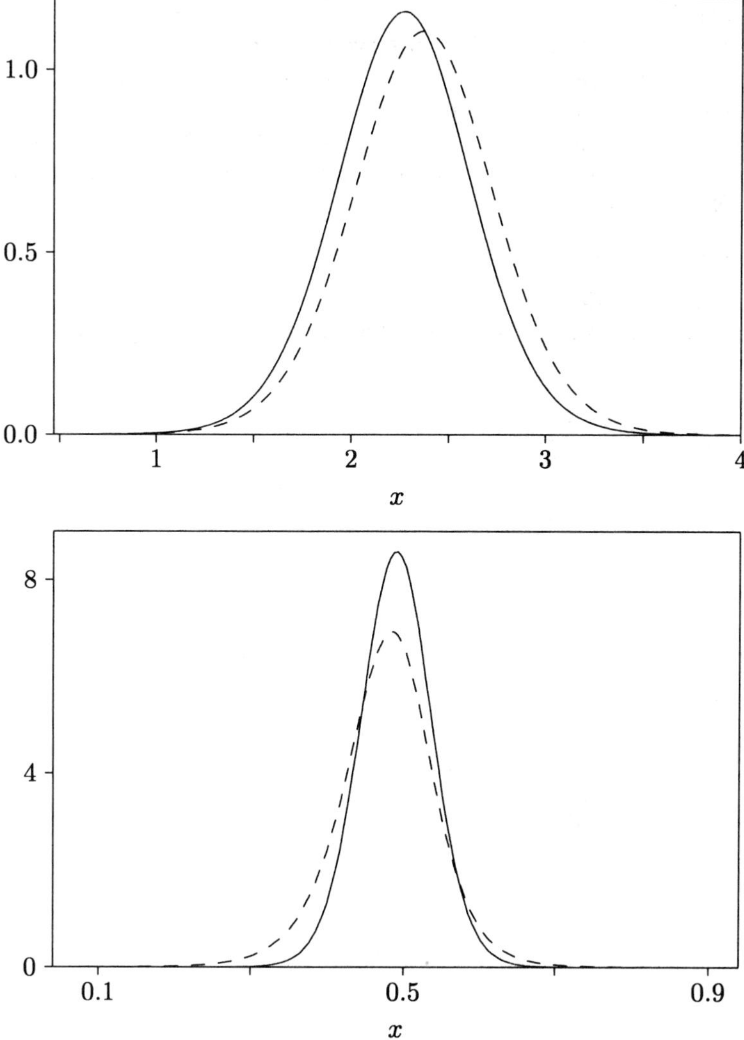

FIGURE 16. Plug-in predictive densities $p(x \mid (\hat{\nu}, \hat{\alpha}))$ and posterior predictive densities $p(x)$ for the same data as in Figure 14. Top figure is for predicting at -10.5 and bottom figure is for -1. Solid line corresponds to plug-in predictive density and dashed line to posterior predictive density under prior recommended by Handcock and Wallis (1994).

on $\mathbb{R}$ with unknown mean μ and spectral density of the form

$$g_{\sigma,\rho}(\omega) = \frac{2^{1/2}\sigma}{\pi\rho\left(\dfrac{2}{\rho^2} + \omega^2\right)},$$

which is just (56) with $d = 1$ and $\nu = \frac{1}{2}$ known. Suppose further that for some $\delta > 0$, Z is observed at δj for $j = 1, \ldots, n$. Use the result of the previous exercise to find an explicit expression for the marginal posterior for σ and ρ for the prior $p(\mu, \sigma, \rho) = 1$. Show that this marginal posterior is not integrable. Hint: the covariance matrix for the observations is explicitly invertible; see, for example, Section 5 of Stein (1990b).

51 Show that (57) implies (58).

52 Using the prior in (58), show that if $n > q$ and the rank of $(\mathbf{m}(\mathbf{x}_1), \ldots, \mathbf{m}(\mathbf{x}_n))$ is q, then the posterior for $(\boldsymbol{\beta}, \boldsymbol{\eta})$ is proper.

Appendix A
Multivariate Normal Distributions

This appendix gives a brief summary of properties of multivariate normal distributions. For proofs and further results, see, for example, Anderson (1984), Muirhead (1982) or Rao (1973, Chapter 8).

The random variable X is said to follow a univariate normal distribution with mean μ and variance $\sigma^2 \geq 0$, written $N(\mu, \sigma^2)$, if it has density

$$p(x) = \frac{1}{(2\pi)^{1/2}\sigma} \exp\left\{-\frac{(x-\mu)^2}{2\sigma^2}\right\}$$

for $\sigma^2 > 0$ and $\Pr(X = \mu) = 1$ for $\sigma^2 = 0$. The random vector $\mathbf{X}$ of length q is said to follow a multivariate normal distribution if $\mathbf{a}^T\mathbf{X}$ follows a univariate normal distribution for every $\mathbf{a} \in \mathbb{R}^q$ (Muirhead 1982, p. 5).

Every multivariate normal distribution has a well-defined mean vector and covariance matrix. Furthermore, if $\mathbf{X}$ is a multivariate normal random vector of length q with $E(\mathbf{X}) = \boldsymbol{\mu}$ and covariance matrix $\text{cov}(\mathbf{X}, \mathbf{X}^T) = \boldsymbol{\Sigma}$, then for any fixed vector $\mathbf{a} \in \mathbb{R}^q$, $\mathbf{a}^T\mathbf{X}$ is $N(\mathbf{a}^T\boldsymbol{\mu}, \mathbf{a}^T\boldsymbol{\Sigma}\mathbf{a})$. More generally, if $\mathbf{A}$ is a matrix with q columns, $\mathbf{AX}$ is $N(\mathbf{A}\boldsymbol{\mu}, \mathbf{A}\boldsymbol{\Sigma}\mathbf{A}^T)$. If $\boldsymbol{\Sigma}$ is positive definite, then $\mathbf{X}$ has density

$$p(\mathbf{x}) = \frac{1}{(2\pi)^{q/2}\{\det(\boldsymbol{\Sigma})\}^{1/2}} \exp\left\{-\tfrac{1}{2}(\mathbf{x}-\boldsymbol{\mu})^T\boldsymbol{\Sigma}^{-1}(\mathbf{x}-\boldsymbol{\mu})\right\}.$$

Suppose the multivariate normal random vector $\mathbf{X}$ is partitioned into two components: $\mathbf{X} = (\mathbf{X}_1^T\ \mathbf{X}_2^T)^T$, where $\mathbf{X}_1$ has q_1 components and $\mathbf{X}_2$ has q_2 components. Then we can write the distribution of $(\mathbf{X}_1^T\ \mathbf{X}_2^T)^T$ as

$$N\left(\begin{pmatrix}\boldsymbol{\mu}_1 \\ \boldsymbol{\mu}_2\end{pmatrix}, \begin{pmatrix}\boldsymbol{\Sigma}_{11} & \boldsymbol{\Sigma}_{12} \\ \boldsymbol{\Sigma}_{21} & \boldsymbol{\Sigma}_{22}\end{pmatrix}\right),$$

where for $i = 1, 2$, $\boldsymbol{\mu}_i$ has length q_i and for $i, j = 1, 2$, $\boldsymbol{\Sigma}_{ij}$ is a $q_i \times q_j$ matrix. Then the conditional distribution of $\mathbf{X}_1$ given $\mathbf{X}_2 = \mathbf{x}_2$ is $N(\boldsymbol{\mu}_1 + \boldsymbol{\Sigma}_{12}\boldsymbol{\Sigma}_{22}^{-}(\mathbf{x}_2 - \boldsymbol{\mu}_2), \boldsymbol{\Sigma}_{11} - \boldsymbol{\Sigma}_{12}\boldsymbol{\Sigma}_{22}^{-}\boldsymbol{\Sigma}_{21})$, where $\boldsymbol{\Sigma}_{22}^{-}$ is any generalized inverse of $\boldsymbol{\Sigma}_{22}$ (for an invertible matrix, the generalized inverse is unique and equals the ordinary inverse).

Finally, if $\mathbf{X} = (X_1, \ldots, X_n)^T$ is $N(\mathbf{0}, \boldsymbol{\Sigma})$ and σ_{ij} is the ijth element of $\boldsymbol{\Sigma}$, then for $i, j, k, \ell = 1, \ldots, n$, $E(X_i X_j X_k) = 0$ and $E(X_i X_j X_k X_\ell) = \sigma_{ij}\sigma_{k\ell} + \sigma_{ik}\sigma_{j\ell} + \sigma_{i\ell}\sigma_{jk}$.

Appendix B
Symbols

Matrices and vectors

$C(\mathbf{M})$	the column space of a matrix $\mathbf{M}$
det	determinant
$\mathbf{h}$	equals $\frac{1}{2}\mathbf{1}$
$\mathbf{I}$	the identity matrix
$\mathbf{M}^T$	the transpose of a matrix $\mathbf{M}$
$\mathbf{1}$	a vector of all ones
tr	trace
$\mathbf{0}$	a vector of zeroes

Sets

A^c	the complement of A
$A \backslash B$	for $B \subset A$, those elements in A that are not in B; equals $A \cap B^c$
$A \circ B$	the symmetric difference of A and B; equals $(A \cup B) \backslash (A \cap B)$
$\mathbb{R}^d$	d-dimensional Euclidean space
$\mathbb{Z}^d$	d-dimensional integer lattice
$\times_{i=1}^d A_i$	for subsets $A_1, \ldots, A_d$ of $\mathbb{R}$, the subset of $\mathbb{R}^d$ whose elements have ith component in A_i for $i = 1, \ldots, d$; also written $A_1 \times \cdots \times A_d$
A^d	for $A \subset \mathbb{R}$, shorthand for $\times_{i=1}^d A$
$B + \mathbf{x}$	for $B \subset \mathbb{R}^d$ and $\mathbf{x} \in \mathbb{R}^d$, the set of points $\mathbf{y}$ for which $\mathbf{y} - \mathbf{x} \in B$
$A_d(r)$	$(-\pi r, \pi r]^d$
$b_d(r)$	the d-dimensional ball of radius r centered at the origin; $b_d = b_d(1)$

$\partial b_d(r)$	the surface of this ball
$\mathcal{B}_m$	$\{-\lfloor\frac{1}{2}(m-1)\rfloor, -\lfloor\frac{1}{2}(m-1)\rfloor+1, \ldots, \lfloor\frac{1}{2}m\rfloor\}^d$, where d is understood by context
c_d	$[-\frac{1}{2}, \frac{1}{2}]^d$
$\mathcal{G}_m$	$\{1, \ldots, m\}^d$, where d is understood by context

Hilbert spaces associated with a random field Z on $\mathbb{R}^d$ with $R \subset \mathbb{R}^d$

$\mathcal{H}_R^0$	the real linear manifold of the random variables $Z(\mathbf{x})$ for $\mathbf{x} \in R$ (generally not a Hilbert space unless R is finite)
$\mathcal{H}_R(m, K)$	the closure of $\mathcal{H}_R^0$ with respect to the inner product defined by the second-order structure (m, K)
$\mathcal{H}_R(F)$	the same as $\mathcal{H}_R(0, K)$ if K is the autocovariance function corresponding to the spectrum F; $\mathcal{H}(F) = \mathcal{H}_{\mathbb{R}^d}(F)$
$\mathcal{L}_R(F)$	the closed real linear manifold of functions of $\boldsymbol{\omega}$ of the form $\exp(i\boldsymbol{\omega}^T\mathbf{x})$ for $\mathbf{x} \in R$ under the inner product defined by F

Probabilities and expectations

corr	correlation
cov	covariance
E	expected value
$G_R(m, K)$	the Gaussian measure for the random field on R with second-order structure (m, K)
$N(\mathbf{m}, \boldsymbol{\Sigma})$	the multivariate normal distribution with mean vector $\mathbf{m}$ and covariance matrix $\boldsymbol{\Sigma}$
P_j	shorthand for $G_R(m_j, K_j)$
Pr	probablility
var	variance

Classes of functions

$\mathcal{D}_d$	all d-dimensional continuous isotropic autocovariance functions
$\mathrm{Lip}(\alpha)$	for $\alpha > 0$, a function f on an interval I is called $\mathrm{Lip}(\alpha)$ on I if there exists finite C such that $\lvert f(s) - f(t)\rvert \leq C\lvert s - t\rvert^\alpha$ for all $s, t \in I$
L^1_{loc}	all real-valued functions on $\mathbb{R}$ that are integrable over all bounded intervals
$\mathcal{Q}^d$	those functions $f : \mathbb{R}^d \to \mathbb{R}$ such that $f(\boldsymbol{\omega}) \asymp \lvert\phi(\boldsymbol{\omega})\rvert^2$ as $\lvert\boldsymbol{\omega}\rvert \to \infty$ for some function ϕ that is the Fourier transform of a square integrable function with bounded support

Functions

cosh	the hyperbolic cosine function, $\cosh x = \frac{1}{2}(e^x + e^{-x})$; other hyperbolic functions used are $\sinh x = \frac{1}{2}(e^x - e^{-x})$, $\tanh x = \sinh x / \cosh x$ and $\operatorname{sech} x = 1/\cosh x$

Γ	the gamma function
δ_{jk}	equals 1 if $j = k$ and 0 otherwise
J_ν	ordinary Bessel function of order ν
$\mathcal{K}_\nu$	modified Bessel function of the second kind of order ν
sinc	for $t \neq 0$, $\operatorname{sinc} t = t^{-1} \sin t$ and $\operatorname{sinc} 0 = 1$
$1\{\cdot\}$	for an event A, $1\{A\} = 1$ if A is true and is 0 otherwise
x^+	the positive part of the real number x; equals x if $x > 0$ and equals 0 otherwise
$\lfloor x \rfloor$	the greatest integer less than or equal to the real number x
$\lvert \mathbf{x} \rvert$	the Euclidean norm of $\mathbf{x} = (x_1, \ldots, x_d)^T$; equals $(x_1^2 + \cdots + x_d^2)^{1/2}$
$\overline{z}$	the complex conjugate of the complex number z

Relationships

$\sim$	for functions f and g on some set R, write $f(t) \sim g(t)$ as $t \to t_0$ if $f(t)/g(t) \to 1$ as $t \to t_0$
$\ll$	for real-valued functions f and g on R, write $f(t) \ll g(t)$ if there exists C finite such that $\lvert f(t) \rvert \leq Cg(t)$ for all $t \in R$; same as $f(t) = O(g(t))$
$\asymp$	for nonnegative functions f and g on R, write $f(t) \asymp g(t)$ if $f(t) \ll g(t)$ and $g(t) \ll f(t)$; write $f(t) \asymp g(t)$ as $t \to t_0$ if, given any sequence $t_1, t_2, \ldots$ such that $t_i \to t_0$, there exists N finite such that $f(t_i) \asymp g(t_i)$ for all $i > N$
$\perp$	orthogonal; can refer either to the orthogonality of two elements in a Hilbert space (have inner product 0) or to the orthogonality of two probability measures
$\equiv$	equivalence for probability measures

Abbreviations

BLP	best linear predictor
BLUP	best linear unbiased predictor
EBLUP	estimated best linear unbiased predictor
IRF	intrinsic random function
LUP	linear unbiased predictor
MLE	maximum likelihood estimator
mse	mean squared error
p.d.	positive definite
REML	restricted maximum likelihood

Miscellaneous

$\sum'_{\mathbf{j}}$	for $\mathbf{j} \in \mathbb{Z}^d$, the sum over all element of Z^d except the origin

References

Abramowitz, M. and Stegun, I. (1965). *Handbook of Mathematical Functions*, ninth ed. Dover, New York.

Akhiezer, N. I. and Glazman, I. M. (1981). *Theory of Linear Operators in Hilbert Space*, trans. E. R. Dawson. Pitman Publishing, Boston.

Anderson, T. W. (1984). *An Introduction to Multivariate Analysis*, second ed. Wiley, New York.

Bailey, D. H. and Swarztrauber, P. N. (1991). The fractional Fourier transform and applications. *SIAM Rev.* **33** 389–404.

Benhenni, K. and Cambanis, S. (1992). Sampling designs for estimating integrals of stochastic processes. *Ann. Statist.* **20** 161–194.

Berger, J. O. (1985). *Statistical Decision Theory and Bayesian Analysis*, second ed. Springer-Verlag, New York.

Billingsley, P. (1968). *Convergence of Probability Measures.* Wiley, New York.

Billingsley, P. (1995). *Probability and Measure*, third ed. Wiley, New York.

Bingham, N. H. (1972). A Tauberian theorem for integral transforms of the Hankel type. *J. London Math. Soc.* **5** 493–503.

Bingham, N. H., Goldie, C. M. and Teugels, J. L. (1987). *Regular Variation.* Cambridge University Press, New York.

Blackwell, D. and Dubins, L. E. (1962). Merging of opinions with increasing information. *Ann. Math. Statist.* **33** 882-886.

Bloomfield, P. (1976). *Fourier Analysis of Time Series: An Introduction.* Wiley, New York.

Bras, R. L. and Rodríguez-Iturbe, I. (1985). *Random Functions and Hydrology.* Addison-Wesley, Reading, MA.

Carr, J. R. (1990). Application of spatial filter theory to kriging. *Math. Geol.* **22** 1063–1079.

Carr, J. R. (1995). *Numerical Analysis for the Geological Sciences.* Prentice-Hall, Englewood Cliffs, NJ.

Carrier, G. F., Krook, M. and Pearson, C. E. (1966). *Functions of a Complex Variable: Theory and Technique.* McGraw-Hill, New York.

Casella, G. and Berger, R. L. (1990). *Statistical Inference.* Brooks/Cole, Pacific Grove, CA.

Chan, G., Hall, P. and Poskitt, D. S. (1995). Periodogram-based estimators of fractal properties. *Ann. Statist.* **23** 1684–1711.

Christakos, G. (1984). On the problem of permissible covariance and variogram models. *Water Resources Research* **20** 251–265.

Christakos, G. (1992). *Random Field Models in Earth Sciences.* Academic Press, San Diego.

Christensen, R. (1991). *Linear Models for Multivariate, Time Series, and Spatial Data.* Springer-Verlag, New York.

Chung, K. L. (1974). *A Course in Probability Theory*, 2nd ed. Academic Press, New York.

Cleveland, W. S. (1971). Projection with the wrong inner product and its application to regression with correlated errors and linear filtering of time series. *Ann. Math. Statist.* **42** 616–624.

Cody, W. J. (1987). SPECFUN—a portable special function package, in *New Computing Environments: Microcomputers in Large-Scale Scientific Computing*, ed. A. Wouk. SIAM, Philadelphia, 1–12.

Cohn, S. E., da Silva, A., Guo, J., Sienkiewicz, M. and Lamich, D. (1998). Assessing the effects of data selection with the DAO Physical-space Statistical Analysis System. *Mon. Wea. Rev.* **126** 2913–2926.

Constantine, A. G. and Hall, P. (1994). Characterizing surface smoothness via estimation of effective fractal dimension. *J. Roy. Statist. Soc. B* **56** 96–113.

Cox, D. R. and Hinkley, D. V. (1974). *Theoretical Statistics.* Chapman & Hall, London.

Cramér, H. and Leadbetter, M. R. (1967). *Stationary and Related Processes: Sample Function Properties and Their Applications.* Wiley, New York.

Cressie, N. (1985). Fitting variogram models by weighted least squares, *J. Internat. Assoc. Math. Geol.* **17** 563–586.

Cressie, N. (1988). Variogram. Entry in *Encyclopedia of Statistical Sciences*, vol. 9, Eds. S. Kotz and N. L. Johnson. Wiley, New York, 489–491.

Cressie, N . (1989). The many faces of spatial prediction, in *Geostatistics*, vol. 1, Ed. M. Armstrong. Kluwer, Dordrecht, 163–176.

Cressie, N. (1990). The origins of kriging. *Math. Geol.* **22** 239–252.

Cressie, N. (1993). *Statistics for Spatial Data*, revised ed. Wiley, New York.

Cressie, N. (1996). Change of support and the modifiable areal unit problem. *Geographical Systems* **3** 159–180.

Cressie, N. and Hawkins, D. M. (1980). Robust estimation of the variogram, I. *J. Internat. Assoc. Math. Geol.* **12** 115–125.

Cressie, N. and Lahiri, S. N. (1993). The asymptotic distribution of REML estimators. *J. Multivariate Anal.* **45** 217–233.

Cressie, N. and Zimmerman, D. L. (1992). On the stability of the geostatistical method. *Math. Geol.* **24** 45–59.

Crowder, M. J. (1976). Maximum likelihood estimation for dependent observations. *J. Roy. Statist. Soc. B* **38** 45–53.

Crum, M. M. (1956). On positive-definite functions. *Proc. London Math. Soc., Third Ser.* **6** 548–560.

Dahlhaus, R. and Künsch, H. (1987). Edge effects and efficient parameter estimation for stationary random fields. *Biometrika* **74** 877–82.

Dalenius, T., Hájek, J. and Zubrzycki, S. (1961). On plane sampling and related geometrical problems. *Proceedings of the Fourth Berkeley Symposium on Mathematical Statistics and Probability* **1** 125–150.

Daley, R. (1991). *Atmospheric Data Analysis.* Cambridge University Press, New York.

Davison, A.C. and Hinkley, D. V. (1997). *Bootstrap Methods and Their Application.* Cambridge University Press, New York.

Dawid, A. P. (1984). Statistical theory: The prequential approach. *J. Roy. Statist. Soc. A* **147** 278–290.

De Oliveira, V., Kedem, B. and Short, D. A. (1997). Bayesian prediction of transformed Gaussian random fields. *J. Amer. Statist. Assoc.* **92** 1422–1433.

Diamond, P. and Armstrong, M. (1984). Robustness of variograms and conditioning of kriging matrices. *J. Internat. Assoc. Math. Geol.* **16** 563–586.

Diggle, P. J., Tawn, J. A. and Moyeed, R. A. (1998). Model-based geostatistics (with discussion). *Appl. Statist.* **47** 299–350.

Efron, B. and Tibshirani, R. J. (1993). *An Introduction to the Bootstrap.* Chapman & Hall, New York.

Feller, W. (1971). *An Introduction to Probability Theory and Its Applications*, vol. II. Wiley, New York.

Ferguson, T. S. (1996). *A Course in Large Sample Theory.* Chapman & Hall, London.

Fuller, W. A. (1996). *Introduction to Statistical Time Series.* Wiley, New York.

Gaspari, G. and Cohn, S. E. (1999). Construction of correlation functions in two and three dimensions. *Q. J. R. Meteorol. Soc.* **125** 723–757.

Gel'fand, I. M. and Vilenkin, N. Ya. (1964). *Generalized Functions*, vol. 4, trans. A. Feinstein. Academic Press, New York.

Gelman, A., Carlin, J. B., Stern, H. S. and Rubin, D. B. (1995). *Bayesian Data Analysis.* Chapman & Hall, London.

Genton, M. G. (1998). Variogram fitting by generalized least squares using an explicit formula for the covariance structure. *Math. Geol.* **30** 323–345.

Gidas, B. and Murua, A. (1997). Estimation and consistency for linear functionals of continuous-time processes from a finite data set, I: Linear predictors. Department of Statistics, University of Chicago Report No. 447.

Gihman, I. I. and Skorohod, A. V. (1974). *The Theory of Stochastic Processes*, vol. 1. Springer-Verlag, Berlin.

Goldberger, A. S. (1962). Best linear unbiased prediction in the generalized linear regression model. *J. Amer. Statist. Assoc.* **57** 369–375.

Golub, G. H. and Van Loan, C. F. (1996). *Matrix Computations*, third ed. Johns Hopkins University Press, Baltimore.

Goovaerts, P. (1997). *Geostatistics for Natural Resources Evaluation.* Oxford University Press, New York.

Gradshteyn, I. S. and Ryzhik, I. M. (1994). *Table of Integrals, Series, and Products*, fifth ed. Academic Press, Orlando.

Guyon, X. (1982). Parameter estimation for a stationary process on a d- dimensional lattice. *Biometrika* **69** 95–105.

Haas, T. C. (1990). Lognormal and moving-window methods of estimating acid deposition. *J. Amer. Statist. Assoc.* **85** 950–963.

Haas, T. C. (1995). Local prediction of a spatio-temporal process with an application to wet sulfate deposition. *J. Amer. Statist. Assoc.* **90** 1189–1199.

Handcock, M. S. (1989). Inference for spatial Gaussian random fields when the objective is prediction. Ph. D. dissertation, Department of Statistics, University of Chicago.

Handcock, M. S. (1991). On cascading Latin hypercube designs and additive models for experiments. *Commun. Statist. A* **20** 417-439.

Handcock, M. S., Meier, K. and Nychka, D. (1994). Discussion of "Kriging and splines: An empirical comparison of their predictive performance" by G. M. Laslett (1994). *J. Amer. Statist. Assoc.* **89** 401-403.

Handcock, M. S. and Stein, M. L. (1993). A Bayesian analysis of kriging. *Technometrics* **35** 403–410.

Handcock, M. S. and Wallis, J. R. (1994). An approach to statistical spatial-temporal modeling of meteorological fields (with discussion). *J. Amer. Statist. Assoc.* **89** 368–390.

Hannan, E. J. (1970). *Multiple Time Series.* Wiley, New York.

Harville, D. A. (1974). Bayesian inference for variance components using only the error contrasts. *Biometrika* **61** 383–385.

Harville, D. A. and Jeske, D. R. (1992). Mean squared error of estimation or prediction under a general linear model. *J. Amer. Statist. Assoc.* **87** 724–731.

Hawkins, D. M. and Cressie, N. (1984). Robust kriging—a proposal. *J. Internat. Assoc. Math. Geol.* **16** 3–18.

Ibragimov, I. A. and Has'minskii, R. Z. (1981). *Statistical Estimation: Asymptotic Theory*, trans. S. Kotz. Springer-Verlag, New York.

Ibragimov, I. A. and Rozanov, Y. A. (1978). *Gaussian Random Processes*, trans. A. B. Aries. Springer-Verlag, New York.

Isaaks, E. H. and Srivastava, R. M. (1989). *An Introduction to Applied Geostatistics.* Oxford University Press, New York.

Istas, J. and Lang, G. (1997). Quadratic variations and estimation of the local Hölder index of a gaussian process. *Ann. Inst. Henri Poincaré* **33** 407–436.

Jeffreys, H. (1938). Science, logic and philosophy. *Nature* **141** 716–719.

Jerri, A. J. (1977). The Shannon sampling theorem—its various extensions and applications: A tutorial review. *Proc. IEEE* **65** 1565–1596.

Journel, A. G. and Huijbregts, C. J. (1978). *Mining Geostatistics.* Academic Press, New York.

Kaluzny, S. P., Vega, S. C., Cardoso, T. P. and Shelly, A. A. (1998). *S+SpatialStats: User's Manual for Windows and Unix.* Springer, New York.

Kent, J. T. and Wood, A. T. A. (1997). Estimating the fractal dimension of a locally self-similar Gaussian process by using increments. *J. Roy. Statist. Soc. B* **59** 679–699.

Kitanidis, P. K. (1983). Statistical estimation of polynomial generalized covariance functions and hydrologic applications. *Water Resources Research* **19** 909–921.

Kitanidis, P. K. (1986). Parameter uncertainty in estimation of spatial functions: Bayesian analysis. *Water Resources Research* **22** 499–507.

Kitanidis, P. K. (1997). *Introduction to Geostatistics: Applications in Hydrogeology.* Cambridge University Press, New York.

Kolmogorov, A. N. (1941). Interpolation und Extrapolation von stationären zufäligen Folgen. *Izv. Akad. Nauk SSSR* **5** 3–14.

Krige, D. G. (1951). A statistical approach to some basic mine valuation problems on the Witwatersrand. *Journal of the Chemical, Metallurgical and Mining Society of South Africa* **52** 119–139.

Kullback, S. (1968). *Information Theory and Statistics.* Dover, Mineola, NY.

Kuo, H. (1975). *Gaussian Measures in Banach Spaces*, Lecture Notes in Mathematics No. **463**. Springer-Verlag, New York.

Laslett, G. M. and McBratney, A. B. (1990). Further comparison of spatial methods for predicting soil pH. *Soil Sci. Am. J.* **54** 1553–1558.

Laslett, G. M., McBratney, A. B., Pahl, P. J. and Hutchinson, M. F. (1987). Comparison of several spatial prediction methods for soil pH. *J. Soil Sci.* **38** 325–341.

Lukacs, E. (1970). *Characteristic Functions.* Griffin, London.

Mandelbrot, B. B. and Van Ness, J. W. (1968). Fractional Brownian motions, fractional noises and applications. *SIAM Rev.* **10** 422–437.

Mardia, K. V. and Marshall, R. J. (1984). Maximum likelihood estimation of models for residual covariance in spatial regression. *Biometrika* **73** 135–146.

Matérn, B. (1960). *Spatial Variation.* Meddelanden från Statens Skogsforskningsinstitut, **49**, No. 5. Almaenna Foerlaget, Stockholm. Second edition (1986), Springer-Verlag, Berlin.

Matheron, G. (1971). *The Theory of Regionalized Variables and its Applications.* Ecole des Mines, Fontainebleau.

Matheron, G. (1973). The intrinsic random functions and their applications. *J. Appl. Probab.* **5** 439–468.

Matheron, G. (1989). *Estimating and Choosing: An Essay on Probability in Practice*, trans. A. M. Hasofer. Springer-Verlag, Berlin.

McCullagh, P. and Nelder, J. A. (1989). *Generalized Linear Models*, second ed. Chapman & Hall, London.

McGilchrist, C. A. (1989). Bias of ML and REML estimators in regression models with ARMA errors. *J. Statist. Comput. Simul.* **32** 127–136.

Micchelli, C. A. (1986). Interpolation of scattered data: distance matrices and conditionally positive definite functions. *Constr. Approx.* **2** 11–22.

Muirhead, R. J. (1982). *Aspects of Multivariate Statistical Theory.* Wiley, New York.

Müller-Gronbach, T. (1998). Hyperbolic cross designs for approximation of random fields. *J. Statist. Plann. Inference* **66** 321–344.

Novak, E. (1988). *Deterministic and Stochastic Error Bounds in Numerical Analysis*, Lecture Notes in Mathematics No. **1349**. Springer-Verlag, Berlin.

Omre, H. (1987). Bayesian kriging—merging observations and qualified guesses in kriging. *Math. Geol.* **19** 25–39.

Pannetier, Y. (1996). *VARIOWIN: Software for Spatial Data Analysis in 2D.* Springer-Verlag, New York.

Papageorgiou, A. and Wasilkowski, G. W. (1990). On the average complexity of multivariate problems. *J. Complexity* **6** 1–23.

Pasenchenko, O. Yu. (1996). Sufficient conditions for the characteristic function of a two-dimensional isotropic distribution. *Theor. Probab. Math. Statist.* **53** 149–152.

Patterson, H. D. and Thompson, R. (1971). Recovery of inter-block information when block sizes are unequal. *Biometrika* **58** 545–554.

Pettitt, A. N. and McBratney, A. B. (1993). Sampling designs for estimating spatial variance components. *Appl. Statist.* **42** 185–209.

Phillips, E. R. (1984). *An Introduction to Analysis and Integration Theory.* Dover, New York.

Pitman, E. J. G. (1968). On the behaviour of the characteristic function of a probability distribution in the neighbourhood of the origin. *J. Austral. Math. Soc. A* **8** 422–443.

Press, W. H., Flannery, B. P., Teukolsky, S. A. and Vetterling, W. T. (1992). *Numerical Recipes*, second ed. Cambridge University Press, New York.

Priestley, M. B. (1981). *Spectral Analysis and Time Series.* Academic Press, London.

Putter, H. and Young, G. A. (1998). On the effect of covariance function estimation on the accuracy of kriging predictors. Manuscript.

Quenouille, M. H. (1949). Problems in plane sampling. *Ann. Math. Statist.* **20** 355–375.

Rao, C. R. (1973). *Linear Statistical Inference and Its Applications*, second ed. Wiley, New York.

Rice, J. A. (1995). *Mathematical Statistics and Data Analysis*, second ed. Wadsworth, Belmont, CA.

Ripley, B. D. (1988). *Statistical Inference for Spatial Processes.* Cambridge University Press, New York.

Ripley, B. D. (1995). Review of *Number-Theoretic Methods in Statistics*, by K.-T. Fang and Y. Wang. *J. Roy. Statist. Soc. A* **158** 189–190.

Ritter, K. (1995). *Average Case Analysis of Numerical Problems.* Unpublished thesis, Erlangen.

Ruhla, C. (1992). *The Physics of Chance: From Blaise Pascal to Neils Bohr*, trans. C. Barton. Oxford University Press, New York.

SAS Institute Inc. (1997). *SAS/STAT Software: Changes and Enhancements Through Release 6.12.* SAS Institute Inc., Cary, NC.

Schoenberg, I. J. (1938). Metric spaces and completely monotone functions. *Ann. Math.* **39** 811–841.

Schowengerdt, R. A. (1983). *Techniques for Image Processing and Classification in Remote Sensing.* Academic Press, New York.

Seber, G. A. F. (1977). *Linear Regression Analysis.* Wiley, New York.

Spivak, M. (1980). *Calculus*, second ed. Publish or Perish, Berkeley, CA.

Starks, T. H. and Sparks, A. R. (1987). Rejoinder to "Comment on 'Estimation of the generalized covariance function. II. A response surface approach' by T. H. Starks and A. R. Sparks." *Math. Geol.* **19** 789-792.

Stein, E. M. and Weiss, G. (1971). *Introduction to Fourier Analysis on Euclidean Spaces.* Princeton University Press, Princeton, NJ.

Stein, M. L. (1986). A modification of minimum norm quadratic estimation of a generalized covariance function for use with large data sets. *Math. Geol.* **18** 625–633.

Stein, M. L. (1988). Asymptotically efficient prediction of a random field with a misspecified covariance function. *Ann. Statist.* **16** 55–63.

Stein, M. L. (1990a). Uniform asymptotic optimality of linear predictions of a random field using an incorrect second-order structure. *Ann. Statist.* **18** 850–872.

Stein, M. L. (1990b). Bounds on the efficiency of linear predictions using an incorrect covariance function. *Ann. Statist.* **18** 1116–1138.

Stein, M. L. (1990c). A comparison of generalized cross validation and modified maximum likelihood for estimating the parameters of a stochastic process. *Ann. Statist.* **18** 1139–1157.

Stein, M. L. (1990d). An application of the theory of equivalence of Gaussian measures to a prediction problem. *IEEE Trans. Inform. Theory* **34** 580–582.

Stein, M. L. (1993a). A simple condition for asymptotic optimality of linear predictions of random fields. *Statist. Probab. Letters* **17** 399–404.

Stein, M. L. (1993b). Spline smoothing with an estimated order parameter. *Ann. Statist.* **21** 1522–1544.

Stein, M. L. (1993c). Asymptotic properties of centered systematic sampling for predicting integrals of spatial processes. *Ann. Appl. Probab.* **3** 874–880.

Stein, M. L. (1995a). Predicting integrals of random fields using observations on a lattice. *Ann. Statist.* **23** 1975–1990.

Stein, M. L. (1995b). Locally lattice sampling designs for isotropic random fields. *Ann. Statist.* **23** 1991–2012.

Stein, M. L. (1995c). Fixed domain asymptotics for spatial periodograms. *J. Amer. Statist. Assoc.* **90** 1277–1288.

Stein, M. L. (1997). Efficiency of linear predictors for periodic processes using an incorrect covariance function. *J. Statist. Plann. Inference* **58** 321–331.

Stein, M. L. (1999). Predicting random fields with increasingly dense observations. *Ann. Appl. Probab.*, to appear.

Stein, M. L. and Handcock, M. S. (1989). Some asymptotic properties of kriging when the covariance function is misspecified. *Math. Geol.* **21** 171–190.

Sweeting, T. J. (1980). Uniform asymptotic normality of the maximum likelihood estimator. *Ann. Statist.* **8** 1375–1381.

Thiébaux, H. J. and Pedder, M. A. (1987). *Spatial Objective Analysis with Applications in Atmospheric Science.* Academic Press, London.

Toyooka, Y. (1982). Prediction in a linear model with estimated parameters. *Biometrika* **69** 453–459.

Traub, J. F., Wasilkowski, G. W. and Woźniakowski, H. (1988). *Information-Based Complexity.* Academic Press, New York.

Trebels, W. (1976). Some necessary conditions for radial Fourier multipliers. *Proc. Amer. Math. Soc.* **58** 97–103.

Tubilla, A. (1975). Error convergence rates for estimates of multidimensional integrals of random functions. Department of Statistics, Stanford University Report No. 72.

Tunicliffe-Wilson, G. (1989). On the use of marginal likelihood in time series model estimation. *J. Roy. Statist. Soc. B* **51** 15–27.

Vanmarcke, E. (1983). *Random Fields.* MIT Press, Cambridge, MA.

Vecchia, A. V. (1988). Estimation and identification for continuous spatial processes. *J. Roy. Statist. Soc. B* **50** 297–312.

Voss, R. F. (1988). Fractals in nature: From characterization to simulation, in *The Science of Fractal Images*, Eds. H. O. Peitgen and D. Saupe. Springer-Verlag, New York.

Wackernagel, H. (1995). *Multivariate Geostatistics.* Springer, Berlin.

Wahba, G. (1990). *Spline Models for Observational Data.* SIAM, Philadelphia.

Wald, A. (1949). Note on the consistency of the maximum likelihood estimate. *Ann. Math. Statist.* **20** 595–601.

Warnes, J. J. and Ripley, B. D. (1987). Problems with likelihood estimation of covariance functions of spatial Gaussian processes. *Biometrika* **74** 640–642.

White, L. V. (1973). An extension of the general equivalence theorem to nonlinear models. *Biometrika* **60** 345–348.

Whittle, P. (1954). On stationary processes in the plane. *Biometrika* **49** 305–314.

Wichura, M. J. (1987). The PiCTeX manual. Department of Statistics, University of Chicago Report No. 205.

Wiener, N. (1949). *Extrapolation, Interpolation, and Smoothing of Stationary Time Series.* MIT Press, Cambridge, MA.

Wilson, P. D. (1988). Maximum likelihood estimation using differences in an autoregressive-1 process. *Comm. Statist. Theory Methods* **17** 17–26.

Woźniakowski, H. (1991). Average case complexity of multivariate integration. *Bull. Amer. Math. Soc.* **24** 185–194.

Yadrenko, M. I. (1983). *Spectral Theory of Random Fields.* Optimization Software, New York.

Yaglom, A. M. (1962). *An Introduction to the Theory of Stationary Random Functions.* Dover, New York.

Yaglom, A. M. (1987a). *Correlation Theory of Stationary and Related Random Functions*, vol. I. Springer-Verlag, New York.

Yaglom, A. M. (1987b). *Correlation Theory of Stationary and Related Random Functions*, vol. II. Springer-Verlag, New York.

Yakowitz, S. J. and Szidarovszky, F. (1985). A comparison of kriging with nonparametric regression methods. *J. Multivariate Anal.* **16** 21–53.

Ying, Z. (1991). Asymptotic properties of a maximum likelihood estimator with data from a Gaussian process. *J. Multivariate Anal.* **36** 280–396.

Ying, Z. (1993). Maximum likelihood estimation of parameters under a spatial sampling scheme. *Ann. Statist.* **21** 1567–1590.

Ylvisaker, D. (1975). Designs on random fields. In *A Survey of Statistical Design and Linear Models*, Ed. J. N. Srivastava, North-Holland, Amsterdam, 593–608.

Zellner, A. (1971). *An Introduction to Bayesian Inference in Econometrics.* Wiley, New York.

Zimmerman, D. L. (1989). Computationally exploitable structure of covariance matrices and generalized covariance matrices in spatial models. *J. Statist. Comput. Simulation* **32** 1–15.

Zimmerman, D. L. and Cressie, N. (1992). Mean squared prediction error in the spatial linear model with estimated covariance parameters. *Ann. Inst. Statist. Math.* **44** 27–43.

Zimmerman, D. L. and Zimmerman, M. B. (1991). A Monte Carlo comparison of spatial semivariogram estimators and corresponding ordinary kriging predictors. *Technometrics* **33** 77–91.

Index

Springer Series in Statistics

(continued from p. ii)

Kotz/Johnson (Eds.): Breakthroughs in Statistics Volume II.
Kotz/Johnson (Eds.): Breakthroughs in Statistics Volume III.
Kres: Statistical Tables for Multivariate Analysis.
Küchler/Sørensen: Exponential Families of Stochastic Processes.
Le Cam: Asymptotic Methods in Statistical Decision Theory.
Le Cam/Yang: Asymptotics in Statistics: Some Basic Concepts.
Longford: Models for Uncertainty in Educational Testing.
Manoukian: Modern Concepts and Theorems of Mathematical Statistics.
Miller, Jr.: Simultaneous Statistical Inference, 2nd edition.
Mosteller/Wallace: Applied Bayesian and Classical Inference: The Case of the Federalist Papers.
Parzen/Tanabe/Kitagawa: Selected Papers of Hirotugu Akaike.
Pollard: Convergence of Stochastic Processes.
Pratt/Gibbons: Concepts of Nonparametric Theory.
Ramsay/Silverman: Functional Data Analysis.
Rao/Toutenburg: Linear Models: Least Squares and Alternatives.
Read/Cressie: Goodness-of-Fit Statistics for Discrete Multivariate Data.
Reinsel: Elements of Multivariate Time Series Analysis, 2nd edition.
Reiss: A Course on Point Processes.
Reiss: Approximate Distributions of Order Statistics: With Applications to Non-parametric Statistics.
Rieder: Robust Asymptotic Statistics.
Rosenbaum: Observational Studies.
Ross: Nonlinear Estimation.
Sachs: Applied Statistics: A Handbook of Techniques, 2nd edition.
Särndal/Swensson/Wretman: Model Assisted Survey Sampling.
Schervish: Theory of Statistics.
Seneta: Non-Negative Matrices and Markov Chains, 2nd edition.
Shao/Tu: The Jackknife and Bootstrap.
Siegmund: Sequential Analysis: Tests and Confidence Intervals.
Simonoff: Smoothing Methods in Statistics.
Singpurwall and Wilson: Statistical Methods in Software Engineering: Reliability and Risk.
Small: The Statistical Theory of Shape.
Stein: Interpolation of Spatial Data: Some Theory for Kriging
Tanner: Tools for Statistical Inference: Methods for the Exploration of Posterior Distributions and Likelihood Functions, 3rd edition.
Tong: The Multivariate Normal Distribution.
van der Vaart/Wellner: Weak Convergence and Empirical Processes: With Applications to Statistics.
Vapnik: Estimation of Dependences Based on Empirical Data.
Weerahandi: Exact Statistical Methods for Data Analysis.
West/Harrison: Bayesian Forecasting and Dynamic Models, 2nd edition.
Wolter: Introduction to Variance Estimation.
Yaglom: Correlation Theory of Stationary and Related Random Functions I: Basic Results.